America's Water and Wastewater Industries: Competition and Privatization

Paul Seidenstat
Michael Nadol
Simon Hakim

PUR Incorporated
Public Utilities Reports, Inc.

Public Utilities Reports Inc.
800-368-5001
www.pur.com

Typography by Type Shoppe II Productions, Ltd., Chestertown, Maryland

First Printing January 2000

Library of Congress Cataloging-in-Publication Data

Seidenstat, Paul.
America's water and wastewater industries : competition and privatization / by Paul Seidenstat, Michael Nadol, and Simon Hakim.
p. cm.
Includes bibliographical references and index.
ISBN 0-910325-80-4
1. Water utilities--Deregulation--United States. 2. Sewage disposal--Economic aspects--United States. 3. Privatization--United States. 4. Competition--United States. I. Nadol, Michael, 1964- II. Hakim, Simon. III. Title.
HD4461 .S44 2000
363.6'1'0973--dc21 99-055744

Printed in Canada

TABLE OF CONTENTS

PREVIEW

Part I offers some concepts and develops a framework for investigating the privatization of water and wastewater systems. David Haarmeyer traces the history of the private sector's participation in the provision of environmental infrastructure services. Emerging from a period of dominance by the public sector, the industry then continues to move toward an ever-greater participation by private providers who have an inherent advantage in both efficiency and water quality.

Janice Landow-Esser and Melissa Manuel, of Holleb & Coff in Chicago, Illinois, explore relevant environmental and legal issues of contracting out. Roger Hartman examines operations and maintenance contracts. He argues that when properly implemented, contracts with professional services management firms can lead to substantial public benefits and better service.

Part II illustrates the principles of privatization through an examination of water and wastewater systems that recently have been privatized. Robert E. Thiele, Jr. of United Water New Jersey details the experience of a private company taking over the water systems of Hoboken and Jersey City, New Jersey. In Hoboken, United Water purchased a "concession" or outright sale of the city's assets, while Jersey City contracted with United Water to provide for the complete operation, maintenance, and administration of the water system.

The City of Franklin, Ohio, was a pioneer in privatization with the sale of its water system. Douglas Herbst and Keith Oldewurtel describe how, in 1996, Franklin also contracted with Earth Tech to design, build, finance, and operate its wastewater system on a long-term contract basis.

Douglas Herbst and Robert Markowitch of Earth Tech, Inc. and Kash Srinivasan of Wilmington, discuss how the water and wastewater systems of Gardner, Massachusetts and Wilmington, Delaware became privatized. The two cities contracted with private companies for the operations and maintenance of their systems, including the provision for capital improvements. Wilmington privatized its wastewater system while Gardner included both its wastewater and water systems. Both entered into 20-year contracts.

Elizabeth Kelly, Paul Reiter, and Scott Haskins of Seattle Public Utilities discuss the development of that city's contract to design, build, and operate its Tolt River water filtration plant, while Remedios del Rosario, David Seader, and John Salo show how Atlanta achieved significant cost savings in its water system with an operations and maintenance contract.

Mayor Kirk Humphreys of Oklahoma City discusses the benefits of a public-private partnership in contracting out the operations of his area's wastewater treatment operations. The city has benefitted greatly with the achievement of cost savings and improved operations, while the private contractor has advanced its technological expertise and has become a more proficient player in the wastewater market.

Part III examines managed competition, which is the process of allowing the government agency to bid against private companies for the right to operate the government facilities. Guru Bose, William Toffey and James Golembeski describe how Philadelphia used managed competition in its biosolids recycling center. Kimberly S. Eagle details how the Charlotte-Mecklenburg Utilities uses managed competition to improve its water and wastewater systems.

Douglas Herbst indicates some major hurdles managed competition systems have to overcome to be effective. Practitioners of the system also should examine its fundamental fairness to private bidders according to Herbst. Alan Manning, of EMA Services, Inc., examines the many techniques utilities can employ to become competitive.

Part IV discusses some key public policy issues involving the privatization of water and wastewater systems. Addressing the recent history of privatizing wastewater systems, Professor Randall Holcombe notes that contracting out in the 1980s primarily was a device to finance major improvements in facilities without placing a greater financial burden on government rather than a method of reducing operating costs. He argues that privatization improves performance only if it injects incentives for efficient production.

Janice Beecher, of Beecher Policy Research, examines two models of privatization: municipal ownership with contracting out and investor-owned water firms. She contends that the latter are subject to public regulation but the former may be exempt. It can be argued that all providers, in a monopoly context, should be subject to regulation but the traditional regulatory role may be outmoded. Beecher suggests a system of structural competition where all providers are subject to a more flexible and innovative system of regulation.

Steve Hanke of Johns Hopkins University and Stephen Walters of Loyola University in Baltimore review the British and French systems of privatization and argue that the franchising device of the French is economically superior to both the British and American schemes of privatization.

PART I

Competition and the Role of the Private Sector

CHAPTER 1

Introduction: Competition and Privatization in the Water and Wastewater Industries

Michael Nadol, Paul Seidenstat, and Simon Hakim

Competition has been slow to come to the United States water and wastewater industries. While many public water systems did begin as private, profit-motivated companies just before and throughout the nineteenth century, devastating outbreaks of typhoid and cholera and major fires in the nation's young urban centers led to a focus on water as more of a public good than a commodity. As local governments increasingly sought to extend water supply beyond those areas that could already afford private supplies, most early water companies could not manage to simultaneously cover the cost to bring service into poorer neighborhoods, maintain reasonable rates for full-paying customers, and still return a profit to their investors. By the turn of the century, more than 200 communities had shifted from private to public ownership (Westerhoff, et al.: 5) and municipally owned water utilities, with monopolistic service territories, had become the dominant national model for service delivery.

Similarly, the U.S. wastewater industry evolved primarily as a governmental response to the public health threat posed by uncontrolled water pollution and wastewater collection; the treatment services today remain overwhelmingly provided for by local governments (US EPA, 1998a). Over most of the twentieth century, these models of public sector monopoly have achieved a high degree of stability, if not efficiency, leading industry managers to take pride in the implicit dependability of providing "silent services" that satisfy critical human needs that are often taken for granted.

During the 1990s, however, this "silence" was broken. In the late 1980s, it was estimated that only 100 to 200 U.S. water and wastewater facilities were under con-

---

Michael Nadol is director of finance for Philadelphia, Paul Seidenstat is an associate professor of economics at Temple University, and Simon Hakim is professor of economics at Temple University.

tracted operation (AMSA/AMWA, 1998). By 1997, a survey of just 14 major industry contract operators found this total to have grown to more than 1,200 facilities in 44 states, including Puerto Rico (Eisenhardt, et al., 1998), and some industry watchers were predicting further growth in operations and management contracting of 20 to 30 percent per year over the next five years (Reason, 1997). Similarly, while there were still only 17 publicly traded U.S. water companies as of 1998, many of these investor-owned utilities are gaining territory through the acquisition of smaller public and private systems. According to one survey of 10 such utilities, total growth in net income increased by 13.5 percent in 1996 alone (Price Waterhouse, 1997).

In an evident defensive reaction, many public sector utilities are increasing their emphasis on cost containment and customer service. Exemplifying this trend, two leading industry groups for predominately public sector utilities, the Association of Metropolitan Sewerage Agencies (AMSA) and the Association of Metropolitan Water Agencies (AMWA), jointly sponsored a national series of workshops entitled "Thinking, Getting, and Staying Competitive." Presented in 1998, these workshops were attended by hundreds of government managers. According to a 1998 survey of 175 municipal leaders conducted by the major environmental engineering firm, Malcolm Pirnie, fully 88 percent of wastewater agency respondents and 93 percent of drinking water agency respondents reported that they were implementing targeted programs to improve effectiveness and efficiency.

This chapter examines this trend of increasing competition within the water and wastewater industries and explores potential strategies for further improvements in performance. It begins with an overview of the economic structure of these industries and an analysis of international models for service delivery. The chapter profiles the current state of these industries within the U.S., exploring the factors now driving change in this market, while examining the industries' initial experiences with the implementation of alternative strategies. Finally, it explores the potential for broader regulatory restructuring, consistent with ongoing deregulation in the U.S. electric industry as the water and wastewater industries approach the twenty-first century.

INJECTING COMPETITION INTO THE FRAMEWORK OF NATURAL MONOPOLY

Classically, the water and wastewater industries have been viewed as natural monopolies. Seen from this textbook perspective, the capital-intensive infrastructure of these industries would render them ill-suited to conventional free market competition. Even relative to other industries with high costs of entry, the need for capital investment to develop water and wastewater treatment plants and pipelines is high. According to a study for the National Council on Public Works, $10 to $12 of capital investment are required for each dollar of revenue generated in the water industry—in contrast to just $3 to $4 for electric utilities, $3 for telephone companies, and $2 for railroads (Haarmeyer, 1994). While others cite a more modest requirement of $4 of capital investment to generate $1 in water or wastewater revenues (Hyman, et al. 1998), there remains no question that this is a capital-intensive enterprise.

Similarly, the most recent U.S. Environmental Protection Agency (EPA) survey data indicates that water systems overall maintain approximately $5 of gross assets for every $1 of revenue, in contrast to a three-to-one asset-to-revenue ratio for investor-owned electric and gas utilities (US EPA, 1997b: 15–16). Moreover, both water and wastewater services feature significant economies of scale, with higher volumes bringing lower costs per unit. After the initial investment in constructing and maintaining a pipeline or treatment plant has been made, the marginal costs of serving additional customers becomes nearly inconsequential. As a result, in an unregulated market, the first water or wastewater company in place would have an overwhelming advantage against any subsequent competitors, and the tremendously inefficient construction of the multiple side-by-side plants and pipelines necessary for competition would occur rarely, if ever.

Despite these tendencies toward monopoly, however, the dominant U.S. model of public sector operation is not only now undergoing change, it is also far from the universal model. As further detailed in several of the chapters within this volume, European countries—most notably France and Great Britain—have organized their entire water and wastewater markets differently from the U.S. Within the developed world, they offer key alternative models for more competitive system management.

In France, water systems are publicly owned, but approximately 75 percent of the French population is provided water service by private companies that operate these systems under contract. Models for such private operation include: concessions, in which private companies are granted long-term contracts to finance, build, and operate all facilities (typically 25 to 30 years); *affermage*, under which the municipality builds and finances the facility, but contracts out all operations and management (usually for 10 to 12 years); and service contracting, under which the municipality retains responsibility for billing and collections, contracting out only specific facility operations (typically for less than 10 years) (Haarmeyer, 1994:47). The theory underlying these varying forms of franchising and contract operation is that competition will be generated among the multiple companies vying for contracts under the multiple options for contract structure. In more than a century of evolution, the French system has led to the growth of multiple large and powerful water companies, widely recognized for technological innovation and high-quality services, but also periodically caught up in allegations of political influence and corruption.

In Britain, under the leadership of former Prime Minister Margaret Thatcher, the United Kingdom sold its ownership interest in the government's 10 former regional water authorities in 1989, creating in their place 10 large, regulated water service companies that both own and operate the water and wastewater infrastructure serving approximately 75 percent of the nation's population (the rest of Britain's population is served by previously existing private water-only utilities) (Jeffery, 1994: 64). Under complex "price-cap" regulation, rates for these 10 major companies are established by a regulatory agency for five-year periods, and are intended to be independent of actual costs. The theory behind this regulatory model is that with fees and profits decoupled from operating expenses, companies will have performance incentives. Since its implementation nearly a decade ago, rate hikes have been steep and profits have been high, generating some substantial initial public criticism within a culture that previously had not been widely metered for water use. On the other hand, significant water quality improvements have been achieved under privatization, with increased capital investments in infrastructure improvements.

In many respects, the French and British models for water system operation are quite dissimilar, reflecting the long and divergent histories of each nation's approach to providing these services. In both models, however, there is a substantially greater emphasis on competition than has existed for most of the twentieth century in the U.S.

OVERVIEW OF U.S. WATER INDUSTRY

In its most recent national survey (US EPA, 1997a), the EPA has estimated that there are more than 180,000 public water systems in the United States. According to EPA and Safe Drinking Water Act (SDWA) definitions, this total includes both "community water systems" serving at least 15 connections or 25 residents year-round, and "noncommunity water systems," such as an individual well serving a school or church, providing water to a nonresidential population at least 60 days out of the year. Water that does not come from a public water system, such as from a well serving one or just a few homes, is considered a private supply excluded from the federal government's statistics. While the nation's 50,289 community water systems in 1995 represented only 28 percent of all identified public water systems, they served an estimated 93 percent of the U.S. population. (See Table 1–1.)

Among these more than 50,000 community systems, only 43 percent were publicly owned, 33 percent were privately owned utilities, and 24 percent were classified as "ancillary systems" (*i.e.*, very small systems, typically privately owned, that provide water as an ancillary service to some other enterprise such as a mobile home park). However, because most private systems are relatively small, an estimated 86 percent of the U.S. population received its water from publicly owned community systems, with only 13 percent relying on private utilities and 1 percent served by ancillary systems.

According to this same EPA survey, total U.S. water industry revenues were estimated to total $25.9 billion as of 1995, primarily from water sales. Of this amount, 86 percent was received by publicly owned systems, mirroring the percentage of the total population served by such systems.

According to the EPA's *1996 Clean Water Needs Survey* report, publicly owned treatment works (POTWs) serve 190 million U.S. residents, approximately 72 percent of the total U.S. population, and the EPA projects this total will grow to 275 million (90 percent of the projected total population) by 2016. Moreover, almost all of the private sector's involvement in wastewater service delivery has historically been limited

TABLE 1–1. U.S. COMMUNITY WATER SYSTEMS BY OWNERSHIP TYPE, POPULATION SERVED, AND ANNUAL REVENUE (1995)

Ownership	Number	% of Total Number	% of Total Population Served	Annual Revenue ($ Billion)
Public	21,789	43%	86%	22.2
Private	16,540	33%	13%	3.7
Ancillary	11,960	24%	1%	N/A

Source: U.S. Environmental Protection Agency, 1997a.

to small subdivisions, trailer parks, and individually owned septic systems. Large-scale wastewater treatment facilities and collection systems continue to be almost entirely publicly owned and while a handful of wastewater systems have been acquired by investor-owned water utilities and environmental services companies, there are no publicly traded U.S. wastewater companies per se.

According to U.S. Bureau of the Census data on government finances (U.S. Bureau of the Census, 1996), total state and local government spending to provide sewerage services totaled nearly $23.6 billion in 1995, with by far the largest share of this expense, over $22.1 billion, supported by local governments. Of this total spending, nearly $8.9 billion combined (and over $8 billion in local dollars) were for capital improvements to treatment and collection systems. (See Table 1–2.)

In evaluating U.S. water and wastewater industries, it should also be noted that there is significant overlap between the two sectors. In many cases, both services are delivered by the same public agency, sometimes even as part of a larger municipal utilities authority or public works department. In addition, both industries are typically served by many of the same major engineering firms, consultants, and contract operators.

FACTORS DRIVING THE U.S. MARKET

Across almost all public services—not at all limited to water and wastewater—there has been a general ideological and management trend toward the increased use of market forces internationally and among U.S. federal, state, and local governments. Privatization, deregulation, and competition have not only become public sector buzzwords, but they have also become a reality in services ranging from transportation, to corrections, energy, and education.

As in virtually all of the public services targeted for privatization, much of this trend has been driven by the promise of overall cost savings and the prospect of cash up front. In the water and wastewater industries, potential cost savings through privatization are often attributed to the greater incentives for efficiency resulting from the private sector's profit motive, and the ability of less bureaucratic private firms to quickly

TABLE 1–2. U.S. WASTEWATER PUBLICLY OWNED TREATMENT AND COLLECTION SYSTEMS IN 1996

Treatment Facilities	% of Total Population Served	Collection Systems	Annual Expenditures, including capital* ($Billion)
16,024	71.8	20,670	$23.6

Sources: U.S. Environmental Protection Agency, 1998c; *U.S. Bureau of the Census, 1996. (Note: because this financial data is drawn from a different survey source in a different year from the other data in this table, any correlation involving this fourth column should be considered approximate).

implement more flexible labor and operating practices. In addition, proponents of privatization cite the increased ability of large firms to leverage technology and state-of-the-art practices because of their sophistication and economies of scale in research and development. Further, corporations unconstrained by public bidding and debt issuance requirements may be able to achieve savings in construction and/or capital financing. Anecdotally, almost every water or wastewater privatization initiative to date purports to have achieved cost savings. However, differences in tax treatment and accounting methodologies, scale of operation, and local operating conditions (*e.g.*, source water, topography, and system age) make more systematic comparisons difficult. In one of the few such efforts, a 1996 study sponsored by the pro-privatization Reason Foundation concluded that California investor-owned water utilities provided "comparable services to consumers at the same price as government-owned water companies even though they pay taxes and do not receive extra nonoperating income" (Neal, et al., 1996). Along similar lines, an analysis of more than 100 water and wastewater utilities during the 1990s by EMA Services, Inc., a major utility management consulting firm, identified an average "competitiveness gap" of 24 percent greater costs for the public sector in comparison to private sector industry leaders (AMSA/AMWA, 1998).

More stringent federal environmental and public health regulatory standards, coupled with technological change, are also contributing to increasing privatization. New drinking water regulations phasing in at the start of the twenty-first century as mandated under 1996 amendments to the SDWA, for example, are toughening standards for guarding against bacteria and other microbial contaminants (the Interim Enhanced Surface Water Treatment Rule), while simultaneously requiring that utilities reduce potentially harmful byproducts of the disinfection process (the Disinfectants and Disinfection Byproducts Rule). To achieve these somewhat conflicting goals, many utilities will need to significantly upgrade their treatment processes, in some cases switching from chlorination to more complex and less familiar technologies, such as ozonation. In total, the EPA estimates that these two new rules alone will cost over $1 billion per year nationwide (US EPA, 1998b). Moreover, these costs are over and above the high price tag already required simply to renew and replace existing treatment facilities and distribution systems. Overall, an EPA Drinking Water Infrastructure Needs Survey study projects that $138.4 billion will be required for drinking water infrastructure nationwide from 1995 through 2015, and many industry officials believe the real cost will be significantly higher. Just for the replacement of water mains and other aging distribution system infrastructure, for example, the largest drinking water industry association, the American Water Works Association (AWWA) estimates 20-year costs of $325 billion—more than four times the EPA's estimate of $77.2 billion (AWWA, 1999).

Similarly, Clean Water Act requirements for wastewater systems are increasing responsibilities and costs not only for plant operation, but also for storm water control and the reduction of pollution overflowing from collection systems. According to the most recent EPA Clean Water Needs Survey of wastewater and storm water infrastructure needs, an additional $135 billion in upgrades are projected to be required from 1996 through 2016.

In the face of such costly requirements for major system upgrades and demands for increased technical and management sophistication, many public decision makers have begun to equate privatization with professionalization. Under a 1998 law passed in the State of Georgia, for example, wastewater plants will be required to privatize when they regularly violate their environmental permit requirements (Hun, 1998). Similarly, at least one of the eight small water systems acquired in 1996 by Philadelphia Suburban Corporation, one of the nation's three largest investor-owned water utilities, was directly motivated by the seller's ability to abandon a groundwater supply threatened by industrial contamination upon sale (Price Waterhouse, 1997).

Globalization has also reached into the U.S. water and wastewater industries, as powerful French water companies and newly privatized British utilities have moved across the seas. In France, private companies have long competed for long-term management contracts and franchises, and have developed into multibillion dollar, highly sophisticated international conglomerates. For example, Vivendi (formerly known as Compagnie Générale des Eaux) took over the Houston-based Professional Services Group in 1980. Through this subsidiary, renamed as Aqua Alliance upon a merger with the Vivendi-owned engineering firm of Metcalf & Eddy in 1998, Vivendi manages approximately 200 wastewater and 170 water facilities in the U.S., including contracts in Newark, New Jersey, Oklahoma City, and New Orleans (Rosenberg, 1999:34). In addition, Vivendi controls several major regional bottled water suppliers and has a minority ownership interest in both the Philadelphia Suburban Corporation and the Maine-based Consumers Water Company, two major water utilities that announced a merger in 1998. Similarly, Suez-Lyonnaise des Eaux, another major international company with origins in the French water industry, has partnered with United Water Resources, one of the three largest investor-owned U.S. water companies, to operate more than 25 North American systems, including contracts for a number of large utilities, such as Milwaukee, Indianapolis, Houston, and Atlanta.

Along with the French, several of the major privatized British water companies are also now active in the United States and elsewhere around the globe. For example, Anglian Water PLC is partnering with the largest U.S. investor-owned water company, American Water Works, Inc., to compete for operations and management contracts as American Anglian Environment Technologies, and is under contract from 1997 until at least 2002 to manage the Buffalo, New York water system, among others. Similarly, United Utilities, PLC has formed a joint venture with the Bechtel Corporation to compete for operations and management contracts as U.S. Water, while Severn Trent, PLC is active through several U.S. contract management firms.

Among U.S. companies, consolidation is also creating new, better financed, and more aggressive players in the water and wastewater industries. In addition to the merger of Philadelphia Suburban and Consumers Water, for example, 1998 also saw the announcement that the American Water Works Company, the nation's largest investor-owned water utility, had reached agreement in principle to acquire National Enterprises, Inc., a privately owned company with subsidiaries providing water service to more than 1.5 million people in four states. This $700 million transaction is the largest U.S. water utility acquisition to date.

Also, moving into and reconfiguring segments of the water and wastewater industries, suppliers of bottled water and point-of-use drinking water treatment devices have grown into major players in their own right. In 1997, bottled water alone represented a $3.9 billion U.S. industry, and was growing three times as fast as the national soft drink market. In itself, this trend is not a major threat to water utilities, given that household water use represents only about 10 percent of total U.S. water consumption, and even of this household water consumption, less than 2.5 percent is estimated to be for drinking and cooking (US EPA, 1997c). Nonetheless, the growth of bottled water sales is creating new opportunities for market differentiation, business development, and service integration. In February 1998, for example, United States Filter Corporation, a $3.5 billion provider of water and wastewater treatment systems, products, and contract management services announced that it was acquiring Culligan Water Technologies, the leading national brand in residential water treatment systems and one of the five largest U.S. bottled water companies. At the same time, the cities of Houston, Memphis, Kansas City, Missouri, and some other municipal water utilities are now beginning to directly market their own bottled water, and there is increasing talk of water utilities offering customers a "market basket" of water-related devices.

Competitive Alternatives

As the U.S. water and wastewater industries become increasingly competitive, different strategies for privatization and restructuring are being used. The following broad categories encompass the major alternative approaches for injecting private sector strengths and competition into this municipally dominated sector.

Asset Sale Under this model, full ownership and operations of a facility or system are transferred to a private enterprise, converting or being folded into a regulated, investor-owned utility. Upon such a sale and transfer, the purchasing company not only pays a sales price to the former public sector owner, but there may also be additional terms and conditions (*e.g.*, limiting rate increases for some period, and/or establishing performance goals) that extend above and beyond the requirements of any general regulatory structure.

Design-Build-Operate (DBO); Build-Own-Operate-Transfer (BOOT) These approaches use a private company to bring a new facility on line. Under a long-term agreement, the private company typically will design, construct, and operate a major facility such as a treatment plant. Where ownership is involved for the private operator (BOOT), the company will also finance the project using private capital. Where the public sector retains ownership (DBO), public debt is used. Under the more limited Design-Build (or "turnkey") variation, a facility newly designed and constructed by a private company is turned over for operation by a public sector utility before it is put into service. At the opposite end of this spectrum, under the Build-Own-Operate (BOO) variation, the facility may remain under private operation and ownership even beyond the term of the initial agreement.

Contract Operations and Maintenance (O&M) Under this approach, the public sector retains ownership of its system, but contracts out operation to a private firm. In full-service contracting (also referred to as a "concession"), all functions are operated by the private vendor. Under narrower contracts, only one or more specific facilities (*e.g.*, a treatment plant) may be privately managed, while other facilities and/or functions (*e.g.*, billing, customer service) may remain under public-sector operation.

Other key distinctions include the degree of responsibility assigned for capital repairs and improvements, and whether or not the operator provides upfront and/or ongoing concession payments to the public sector owner.

Managed Competition In a managed competition, bids are solicited from potential contract operators and compared against a proposal for restructured operation by the public sector's managers and employees. While multiple selection factors may be considered in such managed competitions, this formalized public versus private competition is typically decided primarily on the basis of cost.

Outsourcing In this variation of contract O&M, the non-core functions of a utility are emphasized. As with outsourcing in the private sector, this strategy is generally intended to help an organization focus on its core business functions (*e.g.*, treatment plant operation, or pipeline maintenance) by using a specialized firm to manage a function in which the utility is not as expert or proficient, and/or cannot maximize efficiency and economies of scale (*e.g.*, building and grounds maintenance, billing and collections, meter reading, laboratory services).

Internal Improvement Programs In response to increasing private sector competition, many public sector utilities are seeking to cut costs and improve practices, even without specifically engaging in any formalized competition or use of private firms.

U.S. TRENDS AND EXPERIENCES AT THE END OF THE TWENTIETH CENTURY

While all of the alternatives outlined above are becoming more and more widely used, U.S. experiences at the end of the twentieth century vary significantly from category to category. The following sections briefly outline these trends and highlights selected major cases.

Asset Sales

In the 1990s, asset sales in the U.S. water and wastewater industries have been relatively rare. Most typically, such asset sales have occurred where smaller water systems could be sold to existing, neighboring investor-owned water utilities that offered economies of scale and more sophisticated management. For example, in 1996 alone, United Water Resources acquired two small New Jersey systems to gain approximately 40,000 new customers, Consumers Water Company purchased three local systems, and the Philadelphia Suburban Corporation acquired eight systems to add over 17,000 new customers (Price Waterhouse, 1997). In the wastewater industry, the first sale of a federally subsidized municipal wastewater facility—the Franklin Area Wastewater Plant in southwestern Ohio—took place in 1995, and has not been widely replicated.

A key factor in the limited sale of water and wastewater facilities in the second half of the twentieth century has been the structure of the U.S. tax code, which has implicitly discouraged private ownership in these infrastructure-intensive industries. For many years, significant federal construction grants were available to public utilities, most notably for wastewater treatment facilities built to comply with the Clean Water Act in the late 1970s and 1980s. Not only did such grant programs encourage continued public sector leadership in addressing the environmental goals of that era, but they also made subsequent ownership changes impracticable. Until 1992, municipalities that had received such federal grants were required to make 100 percent repayment upon sale to a private operator. In 1992, however, President George Bush's Executive Order (EO) 12,803 on Infrastructure Privatization gave municipalities new flexibility to sell public infrastructure while only being required to repay any remaining "undepreciated" portion of the grant. Also, in response to EO 12,803, the EPA announced its intent to allow private operators to receive federal assistance. In conjunction with the decreasing availability and significance of such subsidies in the 1990s, this policy change has served to shift the playing field much closer to a level position. Building on these initiatives, President Clinton's EO 12,893 of 1994 on Principles for Federal Infrastructure Investments further encouraged private sector involvement and directed federal agencies to minimize legal and regulatory barriers.

Even with these Executive Orders, however, significant structural barriers remain to the widespread sale of water and wastewater systems. First and foremost, municipal debt continues to be tax-exempt, while privately issued debt remains taxable. As a consequence, public sector entities generally enjoy a lower cost of borrowing, which can lead in turn to significantly lower debt service costs, particularly given the highly capital-intensive nature of these industries.

Beyond this financial consideration, some other government regulations still apply differently to public and private sector utilities. Probably the most significant example involves the environmental regulations applied to wastewater treatment facilities. For a publicly owned treatment works (POTW), discharge standards for the facility and any industrial dischargers to the facility are governed by the permit standards of the National Pollutant Discharge Elimination System (NPDES). A privately owned facility and any industrial dischargers to it, however, could be deemed to be governed by the requirements of the Resource Conservation and Recovery Act (RCRA). According to the EPA, in such a situation, "higher treatment costs may occur if the wastewater treatment facility is designated as a RCRA hazardous waste treatment, storage, or disposal facility." In fact, to avoid just this potentiality, even the much publicized Franklin, Ohio wastewater plant sale featured continued, partial, public ownership under a lease arrangement to retain POTW status (US EPA, 1998c).

In addition, community concerns regarding control over water and wastewater rates and systems have also inhibited the spread of asset sales. Where investor-owned utilities do exist in the U.S., they continue to receive rate hikes under traditional rate of return regulation, ensuring relative stability, but not creating the incentives for increased efficiency sought by the British "price cap" model or the deregulating models of the U.S. energy and telecommunications industries. Beyond whatever near-term restrictions on rate increases may be negotiated as a condition of sale, a community selling its system gives up any local political involvement in rate setting. Along parallel lines, an asset sale involves less long-term local control over such matters as quality and supply of services that involve essential public goods.

For some small communities that happen to be near to major, existing, investor-owned utilities, however, the economies of scale afforded by such larger systems may overwhelm any drawbacks. In addition, other communities may be motivated by the one-time influx of capital that such a major asset sale may bring. For example, after paying off its system debt, Franklin, Ohio's $6.8 million sale generated more than $1 million in proceeds for the county and municipal governments served by the plant (Arrandale, 1995). Similarly, in 1998, the Mayor of Birmingham, Alabama proposed the sale of that city's water and sewer system primarily to capitalize a school construction trust fund, following voter disapproval of a sales tax increase targeted for school improvements (McEntee, 1998). Perhaps exemplifying the difficulty of such a major sale, however, Birmingham voters rejected this proposal in a November 1998 referendum.

Although continued incremental growth in asset sales is likely among communities with small systems adjacent to large and sophisticated investor-owned utilities, widespread private ownership remains many decades away. While some regions, such as Pennsylvania and New Jersey, have begun to see significant use of this strategy, most areas of the nation do not yet have private water or wastewater utilities large enough to pursue this approach. In fact, in certain regions, the era of municipalizing small, poorly coordinated private water utilities is still very much alive. In early 1999, for example, Florida's Brevard, Polk, and Sarasota counties approved an interlocal agreement to form a new Government Utility Authority intended to buy several small water and sewer systems from a private real estate firm in order to strengthen regional water resources management (McEntee, 1999). Absent dramatic government intervention to overhaul the tax code, modify environmental regulations, and generally restructure the market, asset sales face far greater obstacles to implementation than most other competitiveness tools.

Design-Build-Operate; Build-Own-Operate-Transfer

In the water and wastewater industries, most design and construction of facilities have long been performed by private contractors—albeit with municipal direction, coordination, and, upon completion of the project, operation. Typically, a municipal project management group will solicit a private architectural or engineering firm to design a facility and develop bid specifications. Next, a private construction company will be selected based on submission of the lowest qualified bid to meet these specifications. In turn, this construction company (often under the direction of a separately contracted construction manager) will build the facility, turning it over to the public sector utility's employees when ready for operation.

While this approach has successfully brought most of the nation's water and wastewater infrastructure into service, the process has also long been subject to criticism. In this system, project delivery is dependent on multiple, poorly integrated players, who tend to be focused more on their own "deliverable" rather than on the ultimate project outcome. There are rarely any mechanisms or incentives for the designer and the builder to collaborate in order to cut costs. Performance risk that the facility will fail to operate as intended is primarily borne by the utility. Moreover, low-bid construction exacerbates this risk, as minimally qualified but sub-par contractors may underbid more reliable construction firms. Finally, the newest facilities often involve the use of newer treatment technologies, and it can be difficult for some public utility personnel to adapt to the greater demands and complexity of operation (Westerhoff, et al., 1998).

With a DBO approach, in contrast, a single contractor is hired to design, construct, and operate a facility. This approach gives control over a project to a single entity, and focuses them on long-term performance of the facility. As a result of this more interactive and integrated process, some analysts estimate potential savings of 15 to 25 percent of construction costs and 20 to 40 percent of operating costs (Dysard and Callahan, 1998: 62).

Increasingly common in developing nations, a leading example of the DBO approach in the U.S. water industry is the Seattle Public Utilities 1997 Tolt River project, further detailed later in this book. In this initiative, Seattle competitively selected a private contractor to design, build, and provide 25 years of operation for a new, 120 million-gallon-per-day filtration/ozonation water treatment plant. Pursuant to the contract structure negotiated, Seattle Public Utilities provides project financing, retains ownership, and assumes liability for future capital improvements. At a cost of $101 million ($56 million for design and construction and $36 million for 25 years of operation), Seattle estimates that it has saved 40 percent over its estimated total cost of $171 million had it used conventional design, low-bid construction, and city operations (U.S. Conference of Mayors, 1997).

In the BOOT approach, the primary change from the DBO model is that the private sector finances the project under BOOT. Generally, this approach is not advantageous under the current U.S. tax code, given the lower interest rates available to public sector because of the tax-exempt status of municipal debt. When a public entity's debt capacity is constrained, however, BOOT does provide an alternative financing option.

Contract Operations and Maintenance

In the United States, the strongest privatization trend of the 1990s has been in O&M contracting. Perhaps reflecting the widespread involvement of French investors and parent companies among the major contract operators, this rapidly evolving model closely parallels the competitive system of France. At the same time, however, this practice has clearly been spurred on by favorable changes in the U.S. tax code.

Following substantial lobbying from the United States Conference of Mayors, the Department of the Treasury issued new tax regulations in 1997 supporting long-term contracting for operations and management of water and wastewater facilities. Under previous rules, if a facility was under contracted operation for more than five years, it was deemed to be for "private use" and ineligible for tax-exempt capital financing. Under the 1997 modifications, however, contracts of up to 20 years are now permissible without affecting the tax-exempt status of a utility's debt. With longer term contracting now rendered practicable, necessary capital investments can be incorporated into O&M agreements and amortized over a period that makes such agreements more cost-competitive.

Within this general category, contract lengths and provisions have varied significantly. In the water industry, for example, Buffalo, New York adopted a five-year contract (with a five-year renewal option) in 1997 with American Anglian Environment Technologies for full-system operation. Buffalo's contract, which is projected to save $23 million over the first five years (over 30 percent of pre-contract operating costs) (Kinnander, 1998), guarantees no layoffs and continued civil service status for the municipal employees of the system (U.S. Conference of Mayors, 1997). In the wastewater industry, Milwaukee reached a 10-year agreement in 1998 to serve 1.2 million metropolitan customers, projected to save $145 million over its term (Kinnander, 1998). In 1996, Indianapolis, which had entered into a five-year O&M contract for wastewater treatment plant operation in 1994, extended contract operation to its storm water and wastewater collection system (Reason, 1997). Building on estimated savings of 40 percent in operating costs (totaling $22 million) over the first two years of its wastewater treatment contract, Indianapolis projected additional savings of over 25 percent ($13 million over five years) for collection system management. For this initiative, the city selected a team made up of its private treatment plant operator and its local public employee union. Finally, late in 1998, Atlanta gained widespread attention by reaching a 20-year agreement to manage its water system, the largest yet operated under contract. Perhaps reflecting the high degree of private sector interest in such opportunities, Atlanta required interested vendors to pay a nonrefundable $50,000 prequalification fee just to participate in the bidding, and five bidders posted this sum (Perlman, 1998).

Some longer-term privatization initiatives have also been noteworthy for the payment of cash upfront to the municipality. In Cranston, Rhode Island, for example, an existing short-term O&M contract was rebid in 1996 as a 25-year lease that included $48 million in upfront cash for the city and significant capital improvements. As part of this Cranston transaction, current federal regulations permitted the EPA to forgive repayment of over $5 million in undepreciated grant funds (U.S. Conference of Mayors, 1997). In Hawthorne, California, the first long-term (15-year) lease for an existing system in the water industry was reached in 1996, featuring an upfront payment to the city of $6.5 million and $100,000 annual lease payments (Reason, 1997). In North Brunswick, New Jersey, a 20-year franchise was also negotiated in 1996 for both the water and wastewater systems of the township. This agreement included an upfront payment of $30 million, $24 million of which was used to retire existing system debt, and also featured concession fees throughout the contract term (Reason, 1997).

For cash-strapped municipalities, such substantial upfront payments clearly are often attractive. As pointed out in a recent draft of the EPA Guidance on the Privatization of Federally Funded Wastewater Treatment Facilities, however, such upfront payments do not come without a price. Rather, "any payment a local government receives from the sale or lease of a wastewater infrastructure asset is like a loan from the buyer or lessee, which must be repaid with interest by the wastewater users in the form of additional user fees." (US EPA, 1998c). In some cases, the private operator may achieve efficiencies that enable user fees to remain stable (or even decrease) in spite of the cost of this "loan." Even in such cases, however, user fees could have been lowered even further had such a loan not been taken out upfront. Therefore, while such practices may make good sense in the context of an individual community's circumstances, it is more accurate to think of such upfront payments as more of a potential financing tool than as a windfall.

Whether or not such upfront payments are part of the package, however, it is clear that many communities are benefiting financially from O&M contracting.

Moreover, to date, the service quality and environmental records of such arrangements generally have been strong. Absent further change to the current tax code and regulatory structures that present few barriers to such arrangements (particularly relative to asset sale and full private ownership), it is likely that O&M contracting will continue to be the fastest growing strategy for communities privatizing their water and wastewater systems.

Managed Competition

Paralleling the growth of O&M contracting, a strong trend of managed competition has also emerged. Under managed competition, public sector utility managers and employees are encouraged to develop their own restructuring plan and bid for continued operation alongside bids from private operators. While varying accounting methods, budgeting practices, and tax considerations make it difficult to judge such public versus private competitions, there is little question that such processes have helped to spark major cost reductions among utilities that have opted to remain under public operation after their own staff "won" the bid.

In reaction to this trend, some proponents of privatization have challenged the fairness and legitimacy of managed competitions, particularly where public sector bidders do not bear the risk of overruns or failure, and where no allowances are made for the taxes paid by private firms. Such privatization advocates argue that such processes only serve to delay and discourage the industry's transition to a competitive private market that will ultimately prove more cost-effective (Eggers, 1997). In response, proponents of managed competition argue that such processes simply add another layer of beneficial competition that helps to ensure that a municipality receives fair bids from potential private operators. In addition, such managed competitions are often touted as only fair to the people already employed in operating public facilities. At the Charlotte-Mecklenburg Utility Department, for example, a city team developed a proposal to cut costs by nearly 50 percent for two facilities. This public sector bid beat its closest private competitor by 19 percent, resulting in a five-year operating plan projected to save a total of $4.2 million (Eggers, 1997; Westerhoff, et al., 1998). In this book, public sector managers from both Charlotte-Mecklenburg and the Philadelphia Water Department detail their experiences with such initiatives.

Outsourcing

Not all governmental contracting is driven by the pursuit of private sector efficiencies and incentives. Rather, governments and utilities may also contract for many of the same reasons that have made outsourcing a major trend within corporate America in the 1980s and 1990s. By contracting with a specialized firm for specific, non-core functions, an enterprise can leverage their partners' expertise and/or economies of scale, while increasing their own focus on core functions that they may do best. According to one 1995 survey, many of the services most commonly contracted out by local governments are often the same type of non-core functions that many private sector companies have outsourced, including janitorial services (70 percent), building maintenance (42 percent), and security (40 percent) (Wessel, 1995: R8).

In addition to such ancillary functions, water and wastewater utilities may also look at secondary functions such as billing, call center management, and meter reading as potential candidates for outsourcing. In 1997, for example, the Philadelphia Water Department outsourced the installation of a citywide automatic meter reading (AMR) system, as well as long-term operation and maintenance of the system, under a 20-year agreement with a private vendor. Through this arrangement, the city was able to implement a new technology projected to achieve net present value savings of more than $30 million over the 20-year contract term, while placing performance risk on the shoulders of its vendor. No less important, this initiative is helping to sharpen the focus of the utility's revenue collection staff by reducing the size and scope of their direct operations (Blair, et al., 1998).

Internal Improvement Programs

Even among utilities where little or no contracting or managed competition has yet taken place, the effects of privatization within the industry are often evident. As cited above, approximately nine out of ten water and wastewater utility mangers self-report that they have competitiveness programs underway. Among leading industry associations, benchmarking efforts and competitiveness initiatives have also increased steadily. In addition to the "Thinking Getting Staying Competitive" workshops already noted, the largest professional organization in the drinking water industry, the AWWA, is promoting a major self-assessment and peer review program, Qualserve, to promote best management practices. Moreover, multiple engineering and other specialized firms are offering consulting services to public utilities that profess to assist in the implementation of the same strategies that a contract operator would use if the utility were to be privatized. As the generally pro-privatization Reason Foundation reported in a 1998 update on water and wastewater trends, "one of the private sector's most effective advantages, the ability to assimilate technologies and cost-effectively employ expertise at multiple facilities, may be shrinking as a result of these cooperative public sector efforts." (Reason, 1998).

COMPETITION AND DEREGULATION

Another spark for competition has been provided by deregulation throughout other U.S. utilities, particularly the wave of restructuring among electric and gas utilities in the late 1990s. As regulatory change forces competition among these energy utilities, and thereby cost cutting, the water and wastewater industries are being forced to reexamine long-held assumptions about their own status. Among energy utilities, previously integrated services of production, distribution, and marketing are typically "unbundled" to create new areas for competition. While distribution must generally remain a regulated monopoly, structured competition enables multiple firms to compete as either producers and/or retail marketers. In the water and wastewater industries, however, "wheeling" of the product across large distances is not as technologically feasible as in the energy industries. Power costs for pumping through gravity-flow pipelines can be high, drinking water quality can degrade over time and distance, and water from different sources may not react well upon mixing. Similarly, with wastewater, longer distances can increase the risk of flows going septic.

Nonetheless, even though water and wastewater do not travel as well as gas or electrons, the "unbundled" energy utility model may have some applicability. Within certain regions, both conservation successes and demographic shifts from city to suburb have left larger systems with reserve treatment capacity that could provide more cost-effective "production" than the development or upgrade of nearby, smaller plants. Under a different regulatory structure, such economies of scale might be better captured through the encouragement of inter-utility wholesale agreements.

As a result of deregulation, there is also speculation that mergers and strategic alliances may develop between newly competitive energy providers looking for customers and the water and wastewater utilities that still retain large, and largely captive, customer bases. Merging metering, billing, customer service, and back office operations might gain certain efficiencies across multiple utilities. In what may prove to be the first crest of this trend, Enron Corporation, one of the largest U.S. energy companies and a major player in ongoing electric and natural gas deregulation, announced in July 1998 that it is purchasing Wessex Water P.L.C., a British water and sewerage company, for $2.2 billion (AMWA, 1998). Initially, Enron's chairman has stated an intent to concentrate its expansion into the water industry to Europe, Latin America, and Asia. In part, this is attributed to restrictions under the U.S. Public Utility Holding Company Act that constrain expansion of energy firms into other sectors. With several bills in Congress proposing to repeal this Act, however, the potential for future realignment is arising. In New Jersey, for example, Public Service Electric & Gas Co., the state's largest electric utility, announced in 1998 that its non-regulated subsidiary, Energis Resources, would partner with United Water resources to offer an as-yet-unspecified "integrated utility solution" to New Jersey municipalities. Similarly, Philadelphia Suburban stated in 1999 that as many as six electric utilities had initiated discussions regarding potential marketing alliances and/or purchase of an ownership interest in the company (*Philadelphia Inquirer*, 1999).

SUMMARY

In both the U.S. water and wastewater industries, increasing competition, contracted operation, and, to a lesser extent, privatized ownership have become inescapable factors in service delivery. Moreover, the twin engines that have driven most of this change in recent years—tougher environmental regulatory requirements and the imperative to minimize the cost of achieving these mandated improvements—show no signs of winding down.

For the most part, such competition appears to be a positive force. Virtually all water and wastewater systems that have contracted out operations to date are reporting significant cost savings and/or performance improvements. Among public sector utilities that have retained operations in-house following managed competition, similar reports of improvement have been made. As in any business with a tendency toward monopoly, the injection of competitive forces is clearly sparking a newfound drive toward efficiency.

At the same time, however, it is important to remember that the infrastructure-intensive characteristics of these industries—the very factors that created their monopolistic tendencies in the first place—still remain. Absent wise market structuring and regulation, the potential always exists that privatization will simply replace traditional public monopolies with nothing more than newer private monopolies. Looking toward the long-term, the ultimate results of the current trends—however positive so far—have yet to be determined. While competitive private firms offer real advantages in terms of more flexible operating strategies, readier access to capital, and greater incentives for efficiency, these advantages are strong only when competition is actively sustained. In an era of greater contracted operation, it is essential that government retain or obtain sufficient independent expertise—both technical/operational and in management and contracting—to ensure that the public continues to receive good value.

Greater private operation also requires wise regulation to avoid, or at least to minimize, the extent of regulatory capture. Not only is this risk present with regard to ratemaking, but it is also no less a factor in the regulation of environmental and public health externalities. Government-owned utilities are sometimes criticized for opposing more stringent environmental and public health regulations, a common position given the pressure on elected and appointed public officials to minimize the rate increases driven by system improvements. For a private company, however, such system improvements could well appear as welcome opportunities for increasing profits. As a consequence, much as defense contractors have long had an incentive to lobby for greater military investment—whether or not truly needed or cost-effective—private water and wastewater contractors and utilities could have an incentive to exaggerate their need for system improvements. On the other hand, if the incentives provided to private system operators weigh too heavily toward cost containment, there is also a risk that dangerous corners may be cut.

For public policy analysts and academics, these complexities of the water and wastewater industries provide fertile territory for analysis, particularly with so many governance models already in place for comparison, including municipal departments, independent public authorities, and regulated investor-owned utilities. Moreover, with the current period of transition, a host of "natural experiments" are being created for evaluation. At the same time, this sometimes muddled period of transition bears equally careful study and attention for political decision makers, professional utility managers, and the general public. In the evolving water and wastewater industries, transition is giving rise to both the opportunities and the risk that are always inherent in change.

REFERENCES

American Water Works Association (AWWA) (1999). *Waterweek*, AWWA, January 15.

Arrandale, Tom (1995). "A Guide To Clean Water." *Governing*, December.

Association of Metropolitan Sewerage Agencies (AMSA) and Association of Metropolitan Water Agencies (AMWA) (1998). *Thinking, Getting, Staying Competitive: A Public Sector Handbook.* AMSA/AMWA, Washington, DC.

Association of Metropolitan Water Agencies (AMWA) (1998). *Federal Water Review.* Washington, DC, June–July, p. 3.

Blair, Dennis, Bolno, James, Hogan, Michael, and Nadol, Michael (1998). *Philadelphia Rolls Out AMR: Strategy and Implementation.* Annual Conference Proceedings, AWWA Management and Regulations. Dallas,Texas. AWWA.

Dysard, Joe, and Callahan, Neil (1998). "D/B/O Approach Can Save Local Dollars." *American City & County*, May.

Eggers, William D. (1997). "Privatization Killer?" *Infrastructure Finance*, September, pp. 23–25.

Eisenhardt, Paul, and Stocking, Andrew, with Reinhardt, William G. (1998). "Gradual Growth and 17 Long-Term Deals Mark 1997 Water & Wastewater Contract Operations." *Public Works Financing*, March.

Haarmeyer, David (1994). "Privatizing Infrastructure." *Journal AWWA*, March.

Hun, Tara (1998). "Forced Privatization of POTWs Becomes Law in Georgia," *Water Environment and Technology*, August.

Hyman, Leonard S., Hyman, Andrew S., Hyman, Robert C., Meehan, Edward P., Hempstead, James D., and Kochen, Jose V. (1998). *The Water Business*. Vienna, VA: Public Utilities Reports.

Jeffery, Jack (1994). "Privatization in England and Wales," *Journal AWWA*, March.

Kinnander, Ola (1998). "Privatization Trend Cuts Expenditures for Water Agencies," *The Bond Buyer*, April 15.

McEntee, Christopher (1998). "Birmingham Mayor Moves to Sell Water and Sewer Systems," *The Bond Buyer*, August 18.

McEntee, Christopher (1999). "Florida Counties Unite to Buy Private Utility," *The Bond Buyer*, February 4.

Neal, Kathy, Maloney, Patrick J., Marson, Jonas A., and Francis, Tamer E. (1996). *Restructuring America's Water Industry: Comparing Investor-Owned and Government-Owned Water Systems.* Los Angeles, California: Reason Foundation.

Perlman, Ellen (1998). "Contracting by Cross-Fertilization," *Governing*, August.

Philadelphia Inquirer (1999). "Power Companies Courting Phila. Suburban Corp.," February 3.

Price Waterhouse (1997). "The Utility Industry Survey of Developments and Financial Reporting."

Reason Public Policy Institute (1998). "Trends in Water and Wastewater Contracting, *Privatization Watch.* Los Angeles, CA, June.

Reason Public Policy Institute (1997). *Privatization 1997.* Los Angeles.

Rosenberg, Martin (1999). "The Europeans Are Coming!" *Utility Business*, January.

U.S. Bureau of the Census (1996). *State and Local Government Finances by Level of Government: 1994–95.* Washington, DC: U.S. Government Printing Office.

U.S. Conference of Mayors (1997). "Public/Private Partnerships in Municipal Water and Wastewater Systems—Case Studies of Selected Cities." Washington, DC, September.

U.S. Environmental Protection Agency (1997a). *Community Water System Survey.* Washington, DC.

———. (1997b). *Drinking Water Infrastructure Needs Survey.* Washington, DC, January.

———. (1997c). *Water on Tap.* Washington, DC, July.

———. (1998a). *1996 Clean Water Needs Survey Report to Congress.* Washington, DC.

———. (1998b). Drinking Water Priority Rulemaking: Microbial and Disinfection Byproduct Rules. Washington, DC, December.

———. (1998c). Draft Guidance on the Privatization of Federally Funded Wastewater Treatment Facilities. Washington, DC. April.

Wessel, David (1995). "The American Way?" *Wall Street Journal*, October 2.

Westerhoff, Garret P., Gale, Diana, Reiter, Paul D., Haskins, Scott A., and Gilbert, Jerome B. (1998). *The Changing Water Utility*. Denver, CO: American Water Works Association.

CHAPTER 2

Environmental Infrastructure: An Evolving Public-Private Partnership

David Haarmeyer

"Under privatization, the potential exists for the private sector to apply construction time and cost savings, efficient operations, and tax benefits to lower the costs of environmental services and share these savings with the public sector in the form of lower user fees." (Raftelis, 1989: 76).

Privatization starts from the premise that while the public sector is responsible for municipal services and infrastructure, it does not have to provide them. This same idea was put differently by E. S. Savas (1988) in pointing out that the task of government is to steer rather than row and that "privatization is a means of restoring government to its fundamental purpose and relieving it of the burdensome functions for which it is ill-suited."[1]

The economics of privatization rests on competition, specialization, and contractual guarantees. By moving provision of a public service from a monopoly supplier to a system of competitive suppliers, a municipality may secure an infrastructure or utility service (*i.e.*, design, build, and operate) more cost-effectively. The key point is that competition creates powerful incentives for providers to search out innovative and least-cost methods of delivering the required services. In addition, by relying on well-structured contracts, the public sector gains accountability through the assignment of legal obligations to the private sector, which guarantee regulatory compliance for a fixed-price budget.

Defining privatization broadly to mean the use of private sector expertise and resources to meet public needs, indicates that the concept may take several forms, and as a result, better fit the widely divergent needs of different communities.[2] Thus, not

David Haarmeyer is a consultant for Stone and Webster Management Consulting, Inc. in Boston, Mass.

only does the private sector pick up trash, trim trees, and provide janitorial services for city and municipal governments, it increasingly participates through "public-private partnerships" to either construct and operate, through turnkey contracts, or build-own-operate/build-operate-transfer (BOO/BOT) through concession contracts, public prisons, toll roads, and other capital-intensive infrastructure projects.[3] BOO/BOT contracts are unique in that they involve private capital participation. Water and wastewater treatment is one of the faster-growing areas of infrastructure privatization, both in the U.S. and globally.[4]

Private sector participation in some or all phases of the design, construction, and operation of water and wastewater treatment facilities has proved to offer a cost-effective way for local governments both to upgrade existing plants in order to meet more stringent environmental regulations, and to construct needed additional plants. The EPA has estimated the cost of compliance with the Clean Water Act (CWA) to be $137 billion nationwide and compliance with the Safe Drinking Water Act (SDWA) to be estimated at $138 billion (Cook and Rosseel, 1997). While the federal government has replaced its $65 billion wastewater construction grant funding program (1972–1986) with a state revolving loan program, the $3 billion in funding that this program makes available falls short of the $11 billion in total capital investment required on an annual basis.

Even though the private sector operates, on a contract basis, a very small percent of all wastewater facilities and water systems nationwide, this number is growing rapidly (by some estimates, by 30 percent per year).[5] More importantly, new contractual formats are evolving that enable communities to take even greater advantage of private sector resources and expertise. By increasing the scope of private sector responsibility, these formats enable additional operational improvements. Moreover, by securing private financial resources, government finances can be allocated to other more urgent uses.

As a result of legal restrictions and early federal financing, the growth in the private sector's involvement in providing environmental infrastructure services in the U.S. trails the experience of countries overseas. In France, for example, 70 percent of the population's drinking water services and approximately 40 percent of the wastewater services are privately supplied through long-term franchise agreements (Haarmeyer, 1994b). In 1989, Britain privatized its 10 regional water authorities through a public flotation that raised over $8 billion. Many of the private firms operating in these two countries are large, fully integrated companies that today compete to design, build, and operate environmental facilities in the United States and throughout the world. In stark contrast, because the water industry's assets in the U.S. are primarily owned and operated by the government, there are very few large world-class U.S. companies capable of competing in the international marketplace.

EARLY PRIVATE SECTOR PARTICIPATION IN ENVIRONMENTAL INFRASTRUCTURE

Governments have not always been the sole providers of infrastructure and public services, and when they have provided these capital investments, they have rarely been unassisted by the private sector. Beito (1993: 3) has pointed out that during the nineteenth century, "several cities, most notably New Orleans, Phoenix, and Atlantic City, granted franchises for private companies to operate their sewer systems."[6]

Much of the private sector's early involvement in providing roads, water, and sewer systems was displaced by expanding federal, state, and local programs. Regulatory programs, which limited rates private sector firms could charge and the combination of government subsidies for public investment and taxes on private investment, created a significant bias against private sector development and operation of public infrastructure.[7]

After taking over operational responsibilities, governments continued to rely on private firms to procure a wide variety of services and inputs, given the simple fact that the public sector's resources and expertise are limited. These services have generally been support-oriented—repair and servicing of equipment, consulting, and construction, and for a limited duration (*e.g.*, one year or less). What set this era apart was that facilities were fully owned and operated by the public sector and that the services provided by the private sector tended to be discrete and not integrated to minimize cost and create value.

This pattern holds for local government provision of water and wastewater treatment services, which can be broken down to a number of different services. Some of these services include: chemical supply, instrument and equipment supply and maintenance, laboratories for evaluation of water purity, design, construction, training, environmental consulting, and repair and maintenance. Especially for small municipal governments that operate single facilities, providing each of these services would be prohibitively expensive.

THE EVOLUTION TO FULL-CONTRACT OPERATIONS

Today, many municipalities still rely on private sector companies for the provision of these and additional services like sludge disposal, for example, which are mandated through tougher environmental regulations. What is noteworthy is that many of the private firms, especially engineering firms, that in the past provided a narrow range of services to municipalities, today are providing a wide range of integrated services—including operations—under one contractual agreement. Under an operations and maintenance (O&M) contract or full-contract operations, a private contractor agrees to operate and maintain a public facility so that the treated water or wastewater complies with federal environmental quality regulations.

O&M contracts essentially stipulate that for a fixed annual fee, a private company will assume the risk associated with meeting water quality regulations. Contracts generally provide additional "teeth" in the form of requirements that private companies put up performance bonds, and are subject to both early termination and penalties for under performance. Through experience and expertise gained through operating a number of facilities, and a wide array of human, technical, and financial resources, a private company is able to cost-effectively manage commercial risks. O&M contracts provide an efficient method for a municipality to rapidly acquire management skills and technical expertise, which may be especially important in instances when a municipality is facing significant fines from the federal government for water pollution violations.[8]

A number of firms that originally provided water and wastewater design and engineering services are now providing O&M services. Metcalf & Eddy (M&E), CH2M Hill, and James M. Montgomery (JMM) are examples of successful consulting engineering companies that formed new business units to focus on the growing contract operations market. Parsons Corp., one of the world's largest engineering and construction companies, is an example of a company that went from initially providing only narrow engineering and construction services to offering O&M services through its subsidiary, Parsons Municipal Services, Inc.

The trend away from multiple, short-term contracts and towards single, long-term (five to 20 years), full-contract operations was a natural, cost-reducing and value-maximizing response by private-sector companies operating in a competitive marketplace. Important preconditions included: elimination of state and federal legal constraints, proven track record of private companies, and openness by municipal officials. Integrating related service responsibilities under one contract provides a private company with greater opportunity to eliminate waste and add value. Backed by a sound contract, full responsibility provides increased knowledge and flexibility to optimize operations. In addition, full-contract operations enable a government to economize on the transaction costs involved in negotiating a number of separate service agreements.

A chief way private contractors reduce cost and add value is through cost management, especially the chief costs of labor, electricity, and chemicals. As the largest cost component, labor is generally the focal point. Operating a number of plants, for example, allows a private contractor to spread costs more widely and ensure that each plant is staffed properly. Also, it is clear that the flexibility that comes from operating on a larger scale (often on a worldwide basis) allows private firms to offer job opportunities, specialization, advancement, and management assignments that generate work incentives and career potential far beyond those available from a public works department.

Operating a number of facilities also gives private firms the ability to rapidly gain experience, expertise, and develop innovative and efficient plant practices. Predictive and preventive maintenance programs, which ensure water quality standards are consistently met and can extend a plant's useful life, are a by-product of this competitive process and were rarely in place in municipally operated plants. Finally, multiple plant operation allows contractors to realize savings through bulk supply procurement and sharing of common inventory items.

DIVERSIFICATION AND INTERNATIONAL COMPETITION

The growing interest of many municipalities in contracting out the operations of their wastewater and water facilities has encouraged a number of new companies to enter the business. Some companies have entered the market by starting up their own firms, while others entered the market by buying existing environmental services companies. The different strategies of the British water companies are instructive. Severn Trent entered the U.S. market primarily through the acquisition of regional O&M companies. Anglian Water teamed with the largest investor-owned water utility, American Water Works Company. Northwest Water (now United Utilities) entered into a joint venture with U.S. engineering and construction giant Bechtel to go after both the U.S. and international water markets.

Teaming arrangements provide an attractive way to spread risks and combine resources to gain both minimum size and the necessary capabilities. As the French companies have demonstrated, engineering capabilities are important but not sufficient. In addition, a long operating experience is also necessary, as well as project development and financing experience. A fully integrated service portfolio is especially valuable in a highly competitive marketplace.

The French companies are the U.S. and international market leaders primarily because of their long operating experience (both in physically managing systems as well as in negotiating contracts), size, and broad capabilities. The U.S. O&M contract leader is Professional Service Group (PSG), which was originally a subsidiary of the investor-owned water company Philadelphia Suburban. In 1981, PSG was sold to the French company, Vivendi, the largest private water company in France, and also a shareholder of Philadelphia Suburban and Consumers Water Company (another U.S. water company), and Air & Water Technologies Inc. (M&E's parent company). Vivendi was founded in 1853 and is believed to be the largest and most experienced municipal services company in the world. Vivendi is highly diversified, providing, in addition to water services, garbage collection, parking management, cable TV, heating, and construction services.

Head-to-head with Vivendi in both the U.S. and global market is Suez-Lyonnaise des Eaux (LDE), France's second largest water/wastewater company. LDE, established in 1880, is also diversified in a wide range of urban services.[9] LDE owns a U.S. contract operations company, JMM Operational Services, and part of a U.S. investor-owned water utility, United Water. Until 1997, LDE and the U.S. engineering company James Montgomery jointly owned JMM Operational Services. LDE has gained prominence through its participation in two of the world's largest water privatization contracts, located in Buenos Aires and Manila. And like Vivendi, LDE owns equity in British water companies.

The evolving U.S. model for public-private participation has its foundations in France. From the nineteenth century, municipalities in France have relied on three general types of contractual agreements for procuring water/wastewater services. Under a "concession," a private company finances, builds, and operates a municipal facility. Unless negotiated, the lowest winning bid sets what a firm will charge for water service, with the usual contract length of 30 years. In *affermage* (similar to leases) contracts, the municipality is responsible for financing capital investment and contracts generally last no longer than 12 years. In both types of agreements, the operator is remunerated directly by the customer. A "management" contract, the third type, is basically a service contract in which the municipality retains responsibility for billing customers and remunerating the contractor.

Deschamps (1986) and Drouet (1990) note that one of the key attributes of the French system of franchise contracts is its ability to adapt to various local and technical conditions. In addition, as explained by Drouet (1990: 5), the contract formats for water and wastewater management are applicable to other urban services:

> *[T]hese services share a common feature, which is serving urban users through a network system. Similar know how can be applied in customer service, to operate systems. This largely explains the similar priorities chosen by the Lyonnaise and Generale in their diversification moves and widens their competing turf to most urban services. Even funeral services are included. Only gas, electricity, and telephone services, run by national monopolies, remain out of their reach thus far.*

The long history of significant involvement by the private sector in French municipal water/wastewater supply is a key reason why French water companies are internationally recognized as the leading innovators and management experts in the industry and have a large and growing share of the world market. By contrast, U.S. private and investor-owned water companies provide less than 20 percent of the population with water services and have been much less successful in expanding internationally, as well as competing domestically for contract operations.

There are two important distinctions between the U.S. contract operations market and the investor-owned utility market. First, in the contract operations market, companies compete vigorously for the right to manage plant and system operations. In contrast, investor-owned utilities, like municipal utilities that do not contract out their services, are not exposed to this form of competition (although they are exposed to capital market competition). Second, contract operators manage facilities under a fixed-price contract and do not face rate-of-return regulation, while investor-owned utilities do. Hanke and Walters (1987) suggest that rate regulation, which governs U.S. investor-owned water companies (but not French water companies), provides little incentive to innovate and operate efficiently.

NEW PRIVATIZATION: THE EMERGENCE OF A TREND[10]

Starting in the early 1980s, the role of the private sector in wastewater infrastructure expanded from operations and maintenance service provision to include the development and ownership of facilities. At the same time, the demand for private O&M services accelerated. Four factors were particularly important in explaining this trend:

- the Reagan Administration's New Federalism policy that shifted responsibility for financing wastewater treatment facilities from the federal government to states and municipalities;
- stringent environmental standards governing water (*e.g.*, Safe Drinking Water Act of 1974) and sewers (*e.g.*, the CWA);
- tax law changes (*e.g.*, the Economic Recovery Act of 1981) encouraging private capital investment in public environmental projects; and
- growing experience of both private sector and local officials in designing and negotiating contracts.

Thus, while on the one hand the federal government was moving to phase out its construction grant program, which paid between 55 and 75 percent of a plant's total project costs, through the tax code the government increased incentives for private investors to direct capital into these facilities. Before the 1981 Tax Act, the private sector was at a significant tax disadvantage for owning public facilities as a result of local governments not having to pay taxes while also having access to tax-exempt financing (though not to investment incentives offered through the tax code). Under the new tax law, corporate taxes were reduced and private firms could use tax-exempt financing, and accelerated depreciation, and take investment tax credits (ITC).[11]

Particularly attractive to many communities with limited capacity for carrying debt was the ability to finance a project with tax-exempt financing in the form of industrial development bonds (IDBs). By issuing IDBs, which are backed by the revenues of the project, a municipality can retain its debt capacity for other essential services. IDBs were especially popular for communities that were experiencing rapid population growth and that needed additional infrastructure facilities but did not have the income base to support large debt obligations.

As a result of the new financial and regulatory pressures on municipalities and tax law changes, new forms of privatization transactions became attractive. Table 2–1 shows the configurations of public and private responsibility under different forms of project delivery. In terms of this chapter, the two most important are turnkey and full-service contracts. In a turnkey contract, a private partner designs, constructs, and operates (but does not finance) a facility that the government owns. In a full-service contract, or full-privatization contract, a private firm finances, designs, builds, owns, and operates the facility. Because this form of privatization (and lease-back arrangements) involves private equity participation, it is highly sensitive to the availability of tax incentives as created under the 1981 act. In both cases, the private partner generally provides wholesale services to the municipality, who is the retailer.

Ensuing tax legislation in 1986 and 1987 eliminated many of the early tax law's incentives for privatization. The 1986 Tax Reform Act placed a statewide cap on IDBs. The Deficit Reduction Act of 1987 repealed the ITC and significantly reduced the use of accelerated appreciation. While tax incentives in the form of interest and depreciation-expense deductions remained, the overall impact of the tax changes was to substantially increase the tax cost of privatization and generally halt private ownership of wastewater facilities. Holcombe (1990) estimated that after the 1986 changes, the tax cost of privatization was 30 percent of the initial cost of a facility.

TABLE 2–1. DIVISION OF RESPONSIBILITIES OF PUBLIC-PRIVATE PARTNERSHIPS

Project Component	Contract Services	Turnkey Facility	Developer Financing	Privatization	Merchant Facility
Decision	Public	Public	Public	Public	Private
Design	Public	Private	Either	Private	Private
Financing	Public	Private	Private	Private	Private
Ownership	Public	Public	Either	Private	Private
O&M	Private	Public	Either	Private	Private

Source: U.S. Environmental Protection Agency (1989: 6)

During the pre-1986 tax reform period, there were approximately a dozen full privatization transactions where the private sector took an equity interest in municipal wastewater treatment facilities. Since 1986, there have been only a few of these projects, by some accounts only two.[12] Below, two of the earliest privatization agreements are discussed in order to show how these transactions were structured, what motivated them, and what their results were. A later (post-1986) transaction involving a turnkey project is also discussed.

Chandler, Arizona[13]

As a result of rapid population growth, in mid-1983, Chandler, Arizona (a suburb of Phoenix, population 75,000) began considering options for constructing a new five million gallon per day (mgd) wastewater treatment and reclamation facility. To pay for its own financing and operation of a new plant, the city would have had to raise user fees from $3.50 per month to $18.50. The city was reluctant to wait for federal grant assistance and wanted to avoid using its bonding capacity.

The Pasadena, California-based firm Parsons Municipal Services Inc. (PMSI) proposed to finance, build, own, and operate the Chandler plant and keep monthly payments down to only $8.85. To finance the $22.2 million facility, Parsons proposed arranging the sale of 25-year, floating-rate industrial development bonds, backed by Parsons Corp., the parent company of PMSI. Malcolm Pirnie, Inc. was selected to design the new tertiary wastewater treatment plant, which would provide treated effluent for irrigation of a golf course, agricultural land, and residential areas.

The City of Chandler and Parsons negotiated a 25-year service agreement that required Chandler to pay a service fee to PMSI and Parsons Corporation guaranteeing to accept sewage from the city. As a result of a state law prohibiting municipalities from entering into long-term contracts, the actual contract is a one-year service agreement with provisions for the city to automatically renew the contract annually for a total of 25 years. Parsons also agreed to return treated, irrigation-quality water to the city. The facility, which came online in 1985, was $1.5 million below budget. The city estimated its savings at between $950,000 and $1.1 million annually.

Auburn, Alabama[14]

In 1983, officials of the City of Auburn (pop. 37,000) began looking at alternative financing options for two secondary wastewater treatment plants to replace the city's two primary treatment plants, which were 30 years old and operating beyond their capacity. The officials concluded that privatization was the best financial alternative and the quickest way to meet the need for improving wastewater treatment facilities. The city had applied for grant funding but was rated low on the state's priority list. Support for privatization came from both the State Department of Ecology and the regional EPA office, which believed that fines for noncompliance could be assessed against a private contractor easier than against the city.[15]

In December 1984, the city selected M&E to finance, design, build, own, and operate a 5.4-mgd and a 1.6-mgd plant under a 25-year contract.[16] M&E guaranteed the cost for operations and maintenance (subject only to changes in plant flow, inflation, and power costs) which were predetermined over the contract. The city remains responsible for setting and collecting rates from customers. To address legal uncertainties over this unique agreement, the Alabama Privatization Act was passed permitting privatization of water and wastewater facilities.

The privatization law also allowed the city to establish the Governmental Utility Services Corporation as a vehicle to issue industrial revenue bonds for M&E. As a result, Auburn's credit was not used in project financing. The bonds are backed by the project's revenues, while bond insurance and performance guarantees were provided by M&E. The Corporation issued $36,230,000 in floating/fixed rate, tax-exempt revenue bonds, the proceeds of which were loaned to a limited partnership M&E established. M&E's equity share, which was $10 million of this total, was used for bond redemption when the construction was completed.

At the project's onset, it was estimated that over the life of the 25-year contract (which corresponds to amortization term of the debt), privatization would generate savings of $25 million. While sewer rates doubled as a result of the project, under municipal ownership and operation it was estimated that they would have tripled. Under the contract, M&E is paid approximately $640,000 per year for O&M services. M&E hired the six public employees who worked at the retired facilities. Once the plants were up-and-running, they met effluent standards.

Mount Vernon, Illinois[17]

In February 1986, the Illinois Environmental Protection Agency (IEPA) placed Mount Vernon (pop. 17,000) on restricted status and issued a moratorium on sewer hookups as a result of failure by the city's wastewater plant to meet effluent standards. The plant was poorly designed and was operating at overcapacity, a problem that its private operator, St. Louis, Missouri-based Environmental Management Corporation (EMC), was unable to address.

Although an annual deficit of more than $300,000 existed in the city's Sewer Fund, the City Council refused to raise sewer rates. In August 1986, EMC made a proposal to design, build, own, and operate an upgraded and expanded facility at a guaranteed price of $6 million that would meet effluent quality for 20 years. A consultant's feasibility study had estimated that for the city to design and construct the facility it would cost $9 million. Consequently, the Council accepted EMC's proposal.

The proposal for private ownership was subsequently changed as a result of the change in the tax code and the possibility that the city would have to reimburse the federal government for EPA grants used in the older plant's construction. Under the agreement reached in June 1987, the city would be responsible for all financing and retain ownership of the plant while EMC would be responsible for the design, construction, and operations and maintenance for the 20-year life of the service contract.

The total financing cost of $6.64 million was raised through industrial development bonds issued by the City of Mount Vernon in the name of EMC. The tax-exempt bonds were rated AAA because of a letter of credit EMC secured from a Japanese bank. The city invoked a 0.5-percent sales tax to finance the interest cost. EMC is paid an annual fee based on two payments: one for a fixed-fee construction contract and the other for an O&M service contract.

In addition to assuming all effluent quality risks, guaranteeing construction costs, and offering a one-source liability umbrella (for design, construction, and operation), EMC posted an annually renewable performance bond. The plant was completed in June 1988, 12 months from when construction began, and for a 30 percent savings. As a result of the new plant, effluent standards exceeded EPA requirements and IEPA removed the plant from restricted status.

SOURCES OF PRIVATIZATION SAVINGS

The economic advantages of water and wastewater privatization can be traced to the private sector's superior ability to organize and coordinate human, capital, and financial resources. Competition for the right to provide a service drives private companies to continually search for cost-effective methods of organizing resources. The chief areas where private companies have been successful in achieving cost savings include: financing, design and construction, and operations and maintenance. Importantly, the competitive process of identifying and securing cost savings cannot be replicated. For example, in 1993, the consulting company hired to evaluate privatization options for the City of Indianapolis concluded after a detailed study that private sector operations could only reduce costs by 5 percent. The reality was that the winning bidder for the five-year, $72 million contract to manage the city's two advanced wastewater treatment plants brought down the operations costs by 35 percent.[18]

Financing

The advantages of private sector participation in financing new facilities were evident in the pre-1986 privatization transactions. The tax incentives provided by the 10 percent investment tax credit, five-year depreciation, and availability of IDBs were critical in lowering the capital costs through the investment of equity and reduction of debt. Importantly, by equalizing the cost of capital between private and public ownership, these tax incentives enabled governments to realize the construction and operational savings that flowed from an integrated design-build-own-operate approach.

In post-1986 privatization transactions, private companies as tax-paying entities can still take advantage of available tax benefits in the form of interest rate and depreciation expense deductions. As facility financiers, private firms can structure transactions utilizing innovative financing techniques, which by allowing them to bear some of the project's financial risks, can minimize interest rate costs. In the Auburn privatization, for example, the private company assumed the interest-rate risk for the floating rate bonds for the first 10 years of the project. In addition, the participation of an environmental service firm with a large parent company, which has a good credit rating, may also work to lower a project's finance costs.

Design and Construction

With privatization, needed facilities can generally be brought on line more expeditiously and cost-effectively than under traditional public ownership/operation approach. By being responsible for the design, construction, and operation of a facility, a private company has an incentive to secure cooperation and to ensure that the tasks are coordinated in the most cost-effective manner. In doing so, it can obtain savings in construction costs and long-term operation and maintenance costs.

In a comparative study of municipal wastewater treatment plants, Heilman and Johnson (1989) found that privatized plants were designed and constructed in almost one-third the time taken for grant-funded plants. In terms of plant cost-per-gallon of sewage treated, grant-funded plants tended to have higher average costs. Compared to the "centrally regulated but highly fragmented" process for designing and constructing grant-funded plants, the process for privatized plants was "highly integrated and coordinated."[19] Thus, the process for grant-funded plants "encouraged neither a rational commitment to efficiency nor an effective commitment to cooperation"[20] by the involved parties. By contrast, the parties involved in privatized plants "shared both economic interest and affective commitment to efficiency."[21]

As a facility owner and/or operator, a private contractor has both the opportunity and incentive to cost-effectively design and construct the facility. Not bound by administrative regulations that govern federal and state construction projects (for example, wage requirements of the Davis-Bacon Act), private companies have more flexibility to build facilities at a much faster rate than a government authority. For example, unlike a municipality, private companies can begin construction even before plans are complete. A private facility owner/operator will also typically choose plant equipment on the basis of their experience, not by what is required by public bidding laws. Finally, by being fully accountable for a facility's total project costs, a private company is less likely to design and build a "gold-plated" plant that looks better than it functions.

Operation and Maintenance

Lower operating and maintenance costs flow from having the same organization that designs, builds, and owns the facility, operate it. The integration of these tasks in one organization allows important knowledge to be shared on what works and what is needed. The key to how well a company succeeds in integrating these tasks and combining capital, labor, and knowledge is competition, or the ability of a local government to terminate the contract and find an alternative provider. Similarly, by being paid a fixed fee an owner/operator faces powerful incentives to closely scrutinize operating costs.

As mentioned earlier, private environmental services companies emphasize continuous training programs and preventive maintenance that enable companies to save operating and capital costs. Unlike the rigid civil service system of the public sector, private companies also have much more highly structured incentive systems for motivating employee performance. As more flexible organizations, private companies are generally less prone to labor/management conflicts. Finally, private companies can spread overhead costs over a network of facilities, which can lower unit costs of the service to individual communities and provide the company with a broad range of experience.

There have been very few studies comparing private and public operating costs of environmental facilities. Evidence that private firms have been successful in achieving cost savings is reflected in the large and growing number of O&M contracts, and the growing scope in terms of both function and duration of these contracts. Generally, municipalities initiate these forms of partnerships with the private sector because either a contractor offers to operate a facility at less cost than a municipality and/or a contractor can more rapidly address a compliance problem than a municipality can.

In sum, the sources of the private sector's advantages in developing and operating environmental infrastructure projects are competition and economies of scale organized under well-structured contracts. Competitive market forces encourage continuous cost reduction and innovation. A competitively selected firm that is responsible for the design, construction, finance, and operation of a treatment facility is forced to consider a facility's life-cycle costs. Additionally, under a well-structured contract, the firm has the incentive to reduce these costs while at the same time improving environmental quality.

BARRIERS TO PRIVATIZATION

Cost and Noncommercial Culture

The process of considering alternative, nonpublic-sector infrastructure delivery approaches may be costly, both from a community's perspective and a public manager's personal perspective. A community must put aside resources to examine privatization alternatives, debate its merits, and reach a consensus. To obtain sufficient public support for privatization in terms of public understanding and trust, a considerable public relations effort may be required.

More pivotal is the perspective of the public manager, who may be the focal point for initiating and implementing the process. For a public manager with little background or understanding of privatization, considerable time may have to be committed to learning about the process and informing other local government officials. Success or failure of the privatization process may result in a manager's loss of reputation and create disharmony with other public employees who see privatization as a threat. In the face of these costs, the public manager may gain little in the way of personal benefits by successfully designing and implementing a cost-effective infrastructure arrangement with the private sector.

Another considerable barrier to privatization concerns the noncommercial character of public-sector decision-making. Unlike the private sector, which generally uses a net present value analysis to properly weigh the short-term and long-term costs and benefits of a potential project, the public sector has a tendency to concentrate on the short-term costs and benefits. As these costs and benefits tend to reflect election year cycles (*e.g.*, political promises to constituencies), they can emphasize the political merits at the expense of the economic merits. As a result, projects with considerable up-front (political) costs and long-term benefits, like the private development of new infrastructure, may be judged as not viable.

Lack of Information on Cost and Competition

According to the results of the 1992 Apogee Research survey (p. 4), "many states have still not considered the wide range of potential privatization alternatives available to them." While less true today, this may be a good assessment for many smaller municipalities that might have fewer resources to spend on understanding and learning about privatization.

For state and local officials to take advantage of privatization to reduce the cost of service requires that they know how much they currently spend on providing the service and have some methodology for comparing this cost with private-sector delivery. Because it is often not required and requires some work in obtaining, in many cases this information is either not readily available or is inaccurate.

In the 1987 Mercer/Slavin national survey of the contracting practices of 120 cities, counties, and special districts, over 50 percent of the respondents reported that they had no methodology for comparing the cost of government versus the cost of contract service delivery. A positive outcome of the tougher fiscal climate facing local governments is that it encourages a closer look at budgets and service costs.

Perception About Loss of Control

Some municipalities may fear that turning over essential services to the private sector will diminish their control and authority over these services. As mentioned earlier, the government need not provide services itself but ensure that they are provided. Indeed, specialization and conservation of scarce public resources require government to concentrate on those tasks where it has a comparative advantage—enforcing contracts and regulation—not assuming commercial risks. Thus, while there may be a loss in day-to-day operational control, for example, in the detailed engineering and maintenance tasks, there need not be a loss of control in terms of ensuring that the broader objectives, like meeting environmental standards at a minimum cost, are accomplished.

A local government's control over a facility's operation is embodied in the service agreement, a legal contract that is capable of imposing high costs on the private contractor for nonperformance. The challenge for government is to design effective oversight programs and safeguards in the contractual agreement.[22] The disciplinary forces behind a contract can be many and significant. Five provisions stand out:

- performance guarantees;
- fixed-price guarantees;
- penalties for nonperformance;
- performance standards that incentivise contractor to go beyond minimum standards; and
- termination agreement.

These provisions are unique to the private sector (*i.e.*, they do not exist within the public sector). Behind these provision is the key disciplinary force-competition. This takes the form of competition for the contract and acts as an ongoing safeguard, given that the government has the right to terminate the contract at any time. In addition, a reputation for consistent, quality service is highly valuable and fragile in a competitive marketplace. Consequently, firms go to great lengths to maintain a good reputation. As Benjamin M. Rawls (1984: 9), at M&E noted, "A reputable professional-services firm cannot afford poor quality." Indeed, large national environmental service companies often guarantee the performance and obligations of their contract operating subsidiaries.

Finance and Tax Laws

One of the more important reasons why there are relatively few privately owned water and wastewater plants concerns the very different tax treatment for public and private organizations. While interest on corporate debt is taxed, for example, interest on debt issued by a government authority is not taxable. As a result, in terms of raising funds for infrastructure development, a strong bias exists for the project to be financed by the public sector with tax-exempt financing.

In a study on infrastructure privatization in New York State, Steckler and Payson (1992: 189) calculated that the tax-exempt status of government property, income, and debt gives the state and local governments a subsidy of over 40 percent compared to private firms. It is important to note that the subsidy is not free, but is paid for by taxpayers. In terms of tax revenue loss and inefficiencies, the subsidy cost could be significant. This is because by acting as a barrier to private sector participation in financing, the subsidy restricts competition and thus is likely to result in less innovation, less efficient practices, and a more costly service, while at the same time generating less tax revenue.

Federal and State Legal Restrictions

While the EPA, through its Public-Private Partnership (P3) Initiative, is promoting private sector provision of environmental services, the federal government is often a source of barriers to privatization. Besides creating tax disincentives to private development of environmental infrastructure, the federal government also imposes two other important restrictions to privatization.

Circular A-102 from the Office of Management and Budget requires that a municipal grant-financed facility be sold at "fair market value" and a pro-rata share rebated to the U.S. Treasury. Legal analysis (Giraudo, 1991) suggests that there may be leeway around this barrier if state and local governments terminate a federal grant unilaterally, but until this barrier is removed with certainty, it remains a problem. In addition, because privately owned facilities are ineligible for State Revolving Loan Funding (SRLF) support under the CWA, privatization is less financially attractive. Because facilities financed by SRLF must follow many of the same federal procurement requirements that existed under the construction grant program, the savings available through private turnkey projects are restricted.

A number of states do not have laws providing specific legal authority for municipalities to privatize their wastewater treatment facilities. In some states, the law limits municipal contracts with the private sector to one year. This obviously has a negative effect on continuity of service and raises the transaction costs involved in contracting out. Another barrier to privatization at the state level are laws requiring private ownership or operation of public utilities like water and wastewater facilities to be subject to rate regulation by public utility commissions. Especially when a privatization agreement involves financing based on an existing rate structure, unilateral PUC actions may create too much risk and uncertainty for bond purchasers.

CONCLUSION

In the U.S., much like in the rest of the world, the private sector's participation in the provision of environmental infrastructure services has run the gamut. In the early part of the century it provided full-utility service (operation and ownership) until crowded out by government regulation and cheaper finance. Much later, driven in great part by cost and quality pressures faced by local governments, private companies evolved from providing narrow services to a more fully integrated approach of design, build, and operation of facilities under long-term contracts.

The growth of privatization and the increasing scope of private participation are a function of the competition-driven comparative advantage private companies have in coordinating private resources to solve public problems as well as the elimination of legal restrictions that limit participation. Competition, organized through well-designed contracts, encourages private firms to continuously search for cost-reducing and quality-enhancing practices. The rapidly growing international private water industry is testimony of the vibrancy of competition to encourage companies to continually develop resources and expertise that will provide better environmental services for less money.

The billions of dollars of current needs for water and wastewater facilities, combined with the limited technical, economic, and financial resources of many state and local governments, suggests that privatization in the U.S. will continue to grow. Removal of legal and tax restrictions will enable private sector participation to continue to evolve and enable communities to better reap the advantages of competition.

NOTES

[1]Butler (1985: 1) also makes the same point about an "appropriate division of labor between policymakers and the private sector, allowing the former to concentrate on their central role, delegating to the latter the mechanical features of government." Examined this way, privatization indicates an economic reason to narrow the scope of government's mission, an outcome that is often considered to be driven solely by ideological forces.

[2]Understanding and debate about privatization suffers from the concept's multiple meanings; for example, the most narrow interpretation, which implies ownership by the private sector and a perceived loss of public control, raises a strong negative connotation. The advantage of preferring a broader definition is that it more accurately indicates the great variety of ways in which the private sector is capable of assisting the public sector in the delivery of public service or infrastructure. An especially good definition along these lines was provided by Apogee Research (1992: 1): "A variety of techniques and activities to promote more involvement of the private sector in providing traditional government or public services. It enables each party to do what it does best and results in a 'win-win' solution to providing public services."

[3]For surveys on privatization developments in the United States and abroad see the Reason Foundation's *Annual Privatization Reports*.

[4]See Haarmeyer and Mody (1997) for a detailed analysis of risk allocation and financial structuring issues involved in nine international water/wastewater projects with private capital participation.

[5]See Eisenhardt (1997) for survey of U.S. water and wastewater contract operations.

[6]Beito (1993) notes that the private systems were more prevalent in smaller and medium-sized cities and by the turn of the century, almost all cities with populations over 30,000 had publicly operated and owned sewer systems.

[7]The federal government's $65 billion program to subsidize construction of municipal wastewater treatment facilities from 1972–1986 is perhaps the best example of how government crowds out private sector participation. According to Taylor (1990), the construction grant program deterred communities from examining less economically and environmentally costly alternative wastewater treatment technologies.

[8]According to *PW Financing* (May 1998), "Pressures of $100,000 per day fines for water pollution violations and $900 million in urgently needed repairs has forced a political consensus that privatization is the right solution for Atlanta's utility operations."

[9]Like Vivendi, LDE has a number of urban waste collection contracts with municipalities in France and throughout Europe. In a 1989 *Financial Times* article, Jerome Monod, President of LDE, explained the link between water management and waste management: "If you distribute water in a large territory you must also ensure that this territory is clean. You are thus forced to take an interest in industrial and urban waste disposal and treatment."

[10]Heilman and Johnson (1987) describe "new" privatization as capital-intensive, in contrast to "old" privatization, which is labor-intensive. The former involves construction and ownership of facilities in addition to operation and maintenance.

[11]According to Randall Holcombe (1991: 30), who has studied the tax effects on wastewater privatization, "These advantages were never large enough to create a net tax subsidy for privatization, but they did substantially reduce the tax cost of privatization from 1981 to mid-1980s."

[12]H & W Management Science Consultants (1989).

[13]Most of this discussion was drawn from Fixler (1986) and U.S. Environmental Protection Agency (1989).

[14]Most of this discussion is drawn from WATER/Engineering & Management (1987), International City Management Association (1986), and U.S. Environmental Protection Agency (1989).

[15]Often believed to be a crucial factor in the success of privatization, the city had from the beginning an extensive program to generate public understanding and participation in the program. In addition to putting together a "blue ribbon" selection committee of citizens and Auburn University officials, the city held hearings, public meetings, and interviews.

[16]Before selecting M&E, the city had already had an engineering firm prepare plans and specifications for a new treatment plant. A critical decision was made to allow M&E to redesign the plant that they would operate. This would allow the firm not only to better control construction costs, but to be responsible for any operational problems arising from the plant's design.

[17]Much of this discussion is drawn from Bruzan (1989), U.S. Environmental Protection Agency (1989), and discussions with Environmental Management Corp. President Jerry King.

[18]See Haarmeyer (1994a) for more detailed discussion of the Indianapolis contract and reference to a Seattle engineering study that also significantly underestimated cost savings under competition.

[19]Heilman and Johnson (1989: 560).

[20]*Ibid.* p. 564.

[21]*Ibid.* p. 561.

[22]For discussion on O&M contract design with sample contractor evaluation form and request for proposal (RFP) for wastewater treatment facilities, see Sedwick (1984).

REFERENCES

Apogee Research, Inc. (1992). "State Government Privatization: An Updated Opinion Survey of State Governments on Their Use of Privatization." Sponsored by the National Association of State Comptrollers.

Beito, D. (1993). "From Privies to Boulevards: The Private Supply of Infrastructure in the U.S. During the Nineteenth Century." In Jenkins, J. and Sisk, D. E. (editors), *Development by Consent: The Voluntary Supply of Public Goods and Services*. ICS Press, pp. 23–48.

Bruzan, J. (1989). "The Mt. Vernon, Illinois Wastewater Treatment Project Completed." *The Privatization Review*, Spring.

Butler, S. (1985). "Privatizing Federal Spending: A Strategy to Eliminate the Deficit," Washington, D.C.: Heritage Foundation.

Cook, M. and Rosseel, K. (1997). "Securing Improvements in Water Quality: EPA and Water Infrastructure Financing." *Journal of Project Finance 3, No. 3*, Fall: pp 29–36.

Deschamps, J. (1986). "Privatization of Water Systems in France." *Journal of American Water Works Association*, February: pp. 34–40.

Drouet, D. (1990). "The French Water Industry in a Changing European Context." Paris, France: Recherche Developpement International.

Eisenhardt, P. (1997). "U.S. Market Approaches 1,000 Water and Wastewater Facilities." *PWFinancing* (January), pp. 24–28.

Fixler, P. (1986). *Privatization 1986: Annual Report on Privatization of Government Services*. Santa Monica, CA: Reason Foundation.

Giraudo, J. (1991). "Breaking Free of Federal Grant Restrictions: Making Infrastructure Privatization A Real Option." Reason Foundation, Policy Insight, No. 127, February.

Haarmeyer, D. (1994a). "Letting the Private Sector Work." Washington Times, January 13: p. A21.

———. (1994b). "Privatizing Infrastructure: Options for Municipal Water Systems." *Journal of American Water Works*, March: pp. 43–55.

Haarmeyer, D. and A. Mody (1997) *World Water Privatization: Managing Risks in Water and Sanitation*. Singapore: Financial Times Energy Asia Pacific.

Hanke, S. and S. Walters (1987). "Privatizing Waterworks." Hanke, S., editor, *Prospects for Privatization: Proceedings of the Academy of Political Science* 36, No. 3: pp. 104–113.

Heilman, J. and G. Johnson (1987). "Metapolicy Transition and Policy Implementation: New Federalism and Privatization." *Public Administration Review* 47 November/December: pp. 468–478.

———. (1989). "System Process in Capital-Intensive Privatization: A Comparative Case Study of Municipal Wastewater Treatment Works." *Policy Studies Review* 8, No. 3 Spring, pp. 549–572.

Holcombe, R. (1991). "Privatization of Municipal Wastewater Treatment." *Public Budgeting & Finance*, Fall.

———. (1990). "The Tax Cost of Privatization." *Southern Journal of Economics*, January.

H & W Management Science Consultants (1989). Privatization Market Forecast: Opportunities in Municipal Wastewater Treatment, May.

International City Management Association (1986). "Review of Wastewater Treatment Privatization." Clearinghouse Report No. 40030.

Lorenz & Company (1989). Update: Water Pollution Control Industry Outlook, March.

Mercer/Slavin, Inc. (1987). "Findings of a National Survey of Local Government Service Contracting Practices."

Moore, S. (1991). "Wastewater Privatization: On the Comeback?" *Partnership Focus*, March.

"Privatization is Working in Auburn, Alabama."(1987) *Water/Engineering & Management*, January.

PW Financing (1998). "Atlanta Puts Big-City Water into Play." (May): pp. 1–4.

Raftelis, G. (1989). *The Arthur Young Guide to Water and Wastewater Finance and Pricing*. Chelsea, MI: Lewis Publishers.

Rawls, B. (1984). "Privatization: Three Case Studies Demonstrating What We Have Learned." Arthur Young/ENR Privatization Seminar, September/October.

Reason Foundation (1992). *Annual Privatization Reports*. Los Angeles.

Savas, E.S. (1988). "Private Enterprise is Profitable Enterprise." *New York Times*, February 14.

Sedwick, J. (1984). "Contract Services for Wastewater Treatment." In *This Way Up: The Local Official's Handbook for Privatizing and Contracting Out*. R. Armington and W. Ellis, editors. Chicago, Il: Regnery Gateway.

Steckler, S. and L. Payson (1992). "Infrastructure." *Privatization for New York: Competing for a Better Future*. A Report of the New York State Advisory Commission on Privatization. E.S. Savas (editor), January.

Taylor, J. (1990). "Meeting the Challenge of Water Pollution: State Wastewater Recycling Strategies for the 90s." American Legislative Exchange Council, *Issue Analysis*, April.

U.S. Environmental Protection Agency, Administration and Resources Management (1989). *Public-Private Partnership Case Studies*, September.

CHAPTER 3

Environmental and Contracting Issues in Municipal Wastewater Treatment Outsourcing

Janine M. Landow-Esser
and Melissa A. Manuel

Outsourcing, or privatization, is a growing trend with municipalities under pressure to do more with less. Governments have sought to divest themselves of non-core and even core functions ranging from airport and prison management to tax collection (Reason, 1997). Wastewater treatment operations, traditionally the responsibility of the municipal public works department, are also privatization targets. Municipal wastewater treatment outsourcing arrangements can encompass retaining an operations consultant, long-term operations contracts, or outright sale of the treatment facility.

BACKGROUND

Factors Spurring Increased Interest In Municipal Wastewater Treatment Outsourcing

The growing interest in outsourcing of wastewater treatment operations can be traced, in part, to historic financing of municipal treatment plants. The late 1970s and early 1980s were a boom period for the construction and upgrading of municipal wastewater treatment plants. A strong federal construction grant program coupled with investments at the local level provided financing for new construction and improvements. In the ensuing 20 years, plants have aged and many are now in

Janine M. Landow-Esser is an environmental attorney and a partner and Melissa A. Manuel is an environmental attorney with Holleb & Coff in Chicago, Illinois.

need of significant capital improvement.[1] However, the Water Quality Act of 1987 phased out the construction grant program and established the State Revolving Fund (SRF) program, which, in turn, transferred responsibility for wastewater treatment facility financing to the states.[2] Unfortunately, funding available from the SRF has fallen short of funding needs.

At the same time funds were growing scarce, regulatory standards affecting treatment plants were becoming increasingly stringent, forcing municipalities to consider the addition of advanced wastewater treatment technologies in order to maintain compliance with environmental standards. Failure to comply with standards subjects the municipality to potential fines and penalties, adding additional pressure to upgrade the facilities. Finally, industrial and residential development may also cause pressure for expansion or other capital improvements to existing treatment plants.

Traditionally, municipalities would have financed capital improvements by issuing bonds, raising taxes, or increasing user fees. In today's political environment, government officials are called upon to provide better services with a smaller budget. Faced with the unpopular prospect of raising taxes or user fees, municipalities are increasingly looking toward outsourcing as a financing option.

Governmental Support of Outsourcing

The federal government has become a fairly strong proponent of outsourcing, recently outsourcing operation of NASA's space shuttle, as well as services associated with military base closings. In addition to its direct involvement in outsourcing, the federal government has removed some barriers to contracting with private parties, thus making outsourcing a more attractive option to local municipalities.

Executive Order (EO) 12,803 on Infrastructure Privatization, issued by President Bush in 1992, launched a privatization initiative at the federal level.[3] The order required federal agencies and executive departments to review their procedures affecting the management of state or locally owned federally financed infrastructure assets—including wastewater treatment facilities—and to modify those procedures to encourage privatization.[4] In response to that order, the United States Environmental Protection Agency (EPA) announced its intent to consider revising its regulations to allow non-publicly owned treatment works to be eligible for construction grants and loan programs.[5] In addition, the executive order specified the order in which proceeds from the sale of a federally funded asset would be distributed. Under EO 12,803, the sale price must be distributed first to repay the original cost of state and local government investment to date, then to repay the undepreciated portion of the federal investment, with the remaining proceeds going to the state and local government.[6] Thus, a private company need only offer to cover the municipality's original investment in order to make an outsourcing deal attractive.

President Clinton expanded on EO 12,803 in his 1994 EO 12,893 on Principles for Federal Infrastructure Investments.[7] President Clinton's Order instructed federal agencies to seek private sector participation in infrastructure investment, directing agencies to work with state and local governments to "minimize legal and regulatory barriers to private sector participation in the provision of infrastructure facilities and services."[8]

Many municipal outsourcing projects may be financed by issuing tax-exempt bonds. Financing such projects with tax-exempt bonds can generally save approximately 1.5 percent to 2 percent in interest costs. Bonds that constitute private activity bonds (*i.e.*, a form of tax-exempt bonds that meet a private payment and private security tests) must be allocated a volume cap by the appropriate state authority and must meet certain other requirements. A recent revision to Internal Revenue Service guidelines allows public agencies to enter into management contracts with private companies for periods of up to 20 years, an expansion from the previous five-year limit, without resulting in the bonds losing private activity bond status.

Types of Outsourcing Arrangements

In the wastewater treatment plant outsourcing context, there are three general types of outsourcing arrangements. First, a municipality may contract out some or all of its treatment plant operations. Under an operating contract, the municipality maintains ownership of the plant while the outsourced company operates the plant. In 1993, the City of Indianapolis, Indiana entered into a contract with the White River Environmental Partnership (WREP), a consortium of Indianapolis Water Company, Lyonnaise des Eaux and JMM Operational Services, pursuant to which WREP took over operation of the city's two wastewater treatment plants (Goldsmith, 1997). The City of Atlanta also recently announced plans to contract out operation of its largest wastewater treatment plant.[9]

Outsourcing can also involve outright sale of an existing treatment plant. Although President Bush's EO 12,803 should have smoothed the path for asset sales, to date, only one such asset sale has closed. In 1995, the Miami Water Conservancy District finalized a deal whereby it sold its Franklin, Ohio treatment plant to Wheelabrator EOS, Inc. for $6.8 million, the full market value of the plant. In addition to retiring the existing debt on the plant, the participating municipalities received at least $1.5 million as an up-front payment (*Public Works Financing*, 1994). The deal was extremely complex, involving the USEPA, the Office of Management and Budget, and the IRS, in addition to the contracting parties. Although this deal has apparently benefited the municipalities, it is not clear whether asset sales will become commonplace, considering the relative difficulty in structuring such a deal.

The third type of contractual arrangement is commonly known as BOT, for build, operate, and transfer. Under a BOT contract, the outsourcing company designs, builds, owns, and operates the wastewater treatment plant for a specified period. At the end of the contract, which can exceed 20 years, the plant is transferred back to the municipality.

Potential Benefits of Outsourcing

Outsourcing may be attractive to a municipality for several reasons. Successful outsourcing is based on the assumption that a private company skilled in the area of wastewater treatment can operate the system more efficiently, and at a lower cost, than the municipality. Typically, outsourcing is considered primarily as a financing alternative or in response to budgetary constraints. From a financing standpoint, outsourcing may allow a municipality to generate up-front revenues, which then can be used to retire the existing debt on the plant, to fund capital improvements, or to balance the budget. In addition, as the outsourced company will theoretically operate the treatment facility more efficiently, transferring operations to an outsourced company should result in lower user fees, or at least an avoidance of a rate increase.[10]

As wastewater treatment operations are their "core" business, outsourced companies may have more experienced personnel and better access to the latest technologies. In addition, internal information exchange within the outsourced company may allow the municipality to benefit from knowledge gained through operation of a broad spectrum of treatment plants. The ability to tap into this greater pool of knowledge may be particularly attractive to smaller municipalities, which cannot afford high-caliber personnel or cutting-edge technologies. From an environmental standpoint, more experienced personnel may also mean a better environmental compliance record.

Large outsourced companies may also benefit from economies of scale. A private company managing several treatment plants may get better pricing on chemicals, equipment, and supplies and pass such savings on to the municipality. Finally, the municipality may experience a reduction in personnel, resulting in payroll and benefits cost savings.

Possible Barriers to Outsourcing

Labor Issues While a reduction in force may be perceived as a benefit to an outsourcing municipality, the potential loss of jobs may be a significant labor issue. Local public officials are often concerned that cost savings associated with outsourcing will come from lower wages or a decreased workforce. However, at least one comprehensive study suggests that cost savings result from more efficient operations and the use of more advanced technologies rather than from lower wages or fewer employees. Impact on public sector employees can be minimized by requiring the outsourced company to retain existing employees or to provide placement services for employees not retained (Office of the Comptroller, 1996). Other alternatives include reducing the workforce through voluntary means such as early retirement, or transferring existing employees to other areas within the municipality. Municipalities contemplating outsourcing should also be cognizant of possible union issues, which may arise if existing operators are union members.

Potential Loss of "Hidden" Services A municipality must understand the scope of services currently being provided by the public entity in order to determine whether outsourcing is cost-effective. In addition to operation of the wastewater treatment plant, a wastewater utility may provide additional services, including:

- protection against sewer backups, fires, and explosions under the national industrial wastewater pretreatment program;
- technical assistance on wastewater treatment and waste minimization to industrial customers;

- ambient water quality, groundwater, and sediment monitoring;
- provision of reclaimed water supplies;
- septic tank retrofitting and pumping;
- disaster planning and emergency preparedness and response;
- cooperation with neighboring local governments or large institutional users such as universities, military installations, or government complexes;
- public education and outreach programs; and
- participation with state and federal regulatory agencies on issues of policy and rule-making (Association of Metropolitan Sewerage Agencies, 1996).

Where a public employee performs many roles, the municipality must decide whether to continue these roles or whether the additional services will be lost upon outsourcing. Either way, the benefit of outsourcing may be partially offset by the loss of services or cost of continuation.

Potential Loss of Control Over Operations Yet another potentially significant issue is the possibility that the municipality will lose control over the treatment operations once the operations have been contracted out. A private company will be driven by profits and may be unresponsive to political needs or broader community concerns. Depending on the nature of the outsourcing agreement, the municipality may be relegated to the role of the outsourced company's customer. Thus, the municipality may not have the leverage to influence operations of the treatment plant. Issues may also arise should the municipality experience increased development resulting in increased load to the system or should environmental compliance problems become a concern. As discussed further below, the parties to the agreement can attempt to address these issues through the outsourcing contract.

ENVIRONMENTAL LIABILITY ISSUES

Liability Under the Clean Water Act

The National Pollutant Discharge Elimination System Wastewater treatment operations are unique as outsourcing candidates in that there is a significant degree of federal, state and local regulation of treatment plant operations. The Clean Water Act (CWA)[11] establishes a national program for the protection of the nation's waters. An important component of the CWA is the National Pollutant Discharge Elimination System (NPDES) program, which regulates the discharge of pollutants into U.S. waters.[12] Any point source that discharges pollutants into U.S. waters must apply for and obtain an NPDES permit.[13] The permit sets forth the effluent standards that must be met by the point source and also imposes monitoring, record keeping and reporting requirements.[14] Although the NPDES program is a federal program, individual states can and have been delegated authority to administer the NPDES program in lieu of the USEPA.[15] In a delegated state, the discharge permit is issued by the state and the point source is subject to the state's regulatory requirements.

Municipal treatment plants, referred to as Publicly Owned Treatment Works (POTWs) under the NPDES program, are subject to the NPDES permitting requirements because POTWs discharge to U.S. waters. Non-household dischargers to POTWs are further regulated under the industrial pretreatment program,16 which requires industrial dischargers to meet certain pretreatment standards prior to discharging wastewater to the POTW. In addition, industrial users are subject to record keeping and reporting requirements. The pretreatment program is typically administered by the POTW through local ordinance or by permit.17 In order for a state to authorize a POTW to administer the pretreatment program, the POTW must demonstrate that it has the legal authority, including enforcement authority, to administer the program.[18]

NPDES Permitting One of the most significant environmental issues facing a municipality considering outsourcing is NPDES permitting. The federal NPDES regulations state that when a facility is owned by one person but is operated by another person, it is the operator's duty to obtain the permit.[19] Thus, where a municipality has outsourced operation of its treatment plant operations, the outsourced company must apply for the NPDES permit. However, authorized states have the flexibility to require the owner, the operator or both parties to apply for the permit. For example, Indiana requires the operator to apply for the permit while Illinois puts the burden on the owner of the plant. An interesting situation arises where, as in Franklin, Ohio, the outsourced company is both the owner and operator. The Ohio Environmental Protection Agency listed both the host municipalities and Wheelabrator, the outsourced company, as co-permittees.

From the regulatory agency's standpoint, the entity listed as the permittee is responsible for permit compliance. However, even if a municipality is successful in keeping its name off the permit, it is not clear that the municipality will be insulated from liability from permit violations. In *United States v. Smithfield Foods, Inc,*[20] Smithfield Foods, Inc. ("Smithfield") operated a wastewater treatment plant and discharged its treated effluence in accordance with the state equivalent of an NPDES permit. The Smithfield treatment plant treated wastewater generated from Smithfield Packing and Gwaltney, both subsidiaries of Smithfield. However, neither Smithfield Packing nor Gwaltney were listed as permittees on the discharge permit. Permit violations occurred when Smithfield exceeded its discharge limitations and submitted late discharge monitoring reports.[21] In an overfiling action, the United States sought injunctive relief and civil penalties from Smithfield, Smithfield Packing and Gwaltney for the permit violations. In granting the United States' summary judgment motion, the court stated that a party need not be listed on a permit in order to be held liable for a permit violation. The court first noted that the CWA states that 'persons'—not permit holders—are liable for permit violations.[22] The court went on to find that by generating the wastewater, which was treated in the Smithfield treatment plant and discharged to the receiving river, Smithfield Packing and Gwaltney actively caused the permit violations. Thus, both Smithfield Packing and Gwaltney were considered persons liable for violations under the CWA.

In a reverse situation, at least one court has found that the parent company will not be held liable for a subsidiary's CWA violation unless it can be shown that the parent directly caused the violation and was a person under section 309(d).[23] Avatar Holdings, Inc. (Avatar Holdings) was the parent company of Florida Cities Water Company (Florida Cities), a private utility that operated three wastewater treatment plants in Florida. The three treatment plants were found to be in violation of the NPDES permit requirements and the government brought suit against Florida Cities and its parent Avatar Holdings. In refusing to hold Avatar Holdings liable for its subsidiary, the court stated that there was no evidence that Avatar Holdings directed or caused any of the violations of the CWA at issue in this case. Thus, the court found that Avatar Holdings was not a "person who violates" under section 309(d) of the CWA. As in the *Smithfield* case, liability turns on whether the entity is the person who violates the Act. Although these two cases involve related companies, a court could arguably apply the same analysis to determine whether a municipality not listed as an NPDES permittee is liable for permit violations. According to the *Smithfield* court, the test would be whether the municipality caused the violation. Thus, from the standpoint of a municipality that has outsourced its wastewater treatment plant operations, liability may still attach even if the municipality is not listed as the permittee.

It is important to note that permitting agencies have flexibility as to what entity is listed as permittee; an agency may be reluctant to list solely the private outsourced company as the permittee, in part because the outsourced company's role may be transient. More likely is the situation where either the municipality is listed as the permittee or both the municipality and the outsourced company are listed as co-permittees. A municipality that succeeds in excluding itself from the NPDES permit has, to a great degree, distanced itself from environmental compliance responsibilities. However, there is no guarantee that in an enforcement situation, the regulatory agency will not look to the municipality, especially where there is evidence that the municipality's activity, rather than the outsourced company's operations, caused the violation. At this point, existing environmental regulations do not clearly contemplate outsourcing. Therefore, permitting will involve negotiations between the municipality, the outsourced company and the permitting agency.

Authority to Act as the NPDES Permittee Municipal law considerations may also be important in allocating permitting and compliance responsibilities between the municipality and the outsourced company.[24] In *United States v. Michigan,*[25] Wayne and Oakland counties (Counties) operated wastewater treatment plants, which served their constituent communities.[26] The Michigan Department of Natural Resources (MDNR) insisted on issuing the NPDES permits to the Counties exclusively based on the Counties' ownership of the discharge structures. The Counties objected, claiming that they were not the proper permittees. The court agreed and found that the Counties were not authorized by law to perform certain acts such as construction of new treatment facilities or levying user fees. In essence, the Counties had no authority to finance any of the duties required of a permittee and any financing could be accomplished only by voluntary agreement. Because the MDNR sought to impose conditions that were beyond the Counties' authority, the court stated, "MDNR must look beyond the Counties as operators of the discharge structures and also impose the permit conditions on the local communities responsible for the discharge by naming them as joint permittees."[27]

The above case is not directly applicable to the outsourcing discussion in this article as a county, rather than a private company, was the treatment plant operator. However, it does highlight the fact that transferring responsibility for certain governmental duties requires the recipient to have the appropriate authority. While many authorization issues can be addressed through contractual provisions, some traditional government functions, such as enforcement under the pretreatment program, must be retained by the municipality. This calls into question the extent to which the municipality will be able to escape NPDES permit responsibility simply by keeping its name off the NPDES permit.

Biosolids Management The CWA also requires that municipal wastewater treatment sludge be properly managed.[28] If a municipality disposes of its sludge in a municipal solid waste management landfill, the municipality must do so in compliance with the solid waste regulations. This program, authorized pursuant to the federal Resource Conservation and Recovery Act (RCRA),[29] imposes certain duties regarding appropriate waste management. Testing, recordkeeping and reporting requirements may be applicable to such disposal, depending on the state's program, and failure to comply can result in fines and penalties.

Sludge management practices, including surface disposal, incineration and land application, are regulated under EPA's biosolids program.[30] The regulations set forth management practices, monitoring, recordkeeping and reporting requirements applicable to "any person who prepares sewage sludge, applies sewage sludge to the land, or fires sewage sludge in a sewage sludge incinerator and to the owner/operator of a surface disposal site."[31] Furthermore, the "person who prepares sewage sludge" includes "the person who generates sewage sludge during the treatment of domestic sewage in a treatment works. . . ."[32] These requirements can be administered through the NPDES permit, in which case responsibility for compliance can be negotiated during the permit process. However, an authorized state can assign portions of its sludge management program to a local agency.[33] If sludge management is administered separately from the NPDES permit, parties to an outsourcing contract will have to contractually agree to responsibilities for sludge management, and ensure that the agreement is reflected in the appropriate permit.

Criminal Liability

Criminal liability for environmental violations is based on knowledge, which is typically defined by the courts as a general intent. General intent means that a party knew it was dealing with a substance likely to be regulated and knew that it was discharging it. In addition, under the responsible corporate officer doctrine, high ranking officials in a business can be held criminally liable for regulatory and permit violations. Thus, avoidance of criminal liability is important. As with civil violations, neither party can contract away its liability. However, the parties should contemplate the possibility of criminal violations and allocate any defense costs and fines or penalties in the contract.

OSHA/Tort Liability

Occupational, Safety, and Health Administration (OSHA) liability must also be considered in any outsourcing arrangement. OSHA requires an employer to maintain a safe workplace. In the outsourcing context, the issue becomes who is responsible—the outsourced company or the municipality—for maintaining a safe workplace for employees of the outsourced company who work on the municipality's premises, and vice versa? OSHA typically looks to the party in direct control of the workplace and the employees in determining whom to cite for violations. According to Illinois courts, there are a number of factors used to determine who is in charge of the workers at a workplace, including:

- supervision and control of the work;
- retention of the right to supervise and control the work;
- constant participation in ongoing activities at the construction site;
- supervision and coordination of subcontractors;
- responsibility for taking safety precautions at the job site, as well as
- who owns the equipment used on the site; and
- who is most able to ensure safe working conditions.[34]

CONTRACTUAL PROVISIONS

Although the municipality will not likely be able to avoid liability for its acts in violation of the CWA, a municipality should not be held liable for violations resulting solely from actions of the outsourced company, such as improper operation of the system. Similarly, an outsourced company should not be liable for acts of the municipality or upstream dischargers, which cause the treatment facility to violate its permit, *e.g.*, where the wastewater entering the plant exceeds applicable pretreatment limits. While the regulations do not clearly address the outsourcing situation, parties to an outsourcing contract can and should allocate liability through a well-drafted outsourcing contract.

Indemnification

A municipality cannot contract away civil or criminal liability under the environmental statutes. However, the outsourcing contract can include a strong environmental indemnification provision, which shifts the burden to the outsourced company should environmental issues arise. This provision must be carefully drafted to define and cover any claims, causes of action, liabilities, and losses for which the outsourced company is responsible. Liability for defense costs, penalties, and improvements related to enforcement actions can and should be assigned. The implications of process and operational changes on the outsourced company's duties also should be covered. While in some states the municipality may be indemnified against its own negligence as well as violations it causes, this must be clearly stated in the contract. Contracting parties should beware of state law provisions limiting or striking hold harmless or indemnity clauses that protect a party against its own negligence; some states find this void as against public policy.

Performance Standards

The outsourcing contract should include procedures and time frames for cost-effective operations that guarantee effluent water quality. Compliance with all applicable laws, permits, and regulations must be the baseline. Financial penalties, recapture clauses, or termination clauses can be used as additional incentives to properly operate the plant. An outsourcing municipality may also want to consider including performance standards regarding maintenance, quality control, and safety programs.

The outsourced company may also use performance standards to control the quality of the wastewater into the treatment plant. A company contracted to operate the treatment system has little control over the influent to the plant. Although industrial dischargers will be regulated through the pretreatment program, this program is typically administered by the municipality. Thus, the outsourcing company may have little control over enforcement against significant violators or whether variances are granted to industrial dischargers. Even where the POTW does not service industrial users, the outsourced company may not have control over increased development, additional connections, and the resultant increased load on the system. Thus, the outsourced company may want to set forth certain influent criteria in the contract and include a provision whereby exceedences of influent criteria result in payment of a stipulated penalty to the outsourced company. In this manner, the outsourced company can be assured of some degree of control over the wastewater entering the plant.

Pricing

Costs, including user fees and planned increases, should be set out in the contract. All assumptions on which the parties are relying and all contingencies should be stated. If a price increase is to be allowed over the life of the contract, or for significant changes in flow or loadings, the contract should include a specific escalator clause. The parties should also determine who would be responsible for capital improvements and maintenance costs, which may include grounds or fleet maintenance. These responsibilities should be clearly set forth in the contract. In addition, the contract should state which party is responsible for utilities and other related costs.

Improvements

Ownership of any fixtures or improvements added to the facility during the life of the contract should be clearly defined.

Insurance

The contract should allocate responsibility for carrying insurance, specify the amount, and clearly state the scope of coverage required. For example, the municipality may require the outsourced company to carry insurance covering the property and equipment used by the outsourced company and any damage to equipment and facilities owned by the municipality. The outsourced company should name the municipality as an additional insured on the insurance contract and submit proof of insurance before commencing work under the contract. Performance bonds (for example, for the cost of one year of operation and maintenance) and environmental impairment policies should also be considered.

Personnel

As discussed above, labor considerations are an important issue in outsourcing. The obligations of the outsourced company with respect to existing municipal employees should be included in the contract. For example, if the municipality will have to lay off employees when the treatment activities are outsourced, the municipality may want to require the outsourced company to hire the employees or to assist in placing the employees into comparable positions. Existing employment contracts may have to be reviewed before the outsourcing contract is finalized. Finally, other labor law issues may arise if employees are severed. For example, adequate notice and severance pay may be required.

Technology

One reason for contracting with a private company is to take advantage of advanced technology that would otherwise be too expensive. Thus, the contract should require the outsourced company to use reasonably up-to-date technology and to update technology periodically. At a minimum, the outsourced company must meet any governmental mandates.

Reporting

Periodic reporting to the outsourcing municipality may be useful for the municipality if it wishes to remain informed regarding ongoing operations. Monthly and annual reports, maintenance reports, and annual operating, capital, and maintenance budgets are helpful, and reporting requirements should be specified. In addition, the contract should clearly state which party would be responsible for environmental reporting. Invoicing requirements may also be included in the contract.

Oversight

In conjunction with reporting requirements, the municipality may want to include a provision that allows the municipality to monitor and oversee the company to ensure that the outsourced company is fulfilling its contract.

Dispute Resolution

The parties should contractually agree to a dispute resolution process to address unforeseen occurrences. For example, a dispute resolution process may be helpful in determining responsibility for unplanned capital expenditures.

Term

The duration of the contract should be specified, as well as any renewal options. In addition, the contract should provide for phase-in of the outsourced company's system and phase-out at the end of the contract term. It may be desirable to perform beginning and ending inventories.

Recapture

The contract can include a recapture provision that allows the municipality to regain control of some or all of the wastewater treatment process without terminating the entire contract. This will give the municipality some flexibility and will allow it to take over if the outsourced company is not meeting expectations.

Bankruptcy

The contract should set out the parties' rights should either of the parties become bankrupt or insolvent.

Termination

The contract should specify the conditions under which either party may terminate the contract and the length of notice required. It should provide for a smooth transition to the extent possible.

CONCLUSION

Outsourcing of municipal wastewater treatment operations is an increasingly popular trend. However, unlike other municipal services, wastewater treatment operations are strictly regulated under environmental laws. Thus, while a municipality may perceive outsourcing to be beneficial from a financial standpoint, environmental issues must be addressed in order to make an accurate assessment. A well-drafted contract, which clearly defines each party's responsibilities and allocates the risks associated with environmental issues is critical in ensuring a successful partnership.

NOTES

[1]Testimony of Erwin Odeal, AMSA Board Member and Executive Director, Northeast Ohio Regional Sewer District on Behalf of the Association of Metropolitan Sewerage Agencies Before the House Transportation and Infrastructure Committee Water Resources and Environmental Subcommittee, April 23, 1997.

[2]Notice, 57 *Fed. Reg.* 28,867 (1992).

[3]Exec. Order No. 12,803, 57 *Fed. Reg.* 19,063 (1992).

[4]*Ibid.*

[5]Notice, 57 *Fed. Reg.* 28,867 (1992).

[6]*Ibid.*

[7]Exec. Order No. 12,893, 59 *Fed. Reg.* 4,233 (1994).

[8]*Ibid.*

[9]The R.M. Clayton wastewater treatment currently has capacity of 100 million gallons per day (mgd) and is expanding to 116 mgd. City officials will turn over operations to a private contractor. "Atlanta to Shift Treatment Work," *Engineering News-Record*, December 15, 1997.

[10]As a term of the asset sale, Wheelabrator guaranteed lower user fees to the Miami Conservancy District users (*Public Works Financing*, 1994).

[11]Federal Water Pollution Control Act, 33 *U.S.C.A.* §§1251-1387.

[12]40 *C.F.R.* Part 122.

[13]*Ibid.*

[14]For example, monthly discharge monitoring reports documenting the quality of the effluent discharged must be submitted to the regulatory agency. §122.44.

[15]40 *C.F.R.* Part 123.

[16]40 *C.F.R.* Part 403.

[17]*Ibid.*

[18]*Ibid.*

[19]40 *C.F.R.* §122.21(b).

[20]965 *F.Supp.* 769 (E.D.Va. 1997).

[21]*Ibid.* at 781.

[22]*Ibid.*

[23]*U.S. v. Avatar Holdings, Inc.*, No. 93-281-CIV-FTM-21, (M.D.Fla. Aug. 20, 1996).

[24]*United States v. State of Michigan, et al.*, 781 F.Supp. 1230 (E.D.Mich. 1991).

[25]*Ibid.*

[26]*Ibid.* at 1233.

[27]*Ibid.* at 1234.

[28]*CWA* § 405(d).

[29]42 *U.S.C.A.* §6901, *et seq.*

[30]40 *C.F.R.* Part 503.

[31]§ 503.1(b).

[32]§ 503.9(r).

[33]40 *C.F.R.* Part 501.

[34]*Jackson v. Back of the Yards Neighborhood Council*, 608 N.E.2d 124, 128 (Ill. App. Ct. 3d Dist. 1992), *appeal denied*, 612 N.E.2d 513 (1993).

REFERENCES

Association of Metropolitan Sewerage Agencies (1996). *Evaluating Privatization: An AMSA Checklist.*

Engineering News-Record (1997). "Atlanta to Shift Treatment Work." December 15.

Goldsmith, Stephen (1997). "Can Business Really Do Business with Government?" *Harvard Business Review*, May–June.

Odeal, Erwin (1997). Testimony before House Transportation and Infrastructure Committee, Water Resources and Environmental Subcommittee, April 23.

Office of the Comptroller (1996). Privatization of Illinois Municipalities: Competition for City Services; Has the Time Arrived? Springfield, IL.

Public Works Financing (1994). "Case Study: Ohio Wastewater Asset Sale Will Open up Infra-Refinance Market." June, p. 1.

Reason Public Policy Institute (1997). *Annual Report.* Los Angeles.

CHAPTER 4

Professional Service Management and Operations of Municipal Water and Wastewater Utilities

Roger Hartman

Mounting regulatory pressure, competition for competent personnel, and financial realities are leading many communities to consider private sector management and operations and maintenance (O&M) of water and wastewater facilities. When properly implemented, professional services management and O&M contracts can provide greater accountability for operations, allow community leaders to shift the risk of meeting environmental standards to the private sector, and bring needed expertise to the increasingly complex operating environment of water and wastewater systems. Professional services management and O&M often results in a more predictable operations budgeting process, and can offer an opportunity for local governments to realize operational savings that can be shifted to investments in capital facilities (Hartman, 1993).

WHY PROFESSIONAL MANAGEMENT AND OPERATIONS?

Under a professional management and operations service arrangement, the private sector is contracted to manage, maintain, and operate some or all of a publicly owned utility's system components. For water and wastewater utilities, these components might include treatment plants, the sanitary sewer and water distribution systems, laboratory services, meter reading, and billing and collection. Under this contractual arrangement, competition and the profit motive provide the contractor with incentives to reduce costs within the constraints of the contract. But equally important to the municipal owners are protections offered through contractor penalties or termination of the

Roger Hartman is owner and principal of THC, Utility Management Specialists in Boulder, Colorado.

contract for unsatisfactory performance. The contractor takes full responsibility for specific utility functions, operates under a fixed budget, and guarantees plant performance and product quality. The contractor is responsible for payment of all normal and routine costs associated with operation of the specified facilities. But, depending on the type of contractual arrangement, major capital expenditures for expansion and upgrade of facilities generally remain the responsibility of the municipal owner.

Ever-tightening environmental regulations and increasing capital costs for expansion and upgrade of facilities present major challenges for local officials. The 1986 Safe Drinking Water Act (SDWA) amendments represent a significant source of increasing costs for water supply systems. These amendments require establishment of specific limits on a wide range of drinking water contaminants, and set rules for the treatment of surface water and some groundwater supplies. Cost-estimates for compliance have ranged as high as $49 billion through 2000 (USEPA, 1990). The subsequent impact of these added costs on water service rates generally vary with system size. Small systems, lacking economies of scale, tend to be hardest hit by new regulations. The EPA has estimated that systems serving populations less than 10,000 will require annual rate increases of over 35 percent. This compares with an estimated 25 percent annual rate increase for systems serving more than 250,000 people (USEPA, 1990).

For the wastewater services industry, the Clean Water Act (CWA) is of similar interest. Under its predecessor, the Federal Water Pollution Control Act of 1972, provisions were made for federal matching grants of up to 75 percent of capital construction costs. Between 1973 and 1988, over $50 billion was granted to municipalities and districts for construction of wastewater treatment facilities. But, direct grant assistance to local governments for wastewater treatment facility construction ended in 1994, when it was replaced with state revolving loan programs. As a result, local governments are fully responsible for the cost of capital improvements as the next round of wastewater treatment facility construction is undertaken. At the same time, communities also will be faced with renewal and replacement of capital facilities that have reached the end of their design lives, or that have been used more heavily or have deteriorated more rapidly than anticipated. The full cost of capital improvements, renewal and replacement programs, upgrades in response to stricter regulation, and application of new technology will increase water and wastewater service costs.

Over the last two decades, many municipalities have invested heavily in advanced water and wastewater treatment facilities, sometimes with disappointing results. Staffs that have successfully operated less complex facilities sometimes fail when faced with upgraded or more complex equipment and processes. Further, inadequate start-up and ongoing training, often a result of insufficient or misallocated operating funds, can lead to improper operation of new facilities and poorly administered maintenance programs. Fortunately, however, most modern water and wastewater treatment facilities can perform to design capabilities provided they are staffed by experienced operations and management personnel.

THE PROFESSIONAL MANAGEMENT AND O&M INDUSTRY

An increasing number of communities of all sizes are exploring professional services management and operation of their water and wastewater utilities. In 1995, Indianapolis contracted for operation of its two wastewater treatment facilities—a combined capacity of 250 million gallons per day. The city also contracted its wastewater collection system in 1996. New Orleans, Jersey City, and Houston also selected professional services O&M for some or all of their water and wastewater facilities, and other major metropolitan service providers are exploring professional services management for their systems as well. While the number is growing rapidly, there still are fewer than 600 contracts in place nationwide, representing only about 5 percent of the municipal water and wastewater facilities in the U.S. with rated capacities greater than one million gallons per day.

The majority of these management and O&M contracts—85 to 90 percent—are for wastewater treatment facilities. This seeming preference for the contracting of wastewater operations can probably be attributed to budget constraints, regulatory pressure, and the increased application of advancing technology during the federally sponsored construction program of the 1970s and 1980s. In addition, most communities view continued close control of their potable water systems as an issue of public trust and necessity. However, the realities of utility system economics, the need for capital investment to upgrade treatment facilities, and the impact of tighter standards stemming from the SDWA are causing community leaders to reassess the traditional position of water system control.

Currently, a limited number of firms account for most of the management and O&M services nationally. In addition, there is a second tier of smaller firms with a more regional client base. And, as might be expected in a rapidly growing market, consolidation is occurring as the larger national companies acquire the regional firms. At both levels, a number of the professional management and operations firms are subsidiary business units of engineering firms that traditionally provide design services to the municipal water and wastewater industry. Other management and O&M firms are stand-alone companies whose sole business is contract services.

Benefits and Barriers

Pressing capital needs and associated utility rate increases often provide the political impetus for consideration of a professional management and services option. Communities considering professional services operation often cite cost savings as the prime motivation. Cost savings may be accomplished in a variety of ways. Chief among these are reduced staffing levels, backup expertise, energy efficiency, training, maintenance programming, process control, and capital improvements. Clearly, the magnitude of savings is a function of a variety of factors, including the nature of the facility and its specific operating characteristics and problems. While there are numerous documented cases of cost savings as high as 30 percent, there also are cases where operating costs increased under the private contractor. Often, the contractor must shift savings into other deficient areas to improve overall performance. But, the primary goal of the contractor is improved, consistent operating performance, thereby giving the owner greater value for the money spent (Sedwick, 1984).

Contractors generally do not achieve cost savings by cutting back on employee wages, but rather through reduced administrative and overhead costs, and through better application of personnel, technology, and equipment. A review of the 10 largest secondary wastewater treatment facilities in Massachusetts indicated that staffing for three private sector-operated plants was approximately 25 percent lower than that of seven comparable municipal operations (Factor, 1992). This difference in staffing is consistent with personnel reductions observed at other municipal water and wastewater facilities where private sector operation was implemented.

When staff reductions do occur, they often happen over an extended period of time, resulting from attrition in the workforce (Lampkin, 1991). Some affected employees may pursue other public sector opportunities, while others may select early retirement if it is made available. At Indianapolis, 136 of the city's 332 employees were reassigned to other city departments or outplaced through a cooperative program developed by the city and the contractor (Cairo, 1995).

Cost control by the contractor also tends to stabilize operating budgets. This can assist in minimizing overall budget increases and, hence, increases in user charges. In addition, the contractor often improves the quality and consistency of operation through enhanced process control, maintenance management systems, staff training, and technical expertise often unavailable to publicly operated facilities. Finally, by shifting programs and staff to a private contractor, the municipality can recognize reduced administrative burden and cost. This can release limited resources for other priorities and allow more time to plan for the future rather than staying involved with day-to-day utility operations.

Clearly, O&M cost savings can be reapplied to capital needs, but a carefully crafted O&M contract can produce other more immediate sources of capital for the community. A contract in Glencove, New York allowed the city to defer half of its annual contract payments ($1.5 million) for wastewater treatment for the first three years. The city also deferred operating costs for its trash incinerators, which were part of the contract for one year. These provisions provided immediate and needed capital for the community. Further, using a financing authority mechanism, the contractor agreed to finance and implement more than $12 million in other utility-related capital improvements. In return, the contractor received a 20-year operating contract.

Similarly, Farmington, New Mexico secured more than $300,000 in refurbishments to one of its water treatment facilities, and $315,000 in cash through the sale of utility department vehicles and rolling equipment to the O&M contractor. Further, changes and adjustments in system operations allowed deferral of other planned capital improvements. The net result was stabilization of water rates and a 12-percent reduction in wastewater rates. In both of these examples, the contractor was able to offer tangible, immediate relief to vexing capital shortfalls and associated political problems.

In North Brunswick, New Jersey, the city awarded a concession contract for its water and wastewater systems. In return, it received a one-time $6 million concession fee and annual payments of $600,000, increasing to $2 million over the life of its 20-year agreement.

Barriers to contract operations of municipal water and wastewater facilities can take several forms. More often than not, opposition by public employee groups is cited as a major obstacle. More than 80 percent of the respondents to a municipal contracting survey indicated that public employee opposition was a major concern. The same survey cited opposition by elected officials as the second most significant hurdle to contracting of services (Reason Foundation, 1992).

Utility management and employee bargaining groups have generally opposed contracting programs even when present employees were protected or when significant cost savings to the community were identified. Public employee groups often raise issues about the elimination of public jobs, concerns regarding retirement, reduced fringe benefit programs, and contractor accountability. Many of these issues can be overcome in the contracting process or in development of the contract itself.

Most communities that opt for contract O&M require that public employees be the first hired by the new contractor. Further, the contractor may be required to maintain or increase salary levels of employees who move from public to private sector. Generally, contract O&M firms recognize the value of the hands-on knowledge that current employees possess. Offering employment to qualified, knowledgeable, and motivated employees is in the economic interest of the contractor. For example, of the 204 employees now operating the wastewater facilities at Indianapolis, 96 percent are former city employees (Cairo, 1995).

PROFESSIONAL SERVICES O&M—QUALIFICATIONS

While cost savings may be one of the principal reasons municipal officials consider contract O&M, other issues also can motivate municipalities to examine the professional contract O&M option. A shortage of trained personnel, startup of a new facility, ongoing regulatory violations, or personnel and labor issues also may lead to contract operations. Consequently, price alone should not be the sole determinant in selecting a professional management contractor. In addition to a guaranteed price, the contractor is providing professional management, technical and regulatory expertise, and financial controls for water and wastewater operations. A quality contractor with a proven track record of successful operations is the key to achieving maximum benefit from a professional management and O&M contract (Sedwick, 1984).

The selection process for professional O&M services, therefore, should be similar to that used to procure other professional services such as engineering, or financial and legal services. The "best" operations contractor is the one that meets the specific management and operating needs of the community. Because of the range of technical and management skills required for success, and the possibility of access to needed capital, the "low" bidder may not be the "best" contractor to deliver O&M services.

A qualified contractor should have a proven, successful record of operations similar to those that are being contemplated by the municipality. This means that the contractor should have operated facilities that are of similar size, technical and process complexity, and financial transaction volume. The operations contractor also should have demonstrated financial stability, the ability to post a performance bond, and have technical, financial, and management experience, including access to human resources and labor relations capabilities. The ability to offer creative financial arrangements also can enhance the credentials of prospective contractors. By using a prequalified list of potential contractors with these types of credentials, the likelihood of successful contract management and operations should be substantially increased.

COMPARING THE PRICE

Cost is clearly a consideration in evaluating proposals. A valid and defensible assessment of the actual cost is necessary when comparing proposed contractor pricing with the cost of continued public employee operation. The fully allocated cost of continued public sector delivery of the service must account for both direct and indirect costs. Direct costs include employee salaries and benefits, operational costs, supplies, outside maintenance contracts, annual capital costs for equipment, and insurance premiums. Indirect costs include allocated management and administrative support services, and allocated overhead of other executives and staff agencies. From the accounting of these indirect costs, those that are truly avoidable through the contract option are determined. The sum of direct costs and avoidable indirect costs represents the total cost to be avoided by contracting.

For comparative purposes, the annual costs of contract procurement and administration must be added to the proposed contractor price. Procurement costs (*e.g.*, contract and proposal development, bid processing, severance pay and benefit buyouts) are amortized over the life of the proposed contract. Contract administration costs, on the other hand, vary according to the size and complexity of the contract and include contract administration personnel salaries and benefits, services and supplies, and equipment. Again, the sum of the contractor price, amortized procurement costs, and annual contract administration costs represent the costs of contracting. It is this adjusted cost of contracting that should be compared with the total cost avoided by contracting (Martin, 1992).

THE PROFESSIONAL SERVICES O&M CONTRACT

The contract defines the standards of performance for the contractor (Hartman, 1992). Among other things, it may specify that the contractor meet NPDES permit discharge standards for wastewater treatment operations, or applicable SDWA requirements for water treatment and distribution. The contract also may require that the contractor meet these standards while operating the facilities in a cost-effective and professional manner, or that only a specified amount of energy is used in operating the facilities. With regard to facility maintenance, the contract may require that equipment be maintained in accordance with the manufacturers' recommendations, or that corrective maintenance work orders be completed within a specific time period depending on priority or criticality. Within the contract, the owner also can specify routine reporting requirements, financial reconciliation reports, performance summaries from the contractor's maintenance management system, an annual report to the council or board, and an annual audited financial statement.

The service contract should not be overly prescriptive, however. The contract should state what is to be done and how it will be measured. Work methods and procedures generally should be left to the contractor. The reasons for this are twofold. First, the nature of the legal relationship between the owner and the independent contractor must be maintained. If the owner prescribes the means, methods and procedures to be used by the contractor in completing the work, the independent contractor relationship becomes blurred. The owner, in fact, may become responsible for the outcome of the prescribed means, methods and procedures used by the contractor. This negates one of the basic concepts of contracting—to transfer risk for the outcome to the contractor. Second, an overly prescriptive contract may restrict the creativity of the contractor in developing more effective and efficient work methods and procedures. These may differentiate one contractor from another, increase competition, and be reflected in a more competitive price to the owner.

But, for many municipal owners and managers, contractor freedom may smack of loss of control. The owner is placing millions of dollars of assets in the hands of a contractor for management, operation, and maintenance. Yet the owner will continue to be the permit holder and ultimately be responsible for overall performance. Whether loss of control is real or merely perceived is immaterial. What is important is whether the owner and contractor can craft an agreement that affords the owner a level of contractor accountability and liability and fine protection that can substitute for the owner's direct control.

Contract Term and Renewal

If facilities were financed using tax-exempt debt, federal tax laws currently would limit the term of professional management and operations agreements. Under current Internal Revenue Service procedures, such contracts cannot exceed five years, including renewal options, and must be cancelable after three years without cause or penalty to the municipal owner (IRS, 1993).

Often, a contractor may propose to invest in new equipment as a means of improving efficiency, or to automate a procedure to reduce staffing levels. If these investments can be reasonably recovered within the typical five-year term, then such investments are likely to be undertaken with resulting savings reflected in the proposed contract price. If, however, greater financial participation by the contractor is an objective, a five-year contract term may not be sufficient to allow reasonable recovery. Here, a longer-term contract is more desirable. A contractor was awarded a 20-year concession contract for operation of the North Brunswick, New Jersey water and wastewater systems by agreeing to replace the city's outstanding tax-exempt debt with taxable debt. This refinancing approach avoided the term restrictions required by current IRS regulations.

Termination

Termination outlines the conditions under which the contract may be terminated. Generally, there is language for termination with cause, and for emergency takeover by the owner. The principal difference is the time frame for notice and the opportunity for contractor remedy from termination with cause. In some instances, both parties also may negotiate language allowing for termination of the agreement without cause by either party.

Facilities Description

The definition of facilities to be operated provides a physical description of the facilities, including location, major features and appurtenances, function, design parameters and limitations, and size.

Scope of Services

The scope of services describes the services to be provided. If facilities are to be operated and maintained, the scope defines the contractor's work and responsibilities, and how performance will be measured. Terms might require:

- Professional and cost-effective operations while meeting applicable regulations;
- Operations within design limitations;
- Guaranteed effluent or product quality;
- Maintaining the appearance of buildings and grounds;
- Required programs. For example, a computer-supported maintenance management and process control systems, an effective safety program, a laboratory quality assurance and quality control (QA/QC) program, or a public relations program;
- Minimum staffing and certification requirements. Specifying staffing levels may be too prescriptive, but some regulatory agencies do set minimum staffing requirements. In addition, some labor agreements may prescribe certain staffing, training and/or certification as well;
- Specific contractor reporting to the client. This might include: monthly operating reports, quarterly maintenance account reconciliations, and an annual report and audit;
- Annual operating budget, capital acquisition requests, and a maintenance plan by the contractor; and
- Use of local vendors and services when possible and appropriate.

Compensation

Compensation can be addressed in a variety of ways. Basic compensation is the monthly or quarterly payment to the contractor for carrying out the scope of services. This basic payment can be based on a lump sum bid, an incentive-related pricing structure, or a cost-plus arrangement. Generally, there is language that allows adjustment to compensation for cost increases to the contractor, and for significant changes in flows and loadings. This section also can include language on changes in scope, maintenance expenditures, and contractor incentives.

Employee Transition

Employee transition language is common in first-time contracts. The section can cover issues such as job offers to existing employees, retirement arrangements, restrictions on transfers during some specified period, and initial wage increase guarantees. These same issues sometimes are handled in a separate side agreement between the owner and contractor.

Liability and Insurance

Liability and insurance are important issues. The contract specifies the types and amounts of insurance coverages that the contractor must provide, require proof of specified coverages, and, if applicable, require assurances from parent (and grandparent) corporations as well. Typically, the contractor is, at a minimum, obliged to secure workers' compensation, property, and general liability insurance. In some instances, the contractor also may require the owner to carry certain insurance policies as well, *e.g.*, fire insurance on structures.

Some owners have expressed interest in environmental impairment insurance. Such insurance is relatively expensive and not readily available. Contractor's pollution liability insurance, protecting the owner against "sudden and accidental releases" by the contractor, may be more appropriate and more affordable. A performance bond posted by the contractor and backed by the parent corporation, if applicable, also may be required, generally in the amount of one year's O&M price.

Maintenance

Proper corrective, preventive, and predictive maintenance is the first line of defense in protecting the capital investment that communities have in their water and wastewater utility systems. Proper maintenance also assures that facilities and equipment are available and in operating condition when required. But maintenance also is an area where the owner often perceives the most significant potential for loss of control. By including proper monitoring and control features in the contract, and possibly by establishing some contractor incentives, a great deal of control can nevertheless be exercised.

Generally, *normal maintenance* costs include all preventive, repair, and corrective procedure expenditures, exclusive of labor, for an equipment unit up to an agreed upon amount or "deductible." By definition, anything in excess of this amount is termed major maintenance. Under a typical short-term O&M arrangement, the owner pays for maintenance procedures that are in excess of normal, *i.e.*, major maintenance and capital replacements. Assigning this responsibility to the owner and using the concept of a "deductible" are reasonable and necessary accommodations since the contractor has no intimate knowledge of past maintenance practices performed by the owner or others.

The O&M contract also can specify an upper limit budget, or *maintenance ceiling*, for normal maintenance costs. Here, major maintenance, as well as maintenance expenditures in excess of the ceiling, are a risk for the owner. Conversely, the contract generally requires that any unexpended maintenance funds are returned 100 percent to the owner. This is to discourage any tendency an unscrupulous contractor may have to reduce maintenance in an effort to gain additional profit. When a ceiling is specified, the owner should require the submittal of an annual repair/replacement plan with periodic updates of the plan. Monitoring of this plan and associated expenditures should allow the owner to monitor the maintenance function with reasonable confidence.

A third category of maintenance expenditure, *capital replacement*, also is generally the responsibility of the owner under a typical O&M arrangement. These are planned replacements of an entire equipment unit based on a repair/replace evaluation, or cost-effectiveness analysis. These potential added expenses should be contemplated by the owner with a planned list of capital replacements being presented as part of an annual budget plan submitted by the contractor to the owner.

Incentives

Energy Usage Following labor expenses, energy costs may be the second largest cost category of facility operation, often representing as much as 25 percent of O&M expenditures. Energy expenses also present an area that can generate significant savings with careful monitoring and control. The key to realizing these savings is a contract that provides incentive to the contractor to control and reduce energy costs, but recognizes the uncertainties related to changing plant loadings, process requirements or pumping.

Often, the contractor also agrees to a defined energy budget, or "baseline." The baseline is established from analysis of several years of energy usage data, including adjustment for abnormally wet or dry years and changes in plant flows and loadings. In order to provide an incentive for the contractor to reduce energy use, dollar savings resulting from energy use below the baseline may be shared between the client and the contractor. For example, the sharing arrangement may be 75/25 or 50/50, depending on the negotiated contract. In subsequent years, the baseline is adjusted downward to reflect the savings achieved during the preceding year. Using this approach, the greatest opportunity for energy savings and incentive sharing clearly is in the first year of contract operations.

To restrain the contractor from unbridled energy use beyond the baseline, the contractor may be made responsible for any and all use above the energy baseline. A more equitable arrangement may be to limit the contractor's exposure, to say 105 or 110 percent of the baseline after adjustment for increased growth-related loadings. Energy usage beyond that amount presumably would be due to abnormal conditions and could be assigned to the owner, or possibly split between the contractor and the owner.

O&M Performance The contractor is obligated to produce water quality that meets all permitted standards. There are instances, however, when merely meeting permit requirements is not enough. The owner may want the best quality achievable using the available facilities. Or, from a maintenance perspective, the owner may want minimum downtime of critical equipment and rapid turnarounds on all maintenance work orders. Under either of these circumstances, a performance incentive can be made part of the contract. The key is in setting objective standards against which the contractor is to be measured.

CONTRACT ADMINISTRATION AND MONITORING

Loss of owner control often is a key issue in making the contract operations decision. The owner is turning over substantial fixed assets for operation and maintenance by an independent company. All the while, the owner is retaining, in the eyes of regulatory agencies and the utility customer, ultimate responsibility for permit compliance and final performance. Under these conditions, "comfort" is an essential ingredient. An O&M contract that includes the items discussed can give significant "comfort" to the owner through adequate control of the contractor. But it also must be remembered that the contract will require administration by the owner.

The owner should not view any contract as a simple situation, one where the contractor submits a monthly bill and the owner pays it. The more complex the terms and conditions of the contract, the more complex the administration. The owner should anticipate associated ongoing costs for contract administration. For large-dollar and/or more complex arrangements, full-time contract administration is likely to be required. The owner must decide what degree of control (or "comfort") is necessary and weigh that against the cost and effort of contract administration. The cost of that administration must be included in any economic analysis used to evaluate the contract alternative.

CONCLUSION

Water and wastewater service delivery costs are increasing. In part, this is a result of added regulation, but escalating capital and renewal and replacement costs are adding to the increase as well. In addition, the application of more complex technology to water and wastewater treatment is leading to the need for more skilled and experienced operations personnel. In an effort to deal with these issues, more communities are examining professional management and O&M of water and wastewater utilities. The current market for these services is well in excess of $600 million annually, but represents only about 5 percent of the publicly owned water and wastewater facilities in the United States.

Professional management and operational services provide management, financial, and human resources capabilities. As such, a procurement process similar to that used in securing other professional services should be used. It is important that a "low bid" not be the only criterion used in contractor selection. Technical expertise, program management, and demonstrated success under similar circumstances are equally, if not more, important than bottom-line price. Professional operation, efficiency, and performance are the keys to predicting contract success.

The contract sets the standards of performance for the contractor and requires cost-effective and professional operation. The contract may state that equipment be maintained in accordance with the manufacturers' recommendations, or that corrective maintenance work orders be completed within a specific time period depending on priority or criticality. Since loss of control of daily operations is a key issue in contract O&M, it also is important that the owner and contractor develop a contract that affords the owner a level of contractor accountability that can substitute for direct control. If this is accomplished, private operation of municipal water and wastewater facilities is a valuable option for community leaders.

REFERENCES

Cairo, Patrick (1995). "Delegation of Municipal Water and Wastewater Services to Private Companies." Water Environment Federation/American Water Works Association Joint Management Conference, February.

Factor, A. (1992). "Regional Contracting of Wastewater Treatment Plants." Pioneer Institute.

Hartman, R. (1992). "Contracting Out O&M." *American City & County*, September: 38–40.

———. (1993). "Contracting Water and Wastewater Utility Operations." Reason Foundation Privatization Center, How To Guide No. 8, June.

Internal Revenue Service (1993). Revenue Procedure 93–19.

Lampkin, L. (1991). Speaking on "Labor Relations and Public-Private Partnerships." Proceedings of the Privatization Council's Third National Conference; *Privatization Review 4* (Fall): 111–116.

Martin, L. (1992). "A Proposed Methodology for Comparing the Costs of Government versus Contract Service Delivery." *ICMA Municipal Yearbook.*

Office of Technology Assessment (1990). *Rebuilding the Foundations—A Special Report on State and Local Public Works Financing and Management*, OTA-SET-447, March: 23–24.

Reason Foundation (1992). "Contracting Trends in 1991." *Privatization 1992 - Sixth Annual Report on Privatization:* 6–19.

Sedwick, J. (1984). "Contract Services for Wastewater Treatment." In Armington, Raymond Q. and Ellis, William D., editors, *This Way Up: The Local Officials Handbook for Privatization and Contracting Out*, pp: 124–148.

U.S. Environmental Protection Agency (1990). *Administration and Resource Management, A Preliminary Analysis of the Public Costs of Environmental Protection: 1981–2000*, May: 14.

U.S. Environmental Protection Agency (1987). *Contract Operation and Maintenance - The Answer for Your Town?* January.

Water Environment Federation (1992). *Washington Bulletin.* August.

PART II

Case Studies

CHAPTER 5

The Privatization of Two New Jersey Municipal Water Departments

Robert E. Thiele, Jr.

Prior to 1994, privatization of a water or wastewater utility in the United States meant either that the municipal entity was sold to a private company or that portions of the operation were contracted out. The most logical choice for operations contracts were water and wastewater treatment plants. These contracting arrangements originated with engineering companies that first designed and built these plants, then operated them under start-up and warrantee periods. These contracts were known as Operations Assistance Contracts. Then, convincing the client to contract the operation out rather than hire municipal workers to operate the plant expanded the industry to full-contract operations. Conversion of plants from municipal operation became popular as the operating companies developed expertise in contracting and operating these plants.

Meanwhile, the investor-owned water companies concentrated on acquiring municipal water companies as part of their growth strategies. Acquisition of municipal utilities is generally more complicated than procuring operations contracts. Usually, the municipality requires a public referendum in order to sell municipal assets, which intensifies the political issues surrounding either of these options. Once the referendum is passed, the private company needs the approval of the state public utility commission to ensure that the investment being made in municipal assets is prudent and will not be detrimental to existing utility customers. The commission

---

Robert E. Thiele, Jr. is a United Water Co. executive in charge of operations in Hoboken and Jersey City, NJ.

also reviews the rate impacts on the customers acquired. This process is time consuming. In some cases, the regulatory process can take as long as one year. Unlike rate applications before these commissions, which have statutory time limits on discovery and litigation, acquisition cases are generally conducted by commission staff who operate only under the confines of the caseload. If there is another major rate case consuming staff time, then the acquisition case is given lower priority. These delays put added strain on the parties to keep the transaction together. It has been United Waters' experience that the faster the transaction process, the more likely it is that a successful transaction will be consummated. Protracted delays cause the parties to constantly reevaluate the benefits of the deal, and subsequent events may cause one party or the other to revisit their positions on various issues, not the least of which is price.

The risks associated with acquisitions and contract operations are different. For the private company, the risks associated with contract operations are generally larger than those related to acquisitions. The contractor must guarantee its price, and cost overruns are generally borne by the private company. With acquisitions, the acquired municipal operation generally becomes part of a larger existing private utility, while utility and cost recovery of various items becomes a function of their regulatory rate setting process. Generally, upon a showing that the expenditures are prudent, the costs become an ingredient in the revenue allowance set by the utility commission and recovered from all of the utility customers.

For the municipality, the risks are varied. Generally, since this has been a "buyers market," the municipality can set the amount of risk to be borne by the private contractor in the bid specification. One of the prime concerns of municipalities, which drives them to the contract operation option, is the desire to control the rates to the constituents. Since generally only a portion of the utility is contracted out, the municipality retains the obligation to pay debt service on outstanding utility bonds; this means that the rate setting must remain with the municipality in order to provide adequate protection to the bondholders. The issue of the municipal "tax-exempt" bonds has also had a limiting effect on the length of term of these contracts. Until recently, the Internal Revenue Service had rules that limited the terms of the contracts to five years. Violation of this provision could cause the tax-exempt status of the bonds to be revoked, thus placing the bonds in default. In 1997, these rules were relaxed and contracts of up to 20 years were permitted, provided the pricing schedule of the contracts was at least 80 percent of the fee as fixed; the other 20 percent could be incentive-based.

In an acquisition, the municipality generally gives up entire control of the utility, including the rate-setting authority, since the rates become the jurisdiction of the state utility commission. United Water Resources is an investor-owned holding company that owns and operates 30 water and wastewater utilities in 13 states. Since these investor-owned utilities are essentially monopolies in each state in which they operate, a public utility commission regulates them. Thus, in the absence of competition, these commissions control the rates that companies can charge customers and have broad enforcement powers over the quality of service rendered by the utility. Municipally owned utilities generally are not subject to the jurisdiction of these commissions as the rate-setting and other regulations are under the control of the governing body whose members are subject to voter approval.

We concluded a transaction with the Village of Sloatsburg, New York that provided in the contract that rates be "phased in" to the tariffs of the acquiring utility over a 10-year period. While the New York Public Service Commission originally balked at a phase-in over such a long period, it eventually conceded to it at the municipality's insistence. Thus, this municipality did control the effects of new rates on its constituents. Generally, however, this is not the case.

A TALE OF TWO CITIES

This chapter highlights case studies from the neighboring Hudson County cities of Hoboken and Jersey City, New Jersey.* It is interesting to note that the political relationship between these two cities had a part in the decision each city made vis-à-vis privatization options. Hoboken relies on Jersey City for the wholesale supply of its potable water. Hoboken then retails the water to its residents. The Jersey City water system takes water from the Boonton Reservoir, some 23 miles west in Morris County, purifies it and transmits it to these two Hudson County communities by way of an aqueduct system that dates back to the turn of the twentieth century.

Hoboken is a city of almost 40,000 people. Aside from being known as the site of the nation's first baseball game and the birthplace of Frank Sinatra, the common image of Hoboken is that of an old waterfront community with aged wharves, docks, and taverns left from an era in which ship building and manufacturing thrived. In reality, Hoboken today is a vibrant city, undergoing a major renaissance. Stately old brownstone homes are being renovated. Washington Street, the city's main thoroughfare, is also reviving. New restaurants, sidewalk cafes, and clubs are replacing the old taverns that served the longshoreman of yesteryear. With access to New York City in minutes via a subway or a new ferry service, the city is becoming the preferred place of residence for young professionals. Business is again thriving and private sector investments in the city are at new levels. There are five major new projects under construction along the waterfront; the abandoned wharves and docks are being replaced with housing developments, parks and hotels.

Just to the south, Jersey City is also experiencing a revival. As in Hoboken, the waterfront offers an attraction to people who need access to New York City without having to cope with the high cost of living in Manhattan. Most agree that the view of the New York skyline from either Hoboken or Jersey City is the most spectacular in the entire metropolitan area. Jersey City is home to almost 230,000 people. Unlike its neighbor to the north, Jersey City is a vastly diverse community. Over 15 different ethnic communities make Jersey City their home. Unlike Hoboken, the revival is taking a bit longer to accomplish. While the waterfront is experiencing growth, neighborhoods to the west are still in need of new investment.

The common denominator these two cities share is the need for government to become more efficient, reduce its costs, and provide for investment in infrastructure without increasing taxes. While they embarked on different routes, it was this need that encouraged the administrations of both these cities to make a commitment to

*See Figure A–1 in the Appendix.

privatization. I use the word 'commitment' because the successful consummation of these transactions took an enormous commitment on the part of both Mayors Russo (Hoboken) and Schundler (Jersey City).

HOBOKEN

History of the Relationship

In 1882, Hoboken was the largest wholesale customer of the Hackensack Water Company (now United Water New Jersey). Around 1923, the company lost the wholesale water contract to Jersey City, which has been the city's supplier ever since. In 1993, after receiving notice that Jersey City intended to raise its price for wholesale water, the United Water Company was contacted to explore the opportunity to gain back this contract. In the intervening 70 years, many of the connections between the two service areas of Weehawken and Hoboken had ceased to be serviceable and some were most likely destroyed with the construction of the Lincoln Tunnel, which separates the two communities. In any event, to regain a customer with an average demand of about four million gallons per day, investment in new interconnection was feasible at the right price. Many discussions were held, and finally city officials in Hoboken put both Hackensack Water and Jersey City on notice that it intended to receive sealed bids for this service. While the prices were close, Jersey City prevailed in maintaining Hoboken as its customer.

One of the council people active in this process was Anthony Russo. One year later Mr. Russo was elected Mayor of Hoboken. Since he was familiar with United Water, he opened a discussion about a problem he was having with his water department. The department was operating at a loss of about $350,000 per year, with anticipated cost increases, including the cost of wholesale water, set in the bid process a year earlier with Jersey City; the need for huge rate increases was imminent. The total budget of the department was about $3.5 million; the cost of wholesale water was $1.6 million and scheduled to escalate to $2.2 million in five years. In addition, very little in the way of system improvements was being accomplished. Had the city not elected to explore privatization, they would have been looking at a 50 percent increase in rates to close the budget deficit and to provide for system improvements.

The Concept of a Concession

As discussions about what to do went forward, many options evolved, from contract operations to outright sale of the municipal asset. For Hoboken officials, there seemed to be more support for the concept of sale than there was for contract operations. The need for total control of the asset did not seem to be a priority. There were two major issues that worked against a sale of the asset: there was no support for conducting a public referendum on the issue, as required by New Jersey State Law, and the city clearly wanted to maintain control of the water rate. The politics in Hudson County is heavily Democratic and embraces all of the traditional values of the Democratic Party, especially when it comes to issues like labor. It was clear that no transaction would go forward without a clear commitment to the labor force of the water department.

What United Water needed was a transaction encompassing the benefits of a sale without the public referendum and the regulatory review process, while maintaining local control over the rates. Also, the quicker this process could be concluded the better.

In France, there are more types of private contracts for various public services than here in the United States. Funerals, for example, are the responsibility of government, but many communities elect to contract out this service. In the area of water and wastewater utility service, the French utilize two other types of contracts. One is a *concession* and the other is called an *affermage* (which means leasing or farming out). In both of these cases, the end user pays the utility bill to the private contractor. Unlike the investor-owned companies in the U.S., however, in France the physical assets are owned by the municipalities and the rates are not subject to oversight by a state regulatory agency. Rates are set by contract between the private company and the municipality. When these transactions are bid, the bidders often compete on the basis of the rates they intend to charge.

United Water modified features of these two types of contracts to make an offer to the officials of Hoboken. In the case of the *concession,* the contractor is responsible for making all of the capital improvements. This requires critically accurate "due diligence" in determining the capital needs of the system. In this transaction, all of the risk is borne by the private contractor. On the other hand, the *affermage* requires no capital contribution on the part of the private contractor. With the realization that the physical assets of Hoboken are in excess of 100 years old, United Water was unwilling to assume all the risks of a *concession,* as defined by the French. Since United Water would be assuming the entire revenue stream of the water department, funding capital improvements in an *affermage* transaction would require the city to raise taxes to fund bond issues or to place surcharges on bills. Neither of these options were palatable.

The solution was a combination of both concepts. A concession arrangement could be entered into, which limited the exposure of the private contractor for capital improvements. However, the amount of the capital budget would have to be high enough to cover almost all "routine" capital needs of the system. The only risk being retained by the city would be some catastrophic event or a drastic change in governmental regulation that would require massive amounts of capital. In the event of such a catastrophe, it was agreed that either the city would be offered the option of letting the private contractor fund the project and renegotiate the rate, or the city could raise the capital and either raise taxes or surcharge the water bills to fund the bonds. The threshold arrived at was an annual commitment to capital of $300,000 per year. To ensure that this amount was actually spent, there would be an annual accounting of the improvements, and funds not spent at the end of the contract would be returned to the city. Thus, the budget over a 10-year contract would be $3 million.

Legislative Authority

Realizing that this type of transaction was completely innovative, United Water needed to ensure there was not a legislative prohibition on this type of transaction. An extensive legal search revealed that New Jersey was one of the few states with a definitive law on the books that sought to encourage broad types of privatization contracts for water and wastewater systems. However, the process prescribed by these laws was so cumbersome that since they had been enacted in 1985, the law had never been used to consummate a transaction. The law required extensive intervention by both the state utility commission and the state's public advocate, as well as an exhaustive bidding and public hearing processes.

Another piece of state legislation, the Water Supply Act, seemed to give broad powers to municipalities when entering into transactions that ensured safe and adequate water supply to residents. The Water Supply Act would allow public utilities under the jurisdiction of the utility commission to enter into contracts with municipalities. While this legislation looked promising, it did not address the issues of the allowable length of contract or the bidding process. Since both Hoboken and United Water were blazing new trails here, there was a consensus that the parties would negotiate the terms of the contract. It was unclear how this type of transaction could be described in a bid document as the city wanted broad flexibility in determining the terms.

As luck would have it, another New Jersey community, Sayerville, had requested special legislation that allowed relaxation of the provisions of the original privatization legislation. The community of Sayerville wanted flexibility to enter into a contract with an industrial company in acquiring a well field. This transaction was to be concluded without bidding. The legislation was passed with a sunset provision that closed the opportunity to allow exemptions to the original privatization law for six months.

Negotiating the Deal

With a sense of what United Water could accomplish under existing legislation, it worked to conclude the transaction. The major issue of negotiation was what price United Water would pay the city for the right to acquire a concession. The concept of the "concession fee" was initiated. The value of this fee from the private companies' prospective was the present value of the cost savings in the operations, which the private contractor can achieve over the life of the contract. For accounting purposes, the concession fee is treated as an investment in the enterprise and depreciated over the life of the contract. In Hoboken's case, since the cost of purchasing water from Jersey City, which accounts for 50 percent of the total operating budget, was scheduled to increase by 60 percent over five years, rate increases for Hoboken residents were inevitable, regardless of what type of deal was negotiated.

The paradigm then became what "concession fee" or series of "concession fees" could be negotiated that—along with an acceptable series of rate increases—would win political support. The basic concern of the city was to conclude this transaction with the longest amount of rate. The city required no rate change in the first year and only modest rate increases in the next three years. Needless to say, a variety of combinations were accomplished to arrive at the final terms of the deal. Also, as mentioned above, another item that needed to be addressed was any debt outstanding, so as not to violate the Internal Revenue Service's revenue proclamation. Fortunately, the amount of municipal tax-exempt debt related to the water system amounted to only $400,000 and was to be paid off with the proceeds of the concession fee.

The final negotiated terms were as follows: concession fee of $5.5 million to be paid at signing of the contract; a contract term of 10 years; $3 million in capital improvements over the term; the installation in the first year of all new meters that had the capability of being read by telephone; a rate schedule that provided for no rate increases during the first year; increases the following three years of 3 percent, 5 percent, and 6 percent, then three increases of 10 percent per year in years five though 10; rates indexed to inflation thereafter. Thus, the city was spared from increasing its rates by over 50 percent in the first year and gained a financial package of capital infusion, new meters, and a sizable concession fee in the first year.

While the administration in Hoboken can only sponsor a resolution to endorse the contract, it is the city council that ultimately must approve the transaction. The resolution won the City Council's support by a vote of 9 to 0. The final step was to file the contract with the state utility commission. Although this transaction did not specifically require formal approval, since the transaction did involve a company under their jurisdiction, the contract had to be filed with the agency. Impressed by the innovation of the transaction, which demonstrated a benefit to a major New Jersey city with no impact on the other utility customers of United Water, the Board of Public Utilities endorsed the transaction.

Implementation

One of the major considerations in any transaction with a municipality is the fair treatment of city employees. There were 15 employees in Hoboken's water department. While all 15 were offered employment, only five accepted offers by United Water New Jersey. The other employees either decided to take other jobs or remained with the city, filling openings in the department of public works.

United Water was able to obtain project financing from a local bank and, based on the merits of the contract, the financing was completed without specific guarantees by the parent company, United Water Resources. This nonrecourse financing was also a first for such a transaction.

The metering program was completed on time and within the original budget. The meters and telemetry are totally compatible with the company's existing system. Thus, Hoboken's customer service functions are fully integrated with those of United Water New Jersey.

The operations of the distribution system were also integrated into those of United Water New Jersey. Since crews were already dispatched from the Hackensack area to as far south as Weehawken in Hudson County, to assume additional operations for another community in the area was not a problem. The challenge in assuming the operations of these inner-city systems is the inadequacy of information about the facilities. Documentation and mapping of distribution mains, valves, and hydrants often is severely lacking.

United Water has completed documentation of the majority of the valves, and repairs and replacements have been made to those found in disrepair. This effort, more than any other, has improved flows within the city and improved water quality by eliminating sections dead-ended by inoperable valves. Yet, we still find facilities undocumented and continually replace and repair these facilities. The need for documentation is critical in repairing breaks. Knowing the precise location and operation of system valves cuts the amount of time customers are disturbed in the event of water main failures.

The capital budget of $300,000 has been exceeded by almost $100,000 over the first three years of this project. Instead of the city making up this shortfall, United Water has elected to carry it forward. As repairs and improvements are made to the system, capital spending is becoming more predictable and this deficit should be made up in the coming years, fulfilling a $3 million commitment over 10 years.

Contract Extension

As a testament to the success of the contract, in 1996 the city government extended the contract by an additional 10 years, making the contract length 20 years. Additional concession payments of $3 million in 1996 and $2 million in 1997 were made to United Water. With the ability to amortize the total investment over an additional 10 years, United Water was able to hold the rate schedule as originally negotiated. An additional allowance of $250,000 was added to the capital program, but unlike the first $300,000, there was no requirement to refund unused capital funds to the city.

Another capital improvement that was completed as a result of the new wholesale water contract transaction negotiated with Jersey City was a new 24-inch-diameter interconnection between Hoboken and Weehawken. With this new interconnection, water can be delivered to Hoboken either from Jersey City or from United Water New Jersey. Thus, Hoboken has an alternate supply, which will ensure the city has an adequate supply in the event of an emergency by either supplier.

In summary, this transaction is truly a "win-win" situation for both parties. United Water has gained an important customer, which was a key element of its new growth strategy. Hoboken has received needed infusion of new capital that stabilized taxes and has a professional operator of it's utility. Water rates are controlled by contract, complaints no longer go to City Hall, the system undergoes constant improvement, and the waterfront growth is aided by the expertise of United Water crews and engineers in expanding the water infrastructure.

JERSEY CITY

History Of Relationship

Jersey City is a community of about 230,000 people. The history of its water system dates back to the 1700s when a well was constructed. The well was the source of water until 1854 when the city began to take its water from the Passaic River. In 1874, changes were made to the Passaic River, which altered the flow of the river, causing the salt line from Newark Bay to move upstream. Some evidence of sewer flows from the City of Newark were also being detected in Jersey City's water intake, and incidents of typhoid fever began to take their toll.

To overcome these problems, in 1895 the city entered into an agreement with the East Jersey Water Company for a supply of water from the Pequannock River in Morris County. This change resulted in a marked decrease in the incidence of typhoid fever, although the problem continued. In 1906, another improvement took place with the construction of a reservoir on the Rockaway River in Morris County and a 23-mile aqueduct system. This new supply also decreased the incidents of typhoid by another 50 percent. Water company engineers experimented with treatment of the water, including the use of hypochlorite and lime as well as chlorine gas. In 1913, Jersey City became one of the first water systems in the country to use chlorine treatment, which later became the industry staple for the disinfection of water. Another marked decrease in typhoid was realized, and by 1926 the problem was virtually eliminated.

United Water and Jersey City have worked together since 1992 to mutually benefit consumers of water in northeastern New Jersey. In times of drought, the resources of all of the water purveyors are pooled for mutual benefit. In the early 1980s, the Jersey City system was the key part of an emergency plan that "wheeled water" among four other purveyors: the City of Newark, the North Jersey District Water Supply Commission, the Passaic Valley Water Commission, and the Hackensack Water Company (now known as United Water New Jersey).

Procurement Process

In the fall of 1992, United Water was asked to present various plans to Jersey City's administration in an effort to streamline the water department, bring new revenue to the city, and stabilize rates. As with Hoboken, various alternatives were discussed, from the outright sale of the system to various forms of concessions and public-private partnerships. From our success with the City of Hoboken with innovative solutions, United Water began to work with Jersey City to determine exactly what formula would work best.

While a concession arrangement was explored with city officials, it became apparent that they were considerably uncomfortable with this plan, since under IRS regulations the city would have been required to refinance some $80 million to $90 million in existing debt. In addition, there was the ongoing need to finance new improvements. The city was also concerned about losing control of the entire operation. Unlike the administration in Hoboken, Jersey City officials wanted to maintain considerable control over the operation of their water system, as well as over rates and the use of revenue proceeds. As a result, all parties looked at various forms of subcontracting for the utility in total, or for some departments alone, such as billing and customer service.

Ultimately, the city decided to request proposals for the operation and maintenance of the entire water department, including the watershed, the reservoir, the treatment plant, the aqueduct system, the distribution system, laboratory testing, and fleet maintenance, as well as customer billing, collections, and customer services. In August of 1995, the Request for Proposal was issued with a deadline of October 6, 1995.

To manage this process, the city engaged the services of a consulting firm. The latter's responsibility was to publish the bid documents and guide the city through the selection process and contract negotiations. The city had appointed a selection committee representing a cross-section of city government to implement the process and to ensure that the city's objectives were met. Outside consultants also were added as advisors. This cross-section of people was the key to the success of this process. Since all facets of city government were represented, there was total buy-in by these interests when the ultimate proposal and recommendation needed to be made.

The approval process in Jersey City, as in most municipalities, requires that major contracts be approved by the city council. While the mayor and his administration can recommend the implementation of certain contracts, the city council must ultimately approve the transaction, while the mayor does not have a vote. Thus, having the city council directly involved in the process was important not only for keeping the council informed of the developments of the process, but to ensure the support of many of the council members. Inclusion of key personnel of the water department also proved beneficial since their support for the project was achieved early on. All too often these initiatives fail because of the internal politics of the municipality. More often than not, the city administration spearheads these efforts and the city council only sees the final result. If there is any division of motives between the administration and the council, these efforts can fail, especially if the council feels it has been excluded. Obtaining the approval of water department personnel was also important because this ensured that factual and unbiased information was presented to prospective bidders. Their expertise also ensured accuracy of the evaluation of the technical aspects of the various proposals.

Six proposals were received; five were from firms with national reputations and one was from a partnership of two local public agencies. The city had established six objectives it wished to achieve in this process. These objectives were related to economics, employees, operations, maintenance and replacement, capital and implementation, and customer service, billing, and collections.

The economic objectives included a concession fee, reduction in lost and unaccounted-for water, improvements in billing and collections, improved efficiency, and the sale of wholesale water. Employee objectives outlined the desire of the city for the contractor to provide enhanced employment opportunities, a comprehensive plan for transition, the desire to protect pensions and benefits, and giving hiring preference to city residents. The operations objective dealt with operating improvements and efficiencies, and with bringing the city into compliance with requirements under the Safe Drinking Water Act, specifically the lead and copper rule. Maintenance and replacement objectives included the implementation of computerized system records and work orders, a new radio communications system, meter replacement program, and recommendations on disposal of sludge from the plant. Capital planning asked the bidder to assist the city in identifying and implementing long-term capital improvements required for the system. Finally, the customer service, billing and collections objectives were aimed at not only improving billing and collections, but also improving the image of the water department with customers by improving the way people and their complaints are treated and resolved.

By delineating the objectives up front, the bidders had a clear understanding as to what the city expected of the private contractor. The selection committee then used these objectives to evaluate the proposals received. A rating system was devised and a point score was assigned to the various objectives. This enabled the selection committee to make as objective a decision as possible in selecting a contractor to negotiate the final contract.

Another feature of the bid document that the committee relied on was the inclusion of a draft contract. Respondents were asked to mark up this draft with their submissions in order to alert the city to issues that the contractor would want to revise. This also expedited the contract negotiations, since both parties knew from the outset what the contract issues were.

On December 11, 1995, the selection committee announced that it had selected United Water Resources as its finalist. By January 8, 1996, negotiations were concluded and a public hearing was scheduled for January 23. The major features of United Water's proposal included an innovative plan that guaranteed employment for all existing employees for at least one year and a mechanism by which employees could stay on the city payroll and retain their pensions and benefits. United Water would then reimburse the city for the salaries and benefits of these employees. United Water also waived the marketing fee on some seven million gallons per day of additional water that was to be purchased by its affiliate utility company. This was a benefit to the city of $18.5 million over the life of the contract, and guaranteed the city a new customer for 15 years. United Water's fixed fee proposal was projected to save the city another $17.5 million over the life of the contract, while United Water would make a payment of $2.5 million upon signing the contract as a "concession fee." Although this contract was not a concession type arrangement, the bid document referred to an up front payment as a "concession fee." In this case, the fee was for the right to use the city's vehicles and equipment, as well as the right to inventory materials and supplies. In total, the package was estimated to save the city $38.5 million over the first five years.

On February 14, 1996, the city council voted to approve the contract by a margin of six to two, with one abstention. The contract was signed on April 1, 1996 and implementation of the project began on May 1, 1996. One of the major accomplishments of this large procurement was that the original target date for implementation was achieved.

Implication

Labor One of the major accomplishments of this project was the manner in which the labor component was implemented. Three different municipal unions represented the Jersey City Water Department personnel: a supervisors union, a clerical union, and a "blue collar" union. Our task was to transform these people into United Waters' operations without layoffs and in a manner that would not result in reductions in benefits and pensions. As mentioned above, United Water guaranteed employment for all 135 people in the department for the first year, with a goal of downsizing to a staffing level of 80 people going forward. Our transition plan consisted of three basic phases: an orientation phase, followed by an assessment of the personnel, and finally a transition plan to hire as many employees as we found necessary to operate the department.

The orientation phase actually began prior to the implementation of the project. It was crucial that United Water eliminate as much apprehension as possible early on in the process. This was accomplished with open and honest communications. Over a three-day period the company conducted orientation sessions for all water department employees. The purpose of these meetings was to explain what United Water planned and what was expected of the employees. United Water stressed that the city's employees were under no obligation to make any radical decisions on their future during this process, especially since their employment was guaranteed, at least for the first year. The leasing program was explained and it was made clear that people could, at their option, leave the city's employment and come with United Water directly as an employee. The operative word here was "option." Employees were not pressured either way. All questions concerning employee benefits and work rules were answered at the orientation sessions. United Water obtained approval for broad-banded job titles including professional, clerical, technical and unskilled labor. This enabled the company to maximize the utilization of people in various jobs. This was also explained to all employees.

These sessions were followed by an assessment process. United Water used their standard employment application forms as a questionnaire on the background and interests of all employees. This was coupled with a preliminary physical screening that included drug abuse screening and blood pressure checks, as well as a cursory physical exam. This screening enabled United Water to learn the various talents of employees and target higher-profile people for direct employment by the company. The first direct hires included the city's director of operations and the director of accounting and finance. These hires not only gave United Water top-level experience with the city's system, but also sent a signal to the rest of the workforce that high-level people in the department had made a commitment to and were comfortable with the new company.

One of the more innovative features of the transition plan was the downsizing of the work force. As discussed above, United Water was committed to paying the salaries and benefits of all 135 people for the first year. We realized early on that it was beneficial not only to the company but also to the employees to identify those people who would be a part of the ultimate workforce early on, and to work on that transition prior to the end of the first year. The city also recognized the benefit of this plan and accordingly was fully cooperative with these efforts.

The first step was the announcement of an early retirement plan. A system of incentives was offered, paid by United Water from a fund that was created from the difference between its commitment to fund all 135 people and the actual payroll cost incurred after retirements and other attrition. As a result of the initial early retirement program, 14 employees took advantage of the early buyout.

Another factor that worked to reduce the workforce was that in accordance with the provisions of the RFP and the contract, United Water had the ability to relocate people up to 50 miles from the city. United Water's utility customer service center, which services most of our existing customers in Bergen and Hudson Counties, is located in Hackensack, New Jersey. It was our plan to integrate the Jersey City billing and collections effort with our existing operation. This required that the city employees assigned to these tasks would have to relocate. Many of these employees chose not to make this commute and sought jobs either back with the city itself or pursued other employment opportunities. Many employees in other departments also exercised these same options.

By June 1997, United Water had achieved its goal of downsizing to a level of 82 former city employees. Of this workforce, 37 people opted to become direct employees of United Water and 45 employees remained on the city payroll. Other functions are being supplemented by employees of United Water located in the corporate headquarters in Harrington Park, as well as Hackensack. Water testing and quality control are performed at the company's existing laboratory in Haworth, New Jersey.

The operation of the water treatment plant located in Boonton, New Jersey, 23 miles to the west of the city, is operated under the direction of our affiliate company, United Water Services, which has experience and a national reputation as a leader in contract operations of water and waste water treatment plants. Many of the employees at the Boonton plant opted for retirement and most of the remaining people remain as city employees, including the plant manager.

This project demonstrates that the fear many managers of municipal utilities have about losing their jobs from privatization is not grounded in reality. The truth is that the private entity requires the intimate system knowledge and abilities of these people and, by giving them training and access to resources not available from municipalities, improvements are made in quality and efficiency of operations. One major accomplishment brought about by United Water was bringing Jersey City into full compliance with EPA standards for lead and copper. Prior to this contract, Jersey City's water had some of the highest levels of lead among the major cities in the state. Today, water leaving the treatment plant is a higher quality product than ever before.

Billing System Another major accomplishment of this project was to implement a completely new billing system. United Water undertook to have a new system up and running within 90 days after the start of the contract. The city's existing system operated on an outdated mainframe application that was part of the city's overall accounting and tax system. It had many shortcomings compared to a traditional utility billing and customer service system. Many functions required manual input; records were not easily accessible to customer service representatives; and basic operating data, such as a summary of billings and consumption, was unavailable. Standard water utility computations, such as "lost and unaccounted for" water, were unachievable because the billing system could not summarize billed consumption.

The system United Water put in place was a new product developed by Electronic Data Systems (EDS), known as Customer Star II. This system is a state-of-the-art, client server based system, with complete online capabilities for billing and data access. Once implemented, the turnaround time from the date of the meter reading to the date of the billing to the customer was cut from 21 days to three days.

Meter reading was another area in need of improvement. The city's experience in obtaining actual meter readings was about 60 percent of the total billing. A number of factors led to this poor performance record, including the fact that 4,000 meters had broken remote devices. This was coupled with traditional access problems that resulted in a low percentage of actual reads. United Water's solution was to partner with Public Service Electric & Gas Company (PSE&G), which also serves Jersey City, to obtain meter reads. This partnership worked to the benefit of all parties. With the deregulation of the energy industry, utilities like PSE&G were looking for nontraditional ventures to supplement their operations. The sale of meter reads serves this purpose since it can be done at an attractive incremental cost. United Water benefited because it freed up the meter-reading work force acquired from the city. The results of this change ultimately benefited the city, since PSE&G now gets in excess of 80 percent actual reads.

Faster turnaround in billing with Customer Star II and an increase in accuracy in meter reads brought a monumental improvement to the billing process. That was the good news. The bad news for customer relations was that this new wave of accuracy also identified many accounts that had gone on for several years without actual readings and many others that never had received bills at all. Once readings were obtained, these accounts were hit with large increases in billing because of the back-billing of consumption after many years of underestimated bills. It was not uncommon to find accounts that had thousands of dollars in arrears due to inaccurate estimates and undetected leaks over a period of five years or more. For many Jersey City residents on fixed and low incomes, being hit with bills like these was traumatic. This became a challenge for customer service representatives to resolve inquiries from customers, many of whom had paid all their estimated bills on time, who wanted to know why all of a sudden they were now thousands of dollars in arrears. While interest on these accounts was waived and payment plans were worked out, the outrage of these customers was understandable. The magnitude of these "out-of-period" billings was exhibited in the initial increase in the total accounts receivable balance for water and sewer, which escalated from approximately $20 million to over $30 million, or an increase of 30 percent.

To make matters worse, within a month of the implementation of the contract the city began to plan for a sale of its tax liens to third-party investors. All municipal liens, including those for outstanding water and sewer bills, were to be auctioned off. This intensified the pressure experienced by customer service representatives to resolve the customer billing problems United Water had inherited. Naturally, with the new company rendering the bill on behalf of the city, many blamed United Water for the high bills. To address the new level of customer complaints, the company expanded the number of phone lines and personnel assigned to customer service.

To reverse the public's poor perception of United Water, an intense advertising campaign was launched to explain to their customers the new services offered. In addition, company executives accompanied the mayor to local neighborhood town meetings in order to answer any questions related to the new utility billing system. Two new payment centers were made available. The partnership with PSE&G allowed customers to pay water bills at their existing utility payment center, located at Journal Square, a major transportation hub in Jersey City. United Water opened a project headquarters in a new housing and commercial complex located near Liberty State Park for customers to pay bills and meet personally with customer service representatives to resolve problems. Customers still can pay bills at the tax office in city hall by way of a drop box, but personal service is offered at United Water's office.

The company also conducted a customer survey within the city to plan for new initiatives targeted at improving the image of the project. To their surprise, while the volume of complaints received from their customers tainted United Waters' perception of public opinion, the overall perception of the company's performance was somewhat positive. People felt that problems were being resolved and that access to customer representatives had improved. Improvements in service repairs were also evident to customers. All this reinforced the company's desire to continue to communicate to customers in a manner unprecedented in the city.

Today, collections are substantially improved. At the end of the first year of the contract in 1996, collections and increases in the sales of wholesale water had improved by $6 million (or 13.5 percent). There was an additional increase of $660,000 in the second year. In the third year of the contract, United Water increased revenue by another $7 million over the prior year, for a total of $13.7 million in new cash receipts to the city in the first three years of the contract. The improvement in collections is shown in Figure 5–1. Customer inquiries have been substantially reduced over prior years and the company continues to work on reducing the outstanding accounts receivable balance through an intensified collections effort and the write-off of erroneous or questionable back charges.

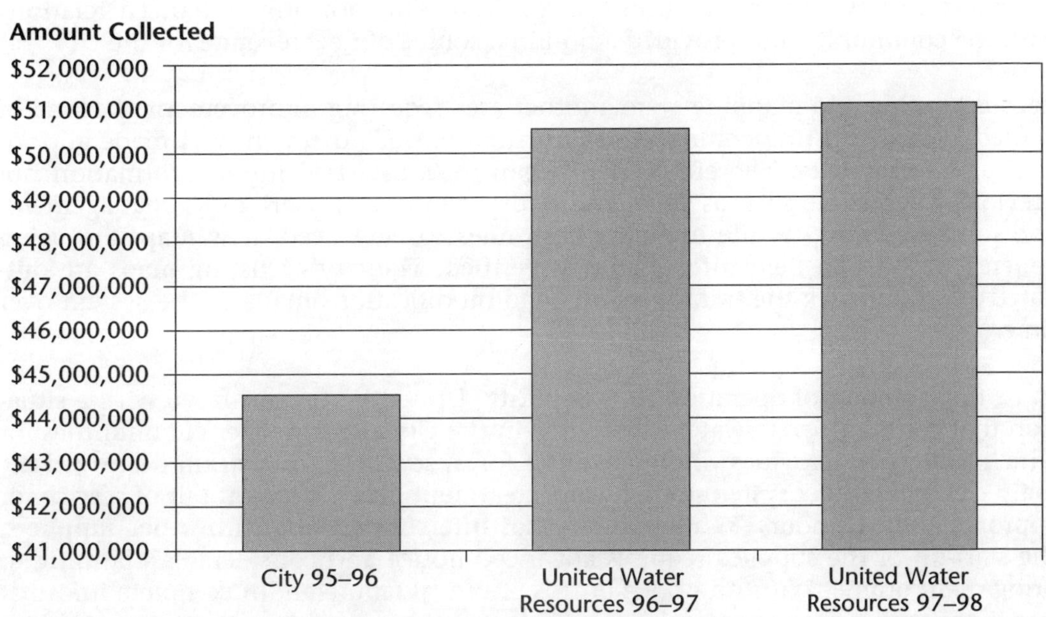

Figure 5–1. Jersey City's Total Water and Sewer Collections

Operations The operations of the water department include the operation and maintenance of 120 square miles of watershed, two reservoirs totaling 11.3 billion gallons (in capacity, including two high-hazard dams), an 80-mgd water treatment plant, a 23.6 mile aqueduct system, 280 miles of distribution mains ranging in size from 4" to 60" in diameter, 5,300 line valves, 3,700 hydrants, and 33,400 service lines and meters.

When United Water took over the Jersey City water department system, some basic service systems needed to be overhauled. The department's aging equipment was under much repair by the city's motor pool. Upon implementation of the contract, United Water placed orders to completely replace the fleet with new leased vehicles. City-owned vehicles were returned to the city for reassignment to other departments. The value of these vehicles exceeded $500,000. With new vehicles, repair costs were reduced and vehicle downtime was virtually eliminated.

In addition to opening a new office in Jersey City, a new construction yard was opened for crew dispatching. This freed up a former water department building, which has since been converted to a new police precinct.

Radio communications was another area that required total renovation. The frequency used by the department was being shared with other departments, and coverage both inside and outside the city was lacking. United Water worked with the city in obtaining a permit to use another city-owned frequency. Through the purchase of new equipment for the vehicles, new base stations, and strategically placed repeater stations, radio communications was enhanced tremendously.

During the first contract, United Water implemented a new recreation program for the city's watershed. Subcontracting out the administration of this program, city residents and nonresidents were issued permits for shoreline fishing, hiking, and bird-watching on city property around the reservoir. This not only enhanced relations with the community, but provided a nominal source of new revenue for the city.

Record keeping and mapping were another area requiring improvement. A clone of United Water's existing computerized distribution record system was made and installed in Jersey City. The effort then began to update and input information not previously accessible such as valve and hydrant locations, work order management, and job scheduling. While mapping continues to be a need, new maps are being generated as actual field information is verified. The city's existing maps are outdated and often lack the improvements and modifications made to the system over time.

In its first summer of operation in Jersey City, United Water was faced with a situation that caused a severe water emergency in the city and for other communities for which Jersey City provides wholesale water for. A severe thunderstorm destroyed the main electrical feeder system to the water treatment plant, taking it out of service for approximately 12 hours. While the city has interconnections with other suppliers, the start-up of the supplies requires advanced notice and considerable planning to bring them online. With no notice of this electrical failure and incomplete information from the power company on the timing of the restoration, there was no alternative but to bypass the plant, allowing nontreated water to enter the aqueduct.

While such a practice does not conform to standards for operation, the city had used this practice during such interruptions in the past. The plant was put back online in about 12 hours, and the city was never without water service. A regional "boil water notice" was placed into effect, and over the next three days the system was tested and flushed with water that was bacteriologically safe. Chlorination was increased to disinfect the system and the "boil water notice" was lifted once the test results confirmed no evidence of bacteria.

This demonstrated an acute deficiency in the city's system, and the lack of stand-by power. While the plant, which was commissioned in the late 1970s, had provisions for a stand-by generator, the unit had been inoperable over the past 20 years. Within one month of this incident, under the direction of United Water, a temporary generator unit was rented and installed at the plant. A permanent unit was in the process of being designed by city engineering but the decision was made not to further delay installation and thus end the city's exposure to this type of event.

Training was another key element in the transition of employees and improvements in the operation. New training programs included safety training and the formulation of a safety committee consisting of a cross-section of people from various water department functions. Safety experts from United Water guided this program. Other training included training on excavation safety, confined space entry, (which impacted people reading meters in vaults and pits), right-to-know procedures, new computer training, and the safe handling of hazardous materials. Other safety improvements included the elimination of the use of highly lethal gaseous chlorine in treatment facilities by conversion to the use of hypochlorite. This change not only improves safety for employees, but for people who reside near these facilities as well.

Automation is another key element of United Water's focus in Jersey City. Through the installation of new technology in one of the gate houses on the aqueduct and more recently at a pump station in the city, United Water will be able to further eliminate labor-intensive functions, freeing up people for other assignments within the organization for implementing new programs for Jersey City.

The latest program to bring quantifiable improvement to water quality is the implementation of a chemical process at the water treatment plant, which reduces the corrosivity of the water. With the latest samples taken, Jersey City has met the requirements of the Federal Safe Drinking Water Act as it relates to lead and copper detected at customer's premises. In the past, Jersey City was in noncompliance with this regulation. As stated above, this was a major objective of the city in the privatization initiative.

The latest achievement was the operation of the system during a shutdown of the aqueduct system. This aqueduct system, built between 1900 and 1910, was scheduled for inspection. Three teams of engineering experts inspected four tunnel sections of the system. While most of these 8.5-foot-diameter tunnels were built with reinforced concrete poured in place, the deeper tunnel sections consisted of brick and mortar construction. The engineers were asked to assess the integrity of these sections and to make recommendations, if needed, on rehabilitation and repair. As mentioned above, bringing other suppliers online takes a considerable amount of planning and coordination. In the past, when this operation was attempted by the city, the shutdown and subsequent start up of the aqueduct resulted in wide-spread incidents of discolored water in the city and among the other wholesale customers. With a careful coordination of all departments and a well-executed plan, the shutdown, which occured in 1998, was accomplished and water was supplied from the Newark and the Passaic Valley Water Commission without a single complaint of discolored water. Start-up of the plant and aqueduct were also accomplished without incident. Not only did the inspection reveal that the integrity of the aqueduct was better than expected but this shutdown procedure, now documented, provided a plan for future successful operations. The company also now knew exactly what procedures were needed in the event of an emergency shutdown.

OVERSIGHT

One of the issues that traditionally comes up in privatization initiatives is how much or how little control a city has in overseeing the performance of the contractor. Some cities, like Hoboken, have complete confidence in United Water's operation and, since collections no longer affect the financial health of the city, they are comfortable relinquishing a great deal of control. In Jersey City's case, the financial health of the city was directly linked to the performance of the contract, and since the operation of the department was more complex and required more interface with the city engineer for construction projects, more control and accountability was desired.

Early on in the contract/bidding process, it was apparent that the oversight meetings needed to be separated into two disciplines, one for operations and the other for billing, collections, and finance. With all of the issues related to customer billing, restoration of customer complaints, establishing a system of reporting, and coordination of the city's auction of its tax liens, half-days are devoted to the financial phase of the project on a monthly basis. While operations are not as complex, the agenda takes almost as long to start on a monthly basis, due to the number of projects being coordinated between the city and United Water.

Since the chief operating officer and the chief financial officer of the water department took positions with United Water, the city felt it was necessary for United Water to hire two consultants to compensate for lost expertise. Both the billing/collection/finance committee and the operations/construction planning committee are very active and meetings continue on a monthly basis.

SUMMARY

While each of these contracts is unique, they both share the distinction of breaking new ground in the municipal water privatization industry. The Hoboken concession-type contract was the first of its kind in the United States. The Jersey City contract, at the time it was implemented, was the first and largest full-service contract where the operations, maintenance, management, and administration—including customer billing and collecting—of an entire municipal water utility were outsourced.

Both of these contracts have been enormously beneficial to the cities. Hoboken, in addition to the $10 million in concession fees, now has a professionally operated water utility that is providing the infrastructure planning and capital improvements required for the city's renaissance. The economic benefit to Jersey City in just two years approaches $21.6 million. This compares favorably to the $35.5 million projected for a full five years. In addition, water quality in the city has improved, customers are getting more accurate bills, and the water system is being upgraded.

Based on the success of these and other projects, the number of municipalities exploring privatization and issuing requests for proposals continues to increase. As our cities age and our water and sewer utilities require major capital improvements, municipalities will continue to look to the private sector to bring efficiencies and improvements to their utility and wastewater customers.

CHAPTER 6

Sale of a Municipal Wastewater System: Franklin, Ohio

Douglas Herbst and Keith Oldewurtel

In 1996, the City of Franklin, Ohio and Earth Tech, Inc. entered into a 20-year service contract whereby the private firm would provide a total package of services for the design, build, finance, operation, and ownership of a new 5-million-gallon-per-day (mgd) water treatment facility. By choosing this innovative public service delivery option, the city gained significant cost savings and performance accountability compared to using conventional approaches involving separate contracts and service providers.

BACKGROUND

The City of Franklin, founded in 1796, occupies 8.1 square miles adjacent to the Great Miami River in the northwest corner of Warren County in southwestern Ohio.* Today, the city has approximately 11,500 residents and an expanding commercial sector, including a large-scale newspaper printing plant.

Prior to privatizing its water treatment plant, Franklin possessed more experience in the privatization of local service delivery than most communities. In 1995, the city was among the first municipalities in the nation to sell a federally funded wastewater treatment plant to a private firm. Franklin and other neighboring jurisdictions approved the sale of the regional facility that served the area. Through this action,

---

Douglas Herbst is senior vice president of Earth Tech, Total Water Management Division and president of the National Council for Public Private Partnerships and Keith Oldewurtel is executive vice-president of Earth Tech, Total Water Management Division.

*See Figure A–2 in the Appendix.

Franklin was able to reduce utility rates by 23 percent during the first year and, with the exception of energy and chemical costs, to limit future increases to the rate of inflation (Reinhardt, 1996).

Stable sewage rates, along with a concerted push for economic development and effective capital planning of water distribution and well field expansions, prompted the entry and expansion of many businesses in the area. Sixteen manufacturing plant expansions or new facilities worth $95 million were started in Franklin between August 1994 (when an agreement was reached with the private firm for the wastewater treatment plant purchase) and December 31, 1995 (Reinhardt, 1996).

The city drilled two new water wells to meet the anticipated water needs of its expanding commercial sector, bringing the total number of wells to six, with a combined peak output of 4-mgd. In order to protect the community's water supply, the city also purchased approximately 300 acres above the underlying aquifer. The city then raised water rates from $1.25 to $2 per 1,000 gallons and qualified for a $5 million loan to build new water lines and storage tanks to support the rapid economic development in the area (Oldewurtel, 1998).

Following the completion of the two new wells, as well as lines and tanks, the city had enough transmission and well field capacity to sign new long-term water/wastewater utility contracts with private industry. However, EPA regulations came into play when the new water supply wells were tapped, requiring the city to begin treating its water by October 15, 1997 in order to meet federal Safe Drinking Water Act regulations.

THE DECISION TO SEEK A PUBLIC-PRIVATE PARTNERSHIP

City leaders estimated a new water treatment facility would cost between $5 million and $15 million. The city had borrowed heavily to build the new water lines and storage tanks and because of this, city leaders determined that a bond issue would have to be floated to pay for the new plant if it was to be procured using conventional methods. The city was concerned about this potential debt burden.

Due to the recent successful privatization of the regional wastewater treatment facility, however, city leaders simultaneously discussed the concept of private sector involvement in the water treatment plant project. Upon further evaluation of its financial and technical resources, the city council determined its goal was not to build or operate or own a new plant, but to provide treated water. Private ownership of the water plant was not a major issue for the city for several reasons. The first reason was that it would be a new facility and there was no existing plant that needed to be replaced. Second, the city had recently been part of the regional wastewater treatment facility that was being sold to a private firm and any issues of private ownership had been resolved at that time.

According to Franklin Mayor Jim Mears, "We ultimately concluded that all we really wanted was treated water, so privatizing the entire process seemed the best answer." (Mears and Uldewurtel, 1997). City leaders also determined that taking a public-private partnership approach would allow Franklin to capitalize on its water assets sooner than if it continued to wait for long-promised federal funding assistance for the design and construction of a new water plant (Reinhardt, 1996).

The status of employees is often an issue when a community moves towards privatization or a public-private partnership; however, since Franklin had no existing water plant, there was no issue concerning an existing workforce. In addition, based on the city's experience with privatizing the wastewater treatment plant, local leaders realized that private firms typically hired from the area to fill new staff positions.

The city decided to pursue a full privatization approach with private sector design-build-finance-operate (DBFO) services with private ownership to develop the facility. By combining these four key components, city leaders concluded that the potential existed for significant cost savings and improved plant performance when compared to a conventional approach in which separate contracts and separate organizations are used to design, build, finance, and operate a facility.

As applied to the Franklin water plant, a DBFO approach would entail a single contract for services with one private sector firm. This firm would then be responsible for designing, building, financing, and operating the plant under a long-term service agreement with the city. The city would pay for services in accordance with a negotiated price schedule, and the proposers would be required to meet rigorous plant and operational specifications. Competition among proposers would lead to selection of the firm with the best competitive price coupled with the highest performance guarantees.

City leaders identified three primary mechanisms as integral to the process of ensuring that Franklin's interests would be protected. These were:

1. **Guarantees.** Performance risk, guaranteed lump-sum pricing, and regulatory compliance and protection from fines would become the responsibility of the private partner.
2. **Demonstrated Capabilities and Qualifications.** The private partner chosen would possess demonstrated technical knowledge, related experience, and financial resources.
3. **Well-Defined Contract Terms and Conditions.** Contract language would clearly outline the requirements and the roles and responsibilities of the city and its private partner.

THE PROCUREMENT PROCESS

With the help of a qualified procurement consulting firm, the city wrote a Request for Qualifications (RFQ), issued on February 9, 1996, to announce the city's intention to provide the private sector with an opportunity to design, build, finance, operate, and own a water treatment facility and related works (termed the "Service Program"). The city also invited the most qualified and experienced firms to respond to the subsequent Request for Proposals (RFP) for the Service Program.

The city's requirements for the successful service provider included the demonstrated ability to provide this service (*i.e.*, design, build, finance, operate, and own the water plant) as a single and complete responsibility proposer without the need to utilize or impact the city's borrowing capacity or ratings. Another requirement was that the drinking water treatment plant be operational and meet all regulatory requirements by November 15, 1997 (City of Franklin, 1996).

The city placed a national advertisement announcing the city's interest in pursuing a public-private partnership for the design, build, financing, operation, and ownership of a new facility and asked for expressions of interest from firms. Twenty-three firms initially expressed interest in the project and the Statement of Qualifications (SOQ) process was used to narrow the field of interested firms to a subset of seven. A list of minimum criteria was established for the firms, including:

- financial analysis (financial wherewithal to provide financing, contract security and guarantees);
- history of facility operations;
- design/build experience; as well as
- assurance that the firms were prepared to complete the program in the timeline that was needed by the city to comply with the consent order.

After evaluating the minimum criteria, the SOQ Evaluation Committee unanimously selected four firms to be invited to participate in the RFP process. Three of the four firms submitted proposals in response to the RFP document (Eisenhardt and Dong, 1996).

Selecting the Private Partner

When going about selecting the private partner, city officials recognized they had an obligation to the community to develop a fair, competitive process that would clearly show that the City of Franklin was getting the best performance for the best price. The city did not simply want the lowest price; it wanted the best price from a qualified firm with the technical and financial resources and experience necessary to successfully design, build, and finance the plant. Upon plant completion, the selected firm also needed the qualifications to operate the facility in a highly professional and efficient manner. To complete the picture, the firm had to have the ability to guarantee its performance and provide the necessary contract security.

City leaders established an evaluation committee comprised of the public works director, city engineer, city manager, a member of the city council, and the procurement consultant. The city's legal counsel and members of the city council provided additional input and review, although to avoid the potential for politics playing a role in the selection process, the city's selection committee did not release the names of the competing firms until a recommendation was in hand. (Eisenhardt, 1999).

To ensure that the proposals were responsive to the city's needs and were evaluated in a consistent and rigorous manner, the procurement consultant worked with the city staff to develop criteria (as outlined in the RFP) against which the proposals would be evaluated. In addition, the evaluation committee assigned scores to each criterion in order to quantitatively evaluate each proposal. The scoring system emphasized the importance of qualitative criteria, such as qualifications, technical approach, and guarantees in relation to cost (see Table 6–1).

TABLE 6–1. PROPOSAL EVALUATION CRITERIA

Score		Criteria
100		**Qualifications**
	50	Corporate qualifications:
		—Clearly demonstrated experience to undertake the proposed project's scope of work: design, construct, own, and operate a water treatment plant using a public-private partnership approach
		—Commitment to the business: number of years in business, firm size, financial resources
		—Demonstrated ability to complete this project in the established time frame and work scope
	50	Qualifications of key individuals
175		**Technical Approach**
	25	Design and construction of water treatment facilities and related works
	25	Expansion capability
	25	Quality of operations and maintenance plan
	25	Suitability of staffing arrangements; interface with city operations
	25	Demonstration of contingency plans—emergency preparedness
	25	Soundness of control and monitoring mechanisms
	25	Schedule
350		**Financing Approach**
	25	Demonstrated financial commitment to the project
	25	Financial capacity of the proposer
	25	Proposer's payment terms and adjustment methodology
	25	Financing of self-supporting infrastructure projects
	25	Buy back
	25	Expansion capability
	200	Cost
100		**Responsiveness to RFP Requirements**
	50	Liability coverages, insurance, guarantees, performance bond, and risk assumption
	25	Responsiveness/timeliness
	25	Responsibility for compliance with local, state, and federal laws and regulations
725		**TOTAL**

Source: Paul Eisenhardt and Anh Dong. "Water Service Project Proposal Evaluation And Recommendation," *Report to the City of Franklin, Ohio*, July 19, 1996, p. 2.

All proposers were invited to present their proposals in person before the evaluation committee. A question and answer session followed the interviews. In addition, the city manager and public works director toured facilities operated by each of the proposers. In each instance, the city representative discussed the facility operations with both the facility owner and operator. The results of these site visits supplemented the reference checks conducted during the SOQ evaluation phase.

Proposals were scored against each criterion on an ordinal scale. In reviewing the proposals and conducting interviews, the evaluation committee worked to balance the trade-off between cost and additional qualitative factors. Therefore, the committee evaluated the proposals both with and without factoring in the cost criterion. In both cases, Earth Tech scored the highest. In addition, Earth Tech was the only firm to offer full equity financing and to assume the cost and scheduling risks associated with regulatory approvals. The equity financing approach allowed Earth Tech to make a strong commitment to financing, with no contingencies as to project debt. Equity financing also represented a very timely financing plan (Eisenhardt and Dong, 1996).

After careful review and evaluation of the proposals, interviews with designated representatives of each of the proposers, site visits, and reference checks, the evaluation committee unanimously scored Earth Tech as the highest-ranking firm and recommended that the city council commence negotiations with the firm for a 20-year service contract to provide DBFO services, and private ownership, for the water treatment plant.

THE EARTH TECH PROPOSAL

Earth Tech is an international provider of total water management services and is a wholly owned subsidiary of Tyco International, a multibillion-dollar international holding company. Earth Tech is a national expert in contract operation, maintenance, and management of municipal water and wastewater treatment plants, providing financing for DBFO projects for municipal public works and private industry. Founded in 1970, the firm employs more than 6,000 staff members in more than 100 offices worldwide.

The firm's proposal to the City of Franklin included the delivery of several key benefits:

- Design/build plant for guaranteed lump-sum price;
- Guarantee to meet plant startup deadline;
- Guarantee of lump-sum O&M price;
- Performance guarantees;
- Provide a facility utilizing proven technology that had received preliminary regulatory approval;
- Provide self-financing—no outside investors; and
- The ability and commitment to finance future improvements.

In its proposal, Earth Tech discussed its sole source accountability for the project. Additionally, the firm would function as the general contractor in total control of construction costs and schedule for the new facility. Financing for the project would come from company cash reserves and the firm would accept all risk and liability associated with the owner/provider of water treatment service in the City of Franklin.

Earth Tech guaranteed to assume all regulatory risks to deliver the project with completion scheduled and water service initiated prior to the state-mandated deadline. Before presenting its proposal to the city, the firm had performed a preliminary design for the plant and submitted it to the Ohio Environmental Protection Agency (OEPA), incorporated the agency's review comments into its proposal, and obtained preliminary approval of a conceptual design. As a result, the facility design was approximately 35 percent complete prior to the firm being selected (Earth Tech, 1996: 47).

Earth Tech accepted all risks associated with gaining regulatory agency approvals for the design of the new facility. In its technical approach, the firm proposed a pressure filtration system with a hydraulic loading rate of 5-gpm/square foot. "Recommended Standards for Water Works," (Ten States' Standards) specifies loading rates of 3-gpm/square foot for both gravity and pressure filters treating groundwater. While this loading rate constitutes best engineering practice for gravity filters, higher loading rates are widely used for pressure filtration systems. Based on Earth Tech's experience, and that of established pressure filter manufacturers with numerous installations in Ohio and the Great Lakes states with groundwater of similar quality as Franklin's, this higher level could be easily achieved and sustained and would provide the Franklin facility reliable service and high water quality at reduced capital costs (Earth Tech, 1996: 23).

Earth Tech accepted the entire regulatory agency approval risk for obtaining OEPA approval for the design loading rate as well as full responsibility for obtaining all required permits and approvals to construct and operate the Franklin water treatment facility. If the facility did not receive OEPA approval, then Earth Tech would absorb all the resulting additional capital and operational costs. The firm also assumed the "timing risk" of initiating construction prior to final OEPA approval in order to meet the construction completion schedule (Earth Tech, 1996: 76).

Earth Tech stated in its proposal that, as a wholly owned subsidiary of a multibillion-dollar holding company, it was well-positioned to secure performance bonding for this project. The firm proposed providing Franklin with a performance bond in the amount of 150 percent of the first year's estimated service fee as security for the faithful performance and payment of all obligations stemming from regular operation and maintenance (O&M) of the facility. In addition, the firm demonstrated bonding capacity and the ability to secure all necessary insurance coverage for the project.

Technical and Operational Approach

Earth Tech evaluated several options to determine the optimum design approach for the Franklin treatment facility. Based on this evaluation, the firm selected pressure filtration using manganese greensand media as the most reliable, cost-effective, and expandable technology to achieve the city's immediate and long-term treatment objectives.

With this design, groundwater is pumped from city wells and treated to remove iron and manganese, ensuring a safe water supply with the addition of chlorine and fluoride. Raw water is pumped directly to the treatment facility where dissolved iron and manganese are turned into solid particles. The water passes through pressure filters for removal of these solid particles. Filtered water then flows through a clearwell where chlorine and fluoride are added. The finished water is pumped to the city's water distribution system.

The technical proposal fully outlined how Earth Tech, under the DBFO delivery mechanism, would combine the traditional design and construction components to create a service that provides for sole-source accountability for all elements of the project, including:

- Design
- Permitting and regulatory approvals
- Construction management, including: subcontracting, equipment purchasing, scheduling, expediting, and onsite inspection and coordination
- Startup
- Long-term O&M service

In contrast to traditional projects where individual elements occur in rigid sequence, the delivery system outlined in the firm's technical proposal would be carefully mapped from the start of design through system startup and long-term operations (Earth Tech, 1996: 37). To ensure that all elements of the project would begin and be completed within the time schedule and that on-time delivery of services would be assured, the proposed delivery system would provide:

- Input by systems operations specialists with field experience during the design process to ensure a truly workable design.
- Designers and schedulers coordinating critical design elements.
- Equipment specifications developed through input from both engineering and field operations specialists.
- Onsite construction activities coordinated to match equipment delivery.
- Bringing the treatment plant online with experienced startup personnel who would then transition to long-term operation as a consistent team.

In its proposal, the firm specifically tailored a work program for the complete O&M of the city's water treatment plant. A comprehensive operation, mobilization and startup schedule detailing the steps to be performed in bringing the new system online was included in the proposal to address the key issues to successfully operate the new facility. The firm also included plans for providing for all personnel and associated wages, salaries and benefits; all services; all materials, including chemicals, fuel, vehicle and equipment expenses, and all utilities and other consumables needed to operate and manage the facility while meeting all applicable laws.

The proposal outlined how Earth Tech would provide services to include process control and maintenance management systems, quality assurance systems, and other risk protection to structures and equipment to ensure efficient service, long-term reliability, and preservation of the capital investment.

In its proposed operating approach, the firm outlined that it would assume responsibility and protect the city from charges relating to faulty and negligent operation or operation not in conformance with applicable laws, including payments of any fines levied by regulatory authorities.

Financial Approach

Franklin approached the financing plans of the proposers as part of the entire package, rather than consider which specific financing plan or combination of debt and equity was more advantageous. Franklin focused on the bottom line: the service charge that the city would be obligated to pay (Eisenhardt, 1999). The firm's proposal resulted in the lowest service fee, as well as the lowest life cycle cost. The equity financing was backed with a strong commitment that if Earth Tech was selected, the financing would be available (provided that acceptable terms and conditions for the service agreement were realized) to meet the time requirements of the RFP. While on the surface the equity financing approach appeared to demand a higher cost of capital, it was determined that this financing approach could be accomplished cost-effectively and in a more timely fashion. The absence of costs of issuance typically associated with a debt-to-equity financing approach, coupled with the relatively small capital needs in relation to issuance costs, made the equity financing approach competitive.

In addition, Franklin evaluated other areas that would impact its bottom-line cost. These included the raw water quality parameters (principally iron and manganese levels) that each proposer would allow by which the guaranteed operating cost and performance obligations would remain in place. Also considered were the service fees and the escalation methodology proposed for the service fees. Sensitivity analyses were run on different quality raw water and the service impacts were evaluated (Eisenhardt, 1999).

Terms and conditions in Earth Tech's proposed finance package were structured to minimize the city's financial commitment and risk and to ensure that the city's borrowing capacity/ratings would not be utilized or impacted. Earth Tech proposed the establishment of a monthly service fee to Franklin consisting of two components: 1) a facility charge and 2) an O&M charge. The service fee arrangement would provide guaranteed costs to the city; the private firm would bear the risk for performance with adjustment only provided for Cost of Living Allowances (COLAs) and price increases for certain variable costs in the O&M component (Earth Tech, 1996: 69).

Facility Charge

The design/build of the water treatment plant was determined as a lump sum, no change order contract. Earth Tech guaranteed delivery of the water plant at a fixed price. The design/build fee, together with finance charges, results in the Facility Charge.

O&M Charge

O&M service consists of two components: 1) fixed cost—a guaranteed lump-sum cost that will only change in accordance with annual COLAs; and 2) variable cost—a consumption-based fee changed annually to account for actual unit cost increase/decrease (see Table 6–2 for a list of items included in each component of the O&M charge).

Options for the city to purchase the facility at the end of the Service Agreement or extension of the Service Agreements were provided, as well as an option to purchase at any time upon cancellation of the Service Agreement.

CONTRACT NEGOTIATIONS

Once Earth Tech was selected, the contract was fully negotiated in less than 35 days. The negotiation process was greatly expedited by use of the term sheet, a clearly written "principles of agreement" (without the "legalese") that had been developed at the beginning of the procurement process and adjusted throughout this period.

TABLE 6–2. ITEMS INCLUDED IN FIXED COSTS AND VARIABLE COSTS

Fixed Costs	**Personnel Services** • All onsite staff salaries • Earth Tech administrative staff • Overtime, holiday pay, meal allowances • All health, welfare, and insurance benefits **Equipment** • Office equipment • Lab equipment • Safety equipment • Tools • Communication equipment • Maintenance equipment **Materials and Supplies** • Cleaning and custodial • Lab supplies and chemicals • Uniforms • Office supplies **Other** • Performance Bond—(Full Service Fee—Facility Charge and O&M)	**Outside Services** • Electrical instrumentation • Groundskeeping • Major mechanical repair • Diagnostic services • Laboratory analyses • Equipment rentals • Insurance • Other professional services **Maintenance and Repair Costs** • Buildings, HVAC • Small mechanical repair • Spare parts • Equipment rental • Lumber and paint • Offsite Earth Tech mechanical specialists **Utility Costs** • Water • Telephone • Heating fuel • Refuse • Telemetry
Variable Costs	**Utility Costs** • Electric **Chemicals** • Polymer • Potassium permanganate • Chlorine • Fluoride	**Residuals Management** • Removal • Disposal • Compliance testing

Source: Earth Tech, "Proposal For 5-mgd Water Treatment Service To The City of Franklin, Ohio," July 24, 1996, pp. 70–71.

The term sheet specified the scope of services, liability assumption, guarantees, and the responsibilities of both parties. When interviewing each firm during the selection process, the committee further clarified how each firm's proposal lined up to the city's term sheet. Because both the city and the private firm agreed on the principles in the term sheet prior to contract negotiations, the negotiations consisted primarily of both parties' lawyers clarifying the "legalese" (Eisenhardt, 1999).

FRANKLIN SERVICE AGREEMENT SUMMARY

The Service Agreement is a typical service agreement between a private sector vendor and a public sector partner for the design/build/finance/operation and ownership of a water treatment facility (see Table 6–3).

Design/Construction

Earth Tech's obligation was to provide for all the design engineering, construction, inspection, labor, materials, tools, testing equipment, and subcontracted items necessary for the execution and completion of all structures and equipment required for the design, construction, start-up, and acceptance testing of a water treatment facility to produce 5-mgd of finished water meeting state, local and federal drinking water requirements.

Earth Tech secured all permits and approvals with Franklin's assistance. The facility design specifications were included in an appendix and contained the parameters and corresponding values of these parameters for contractual compliance as well as the USEPA Drinking Water Standards the facility had to meet. The contract allowed for Franklin to perform oversight and monitoring activities of the design and construction. Franklin had the right to request reasonable changes to the project with any resulting adjustments to the service fee being made.

Earth Tech provided 100 percent Performance and Payment Bonds. Franklin provided the facility site, which will be governed by a long-term lease that was also appended to the contract. Earth Tech guaranteed to have the facility operational by a date certain with allowable delays only for changes requested by Franklin and uncontrollable/force majeure circumstances. Acceptance testing protocol and procedures were appended to the contract; they allowed for Franklin to participate in the testing to assure that the facility performed as expected.

Operations

Earth Tech's obligation is to operate and maintain in good working order and repair the facility in accordance with generally accepted industry practices within the design specifications to produce finished water in accordance with the contract. Earth Tech applied for and obtained all permits and approvals, with assistance as required by Franklin. Franklin is required to deliver raw water as defined in the contract and, so long as that occurs, Earth Tech's guarantee as to finished water quality and quantity is in place. If raw water is not as contemplated by the parties, relief was afforded

TABLE 6–3. INDEX OF AGREEMENT FOR WATER TREATMENT SERVICE

Article 1	**Construction of the Facility**
1.01	Earth Tech's obligations
1.02	Approvals
1.03	Responsibilities with respect to construction
1.04	Royalties and patents
1.05	Performance and payment bond
1.06	Franklin's responsibilities
1.07	Subcontracts
1.08	Contract time schedule
1.09	Acceptance testing
Article 2	**Services**
2.01	Operations and maintenance
2.02	Residual disposal
2.03	Additional services
2.04	Increasing water sales
2.05	Cost savings improvements
2.06	Franklin's responsibilities
2.07	User charge system
2.08	Performance bond
Article 3	**Facility Price and Annual Service Fee**
3.01	Facility price
3.02	Annual service fee billing procedures
3.03	Calculation of annual service fee
3.04	Additional
Article 4	**Breach**
4.01	Damages
Article 5	**Termination**
5.01	Termination by Franklin
5.02	Termination by Earth Tech
5.03	Termination in case of certain events
5.04	Transfer of title
5.05	Delayed payment
Article 6	**Liability and Indemnity**
6.01	Indemnification by Earth Tech
6.02	Indemnification by Franklin
6.03	Right to defend
6.04	Authority to contest
6.05	Limitation on scope of liability
Article 7	**Uncontrollable Circumstances, Force Majeure Occurence**
Article 8	**Term/Conditions**
8.01	Term
8.02	Conditions
Article 9	**Purchase/Renewal**
9.01	Purchase option
9.02	Renewal
9.03	Special early purchase option
Article 10	**DisputeResolution**
10.01	Scope
10.02	Covenant to continue work
Article 11	**Insurance and Waiver of Subrogation**
11.01	Earth Tech's liability insurance
11.02	Comprehensive general liability insurance
11.03	Franklin's liability insurance
11.04	Insurance to protect project
11.05	Waiver of subrogation
Article 12	**Representations**
12.01	Earth Tech representations
12.02	Franklin representations
Article 13	**Miscellaneous**
13.01	Reports and records
13.02	Modification and waiver
13.03	Notices
13.03	Headings
13.04	Third-party rights
13.05	Entire agreement
13.06	Counterparts
13.07	Choice of law
13.08	Assignment
13.09	Set-off
13.10	Construction
13.11	Payment and interest
13.12	Severability
13.13	Further assurances

Schedule I — Definitions
Schedule II — Dispute resolution procedure
Appendix A — Facility design specifications
Appendix B — Acceptance test procedures
Appendix C — Sample service fee adjustment calculations
Appendix D — Facility site lease

Source: Earth Tech, "Agreement With The City of Franklin, Ohio, For 5-mgd Water Treatment Service," October 21, 1996, pp. ii–iii, index.

Earth Tech for both cost and performance. However, Earth Tech is under a best efforts obligation to mitigate impacts to both cost and performance. Other Earth Tech obligations include:

- Performance of sampling and analyses.
- Metering at three key locations.
- Inspection and testing of meters with Franklin representation at inspections and testing.
- Full responsibility for facility staffing.
- Compliance with all local, state, and federal regulatory requirements.
- Maintenance of records and preparation of reports with rights for Franklin to inspect records and receive reports.
- Develop Preventive Maintenance and QA/QC programs including computerized preventive maintenance programs, inventory control systems, process control programs, and laboratory quality assurance/quality control.
- Obligations to dispose of backwash sludge.
- Develop O&M Manuals.
- Assist Franklin with water sales to new service areas and neighboring communities.

The parties also agreed to allow for cost savings improvements to be made during the term of the agreement. Earth Tech will make the improvements and, after the costs are recovered from the savings, the parties will equally split further and future savings. Earth Tech provided a performance bond for 150 percent of the annual O&M fee as contract security.

Facility Price and Service Fee

Earth Tech guaranteed a fixed price for the facility. The only adjustments that were allowable were for changes Franklin requested or uncontrollable/force majeure circumstances. One risk as stated elsewhere was that the risk for additional design/build and O&M costs due to filtration rates was taken by Earth Tech.

The service fee is paid monthly and made up of two components: the facility charge and the O&M charge. The facility charge is a fixed monthly charge that reflects the amortization of the financing of the facility. Adjustments are allowed to this monthly amount to reflect additional capital improvements made and financed by Earth Tech during the term of the agreement for various reasons. The O&M component consists of a fixed and a variable amount. The variable amount includes consideration for electricity, chemicals, and actual charges to the cost of residuals management.

All other O&M costs are included in the fixed monthly portion. Additionally, the variable component can be adjusted if the raw water quantity and quality exceed the parameters defined for raw water. There are other service fee adjustments in the event of damages, fines, and penalties or for the outcome of disputes.

Breach/Termination

The contract contains typical and standard language, which includes events of breach and notice and cure periods prior to termination. If Earth Tech fails to properly treat raw water, Franklin has the right to immediately assume possession and operation, pending termination. The contract deals with termination by Franklin prior to construction being complete and during the operations period.

Prior to construction completion, Franklin could take possession and finish the work by whatever reasonable means Franklin deems appropriate and Earth Tech would not have been entitled to any further payment until the facility was built and in commercial operation. At that point, Earth Tech would have been paid for costs incurred for work up to date of termination including the amount to pay or defease outstanding indebtedness to finance the guaranteed facility price. In other words, Earth Tech would have gotten back its investment up to the point of termination.

For termination by Franklin, due to a breach by Earth Tech, the firm would have to pay Franklin (based on methodology defined in the contract), a one-time payment for the excess of O&M costs incurred if Franklin or another O&M service provider were to operate the facility for the remainder of the term plus third-party costs incurred by Franklin in getting a replacement operator on board. There was a cap of $500,000 for this liability. If the termination occurred in the operational period, Franklin would pay Earth Tech the amount to redeem or defease all outstanding indebtedness to finance the guaranteed facility price and an amount equal to the unamortized facility price not financed by debt as of the date of termination. Also, if additional capital improvements were made prior to the termination, Earth Tech would have been entitled to be repaid the outstanding indebtedness and/or unrecovered capital (that is not debt) on those capital improvements for which Franklin was responsible.

If Earth Tech terminated the agreement, the firm would have received the outstanding indebtedness and/or capital recovery for the facility price and all other additional capital improvements made up to the termination by Earth Tech, in addition to consideration for demobilization and lost profit as determined by contract methodology, not to exceed $500,000. With a termination the transfer of title and assignments of permits and contracts were provided for.

Liability and Indemnity

There is reciprocal indemnification in the contract. Earth Tech indemnifies against its intentional or willful misconduct, negligence, or breach. Franklin indemnifies against intentional or willful misconduct, negligence or breach and any uncontrollable/force majeure circumstance, and receipt of raw water outside the contractual definition. There is an agreement that there will be no liability for special, consequential, indirect, incidental, or similar damages.

Uncontrollable/Force Majeure Circumstances

The section provides for performance and cost relief for Earth Tech in such an event. Earth Tech is under best efforts to mitigate impacts and resume performance as soon as possible after such event.

Term/Conditions

The term of the agreement is for 20 years. The conditions precedent that had to be satisfied by a date certain are the standard and normal conditions as representations are true and correct, no pending suits, actions or investigations, and authority to execute.

Purchase/Renewal

At the end of the term, Franklin has the option to purchase the facility from Earth Tech for the greater of $1 or the fair market value or renew the agreement for a term not less than five years. The fair market value at the end of the term is a specified amount and it approximates 15 percent of the original design/build cost. The purchase price also includes any outstanding indebtedness or unrecovered capital from capital improvements that Franklin was responsible for and made during the term of the agreement.

Franklin also has a purchase option in the event expansion capital improvements are needed and the city is eligible and has a commitment of governmental funding that is less expensive than Earth Tech financing. This would involve a purchase price of outstanding indebtedness and/or unrecovered capital for the initial new facility and any capital improvements made for which Franklin was responsible for up to the purchase date. Also, the service agreement would remain (*i.e.*, Earth Tech would still operate the facility) and the service fee would be adjusted to eliminate the facility charge component. Earth Tech would also be responsible for the design and construction of the expansion.

Dispute Resolution

Dispute resolution language was negotiated in an attempt to bring about a quick and efficient resolution of disputes. There are two levels of dispute resolution involving representatives of the parties. These parties are obligated to attempt to resolve the dispute within five working days. If level 1 cannot successfully resolve the dispute, it is sent to level 2. Thereafter, it is resolved through arbitration. Both parties are also obligated to continue work and to perform during dispute resolution.

Insurance

Earth Tech is required to purchase and maintain certain insurance. Liability insurance is required for protection from claims that arise from operations whether by Earth Tech, subcontractors, or anyone directly or indirectly employed by any of them. Insurance is typical and standard for this type of transaction. This was handled through comprehensive general and automobile liability insurance. Insurance was also obtained to protect the facility. Earth Tech has property insurance for the full cost replacement at the time of any loss. If the facility is destroyed by casualty as defined by the coverage for fire, extended coverage, and all risk insurance for physical loss, Earth Tech will rebuild the facility to a condition comparable to the precasualty condition.

Franklin also is required to maintain certain insurance. The parties agreed to waive all rights against each other and subcontractors for damages covered by insurance. All subcontractors are required to give similar waivers. Franklin also granted a waiver of subrogation against Earth Tech and subcontractors. Earth Tech is required to demonstrate that such insurance is in force with certificate filings to Franklin.

Representations

Standard and typical representations have been given by the parties as to being duly organized and valid and with full power and authority, and that the agreement is duly authorized and executed and enforceable, and that there are no suits or proceedings before a court pending or threatened that would affect the performance or the agreement.

Key Definitions

Certain definitions exist that are key to performance and cost, and that govern when performance and cost relief are granted. These include: Change in Law, Force Majeure, and Uncontrollable Circumstances.

1. **Change in Law:** after effective date of agreement and that which impacts the performance and cost of the parties obligations; 1) enactment, adoption, repeal, or change in interpretation of federal, state, city, or other local law, 2) issuance of order, decree, or judgment of federal, state or local court (provided not a result of negligent or willful action or failure to act of the party relying on such), 3) suspension, termination interruption, failure of renewal of any permit or approval essential to construction or operation of the facility (same proviso as in 2) above).
2. **Force Majeure:** the standard definition—abnormal weather conditions, landslide, lightning, fire, earthquake, tornadoes, and flood.
3. **Uncontrollable Circumstances:** an event or condition beyond the reasonable control of Earth Tech. These shall include:
   a. change in law on or after date of agreement,
   b. labor disputes other than ones involving Earth Tech employees,
   c. loss or inability of utility services,
   d. act or omission of Franklin, and
   e. concealed or unknown site conditions, including hazardous materials.

FAST-TRACK CONSTRUCTION SCHEDULE

Earth Tech was awarded the contract on July 23, 1996; the contract negotiations were concluded and notice to proceed on the project was received by the firm in September. After obtaining state and local approvals, construction on the site was initiated in June 1997 and was completed on a fast-track schedule, with the facility functional in five months. The facility began producing potable water on November 1, 1997 (Oldenwurtel, 1998).

The entire process, from the time the decision was made to build the plant and through its completion, including procurement, took approximately 18 months. Often, municipal projects of similar size and scope take four to five years to complete under conventional approaches utilizing separate contracts and service providers.

The firm fulfilled the role of general contractor for the project and was responsible for the procurement of all process equipment and for securing and coordinating trade subcontractors and utilities. In addition, Earth Tech utilized in-house resources to complete the site's civil and utility work. As both the designer and builder of the project, Earth Tech was able to initiate the procurement and fabrication of major process equipment before trade subcontractors were selected. This, combined with the sizable up-front investment in the plant's detailed design during the proposal stage of the procurement, as well as the firm's willingness to commit to a strict schedule with financial penalties for failure to perform, effectively reduced the field construction period for the project.

The city has retained all customer billing and collection responsibilities. In addition, a unique Supervisory, Control, and Data Acquisition (SCADA) system enables personnel from the city's public works department to access live operating data about the plant (*e.g.*, amount of water in storage tanks, which pumps are operating, etc.) from their offices and other remote locations (Oldewurtel, 1999).

FRANKLIN BENEFITS

By adopting the public-private partnership approach to service delivery, Franklin was able to secure a total package of services needed to design, build, finance, and operate a new water treatment facility—thus relying on a single-source provider for all of the city's water needs.

Franklin has not only fulfilled its legal obligations but also has improved essential services to its citizens. Residents are being supplied with higher quality water with efficient long-term plant operations at a lower cost than it could have through the traditional multivendor approach. Local leaders estimate that the sole-source approach to water treatment created a 30 percent cost reduction compared to a more traditional approach (Mears and Oldewurtel, 1997). Such an approach would have entailed letting separate contracts to individual firms, each handling the separate functions of financing, engineering, construction, and contract operations. Forty percent cost reductions were estimated from taking the sole-source approach. Highlights of the benefits to Franklin of adopting the DBFO approach include:

- **Cost Guarantees.** Engineering and construction were performed on a lump-sum, no-change-order basis, and all project costs were identified and paid for directly by Earth Tech. The private firm assumed all the risks for construction costs and provided a guaranteed lump-sum price for O&M costs.
- **Performance Guarantees.** The service fee is structured for the private firm to assume performance risk. Terms of default have been established that shift risk for performance to Earth Tech.
- **Responsibility for Compliance.** All permits, regulatory approvals, and inspections required were provided and paid for by Earth Tech and the firm accepted all risks associated with gaining regulatory agency approval for the design of the new facility. The "timing risk" of initiating construction prior to final OEPA approval in order to meet the construction completion schedule was also borne by the private firm. The private firm is responsible for operation of the water treatment facility in accordance with all OEPA and other regulatory agency requirements.

- **Risk Sharing.** Changes in cost due to reasonably expected changes in water quality are limited to additional O&M cost for chemicals. Changes in cost for flow are addressed through the service fee, which is comprised of fixed and variable components. (Up to 5-mgd capacity, no changes in the fixed components will result.)
- **Expansion.** Water treatment plant expansion needed for new industry/economic development is provided for in the contract.

SUMMARY

The City of Franklin entered into a public-private partnership whereby a private firm, Earth Tech, is under contract to provide 5-mgd of drinking water for the next 20 years. Under the agreement, the private firm provided the design, construction, and financing of a new $3.5 million water treatment plant. In addition, the private firm provided full equity financing, guaranteed a lump-sum cost for the facility, and assumed all cost and scheduling risks associated with regulatory approvals.

During the proposal phase and prior to the awarding of the contract, Earth Tech evaluated all feasible treatment alternatives and selected the optimum treatment system for the Franklin water treatment facility. A preliminary design was submitted to OEPA, incorporating review comments, and preliminary approval of the conceptual design was obtained. Thereby, the facility design was approximately 35 percent complete prior to the firm being selected by the city.

Using the DBFO approach, the city was able to conserve its capital for future expenses. The contract also includes an option to transfer the water plant to city ownership after 20 years and, at the outset, the city set minimum design standards to ensure that all proposals would result in a plant that meets quality and longevity standards. Detailed performance specifications in the contract address ongoing maintenance and operation requirements of the facilities so that the city's investment is protected.

By placing the responsibility for facility development and operation into a single commitment between the proposer and the city, Franklin has achieved considerable savings and efficiencies. The private firm provided all the technical, financial, and operating components under one corporate guarantee. Combining these functions into one contract places the responsibility for plant performance directly on the firm that will be operating the facility for the next 20 years.

REFERENCES

City of Franklin, Ohio (1996). "Summary of Request for Qualifications and Request for Proposal." February 9, p. 2.

Earth Tech (1996). "Proposal for 5-mgd Water Treatment Service to the City of Franklin, Ohio." July 24.

Eisenhardt, Paul (1999). Telephone Interview, January 21.

Eisenhardt, Paul, and Dong, Anh (1996). "Water Service Project Proposal Evaluation and Recommendation." *Report To The City Of Franklin, Ohio*, July 19.

Mears, Jim, and Oldewurtel, Keith (1997). "City Finds Private Partner to Help Meet Its Water Needs." *American City & County* 112, November, p. 34.

Oldewurtel, Keith (1998). "Privatization and the Public Works Professional." *Public Works* 129, April, p. 33.

Reinhardt, William (1996). "Franklin, Ohio's Privatization Bonanza." *Public Works Financing* 117, January, p. 6.

CHAPTER 7

Public-Private Partnership: Gardner, Massachusetts

Douglas Herbst and Robert Markowitch

In 1998, Gardner, Massachusetts, entered into a 20-year public-private partnership agreement with a private firm to provide day-to-day operation, maintenance, and management (OM&M) of the city's water and wastewater services and to carry out a series of capital improvements. The capital improvements included the design and construction of a new 3-million-gallon-per-day (mgd) water filtration facility, the installation of approximately 10 miles of water main, and the replacement of more than 5,000 water meters. Additionally, the agreement calls for capital improvements to be completed in the wastewater system, including refurbishing of one primary clarifier and one intermediate clarifier at the wastewater facility and rehabilitation of three sewage pumping stations.

Cost estimates by the city for the design, construction, and operation service functions exceeded $52 million under conventional design and public works bidding procedures. However, by taking the design-build-operate service delivery approach, city leaders estimated Gardner would save $19 million (Manca, 1998). Although private financing alternatives were both requested and proposed, Gardner was able to secure a zero-percent-interest, state-revolving loan to fund the necessary capital improvements.

BACKGROUND

Gardner, founded as a town in 1785 and as a city in 1923, occupies 23 square miles in north central Massachusetts, 59 miles northwest of Boston.* Although the industrial

Douglas Herbst is senior vice president of Earth Tech, Total Water Management Division and president of the National Council for Public Private Partnerships and Robert Markowitch is a regional director for Earth Tech.

*See Figure A–3 in the Appendix.

aspects of Gardner have diminished over the last 40 years, the area is currently experiencing a renewal as a bedroom community and today has a population of approximately 21,000 people.

The city administration has been aware of and planning for the need to upgrade its infrastructure. In 1986, Gardner made a major investment in the upgrade and expansion of its wastewater treatment plant. Since that time, the city has contracted out the operation, maintenance, and management of the wastewater treatment facility to a private firm.

Recent changes in law have made it necessary for the community to address specific issues concerning its public drinking water supply and quality. Dual 2-million-gallon pre-stressed concrete water tanks were constructed in 1989 and a 750,000 gallon elevated storage tank was constructed in 1990. A program to replace, or clean and line unlined cast iron pipe installed prior to 1927 began in 1977 and is projected to continue through 2012. In addition, expansion of the sewer and water collection and distribution systems is underway. These major improvements to the water storage and distribution facilities have necessitated a large increase in water rates during recent years—from $0.98/hundred cubic feet (hcf) prior to 1990 to $1.87/hcf in 1995 and $2.24/hcf in 1996 and 1997 (Gardner, 1997). In addition, construction of a new water treatment facility is required by May 2000 in order for the city to comply with the Final Surface Water Treatment Rule of the Safe Drinking Water Act (SDWA) Amendments of 1996 that calls for filtration of its water source.

THE DECISION TO SEEK A PUBLIC-PRIVATE PARTNERSHIP

Facing the deadline imposed by the Massachusetts Department of Environmental Protection (MA DEP), the city began investigating ways to procure a new water filtration plant. Financing presented a challenge because the city's recent construction program for schools and other infrastructure had burdened the city's bonding capacity. As of November 20, 1997, the amounts of debt outstanding were $7,404,250 and $3,500,000 for the city's Water Fund and Sewer Fund, respectively. The city had recently issued $1,500,000 of General Obligation debt to finance expansion of the sewer and water collection and distribution systems, and additional capital totaling $3.8 million came from federal grants and from the Massachusetts Water Pollution Abatement Trust. With a November 1998 bond issue, the total financing for the sewer and water collection and distribution systems expansion project was $6.9 million (Gardner, 1997).

City leaders began investigating the design-build-finance-operate (DBFO) service delivery concept as a solution to Gardner's need for a new water filtration facility as well as other capital needs for its water and wastewater system. Ultimately, Gardner officials determined the design-build-operate approach had merit and could satisfy a number of key city objectives, including:

- **Minimizing costs.** By combining the three key components of designing, building, and operation of the necessary water filtration facility, the potential existed for significant cost savings and improved plant performance when compared to using traditional methods with separate contracts and multiple responsible parties.

- **Securing system reliability and performance.** To meet the terms of a public-private partnership, the service provider must preserve the city's and its own capital investment and assure the long-term reliability and efficiency of the systems.
- **Limiting the city's economic exposure for meeting DEP water permit requirements and for DEP noncompliance.** The service provider would be required to achieve continuous compliance with DEP water permit requirements and consent to order timetables. The service provider would also be required to provide high quality, uninterrupted, economical operation of the water treatment system in a manner that meets all applicable Massachusetts and federal laws. The service provider would also be required to maintain continuous compliance with NPDES permit limits at the city's wastewater treatment plant.

Protection of existing city workers' jobs was an important issue for the Gardner officials. Through its experience with contracting out the OM&M of its wastewater treatment facility in 1986, city government realized that protection of jobs, wages, and benefits for designated, existing city employees could be thoroughly addressed in the Request for Proposals (RFP) and ultimately guaranteed in the partnership agreement.

After weighing its options, the city decided to pursue the DBFO public-private partnership arrangement for the city's new water filtration plant and the operation, maintenance, and management of its wastewater treatment plant. The city did not, however, opt for private ownership of either the water or wastewater facilities. Unlike Franklin, Ohio, for example, which entered into a public-private partnership for the DBFO of its water plant, including private ownership, Gardner was not faced with IRS restrictions on the length of the term for service agreements.

On January 16, 1997, the IRS revised regulations regarding management of service agreements. Prior to these changes, service contracts were limited to five-year terms, with a termination for convenience by the public sector partner (without penalty or cause) at the end of the third year. In the case of Franklin, Ohio, private ownership allowed that partnership to be long-term and unaffected by this earlier IRS restriction.

The new IRS rules provide for permissible contract terms of up to 20 years. With the rule change, public sector partners can now enter into long-term service agreements for OM&M and include design, build, and finance obligations without relinquishing ownership. This represents a major market shift and should allow for the increased use of public-private partnerships to meet municipal water and wastewater needs.

The rule change has increased the benefits achievable by local governments like Gardner, such as the realization of necessary capital improvements with private sector responsibility for design-build-operate and cost and performance guarantees. It also includes the benefit of a private sector financing option for needed capital improvements and/or up-front contract payments (typically referred to as concession fees) that provide municipalities with funds to meet other permissible public needs (Herbst, 1997). Long-term contracts also give the private sector the ability and incentive to invest its own capital to increase efficiencies and reduce costs.

"DUAL PATH" PROCUREMENT PROCESS

It was imperative that Gardner satisfy the State Administrative Consent Order (ACO) to build a new water filtration plant and that it meet all requirements of the Safe Drinking Water Act by May 2000. In addition, before Gardner could fully pursue a DBFO service delivery option, special state legislation was required. The Commonwealth of Massachusetts requires that special legislation be passed before a city can enter into a long-term service agreement with a private entity.

Because of the impending ACO deadline, as well as the requirement for new legislation for the DBFO process, the city felt it prudent to conduct a "dual path" procurement process. The city hired an engineering consultant to conduct a study and prepare a preliminary design for a water treatment facility. The engineering consultant was also authorized to hire a procurement subcontractor familiar with the DBFO concept and that had recent experience in crafting the special legislation required for this type of delivery system to proceed (Gardner, 1997).

The special state legislation allowing the long-term service agreement for Gardner's water and wastewater system passed in the Massachusetts House of Representatives on June 24, 1997 and in the state Senate on June 26, 1997.

THE PROCUREMENT PROCESS

Once the special state legislation was passed, an RFP was crafted for the design-build-finance of a new water filtration plant and other capital improvements to the water and wastewater system, and for the operation, maintenance, and management of the city's water and wastewater facilities under a 20-year contract. In its RFP, the city stated its objectives as:

- Private sector finance and design/construction of new capital facilities and improvements.
- Efficiency and effectiveness of operation and maintenance (O&M) services.
- Sustaining facilities' performance to meet appropriate standards and permit conditions.
- Limiting of economic exposure via long-term, stable, less costly rates.

The RFP sought proposals from private firms to:

1. Finance, redesign, permit, and construct the new Crystal Lake Water Treatment Facility (including a new intake structure).
2. Provide long-term (20-year) contract operation and maintenance services for the Crystal Lake Water Treatment Facility and Snake Pond Groundwater Well and Treatment Facility.
3. Provide long-term (20-year) contract for O&M services for the water distribution system (this service was requested in the RFP, but stated that selection would be at the city's option).
4. Provide long-term (20-year) contract maintenance of the grounds of the water supply system reservoirs and dams.

5. Provide long-term (20 years) contract O&M services for the wastewater treatment plant, siphons, pump stations, and sludge landfill.
6. Provide long-term (20-year) contract O&M services for the wastewater collection system (this service was requested in the RFP, but stated that selection would be at the city's option).
7. Finance, design, and construct capital improvements on both the water and wastewater treatment facilities (the city would maintain responsibility for capital improvements to the wastewater collection system).

City responsibilities would include:

- Maintain ownership of all facilities;
- Design and construct the new Snake Pond Groundwater Well and Treatment Facility;
- Provide certain specialized vehicles for use by the contractor on the wastewater system; and
- Perform billing and collection functions for both the water and sewer systems.

Proposers were required to bid on the construction, financing, and operation of the "base design" of the water filtration facility already prepared by Gardner's engineering consultant as part of the city's "dual path" procurement strategy. Proposers were also allowed to submit alternative designs to meet the city's water treatment needs (Gardner, 1997).

Studies that were conducted after the "base design" for a 4.0-mgd water filtration facility had been completed found that the safe yield of the surface water supply system was 1.58-mgd, while the average daily water consumption was 2.1-mgd. Average daily demand in the year 2020 is estimated to be 2.5-mgd. To eliminate the deficit, the city received approval from the MA DEP to develop a groundwater source near Snake Pond as a 1.3-mgd water supply well to supplement the existing Crystal Lake water supply system. With the approval of the 1.3-mgd groundwater source, Gardner decided to consider downsizing the necessary Crystal Lake plant from a 4.0-mgd capacity of the "base design" to a 3.0-mgd facility. The city stated in its RFP that the proposers were to bid on the design, construction, finance, and operation of the Crystal Lake facility, and the city would be responsible for the design and construction of the 1.3-mgd groundwater well and treatment facility.

The scope of services outlined in the RFP also included:

- water main replacement,
- water meter replacement,
- operation of the wastewater treatment facility, and
- optional operation of the water distribution system, and the wastewater collection system.

Two selection committees were appointed to review and rate the technical and price proposals that had been submitted. The technical review committee evaluated the technical proposals first to ascertain if the proposer was "responsible and responsive." This was carried out by determining whether the proposal met the minimum rating for each criterion in the city's prepared rating sheet, which entailed noncost factors relating to technical, business, and schedule considerations; proposed service; proposer experience and financial capability; and equal employment opportunity.

Technical proposals deemed responsible and responsive were then ranked as "advantageous" or "highly advantageous" in accordance with the criteria set forth in the city's rating sheet. Following this evaluation, each technical proposal was assigned a composite rating of either "minimum," "advantageous," or "highly advantageous," again, in accordance with predetermined criteria. Once the technical proposals received a composite rating, the content of each firm's cost proposal was revealed by the financial review committee to the technical selection committee.

Earth Tech's proposal was rated both as highly advantageous by the technical committee and as the lowest price by the financial committee. As a result, the firm was selected to enter into contract negotiations with the city. The entire selection process took approximately six months after the proposals were submitted. The entire procurement process, from issuance of the RFP to service agreement execution, took 11 months.

EARTH TECH'S PROPOSAL

In its proposal, Earth Tech submitted proposed costs for the base design of the water filtration plant required by the city in accordance with the RFP, and also provided for alternative designs that would be more cost-effective to construct and operate. The firm also outlined how it planned to accelerate the program for installing 5,100 water meters in order to greatly reduce unaccounted water, increase revenues, and improve cash flow for the city by as much as $500,000 per year (Earth Tech, 1998).

In its cost proposal, Earth Tech proposed to finance the water treatment facility and the capital improvement program with a combination of tax-exempt (private activity bonds) debt and Earth Tech's equity. The private firm proposed to stand behind its proposed financing program with a commitment to fund the capital costs with corporate financing if tax-exempt financing was not feasible at the time financing was required. All current city employees would be offered employment at equivalent compensation and benefit levels. No existing employee would be transferred or terminated without cause as provided in the RFP.

Earth Tech accepted the entire regulatory agency approval risk for obtaining MA DEP approval for the alternative design of the water filtration facility; the firm also accepted full responsibility for obtaining all permits and approvals as required to construct and operate the plant. If the facility did not receive MA DEP approval, then all the resulting additional capital and operational costs would be absorbed by Earth Tech. In addition, the firm provided a guaranteed lump-sum price for the new water filtration plant and guaranteed to meet the plant startup deadline. The proposal also included a guaranteed lump-sum operation and maintenance price.

TECHNICAL AND OPERATIONAL APPROACH

The technology of the base design incorporated ultrafiltration units utilizing buoyant filter media. Earth Tech provided an alternative 3-mgd water filtration plant design that it claimed would be more cost-effective to construct and operate, utilizing state-of-the-art, proven technologies that would not sacrifice the quality of service. The technology of the alternative design consists of ultrafiltration units utilizing nonbuoyant filter media. Some of the advantages offered to the city by Earth Tech's alternative design were:

1. By relocation of the clearwell within the treatment facility itself, the construction of a precast tank and an entire pumping system was eliminated, producing construction and power cost savings.
2. The architecture of the building is the same as that of the original design. However, by utilizing space more efficiently, the square footage of the treatment plant was reduced without sacrificing interior space for operation and maintenance and equipment redundancy.
3. Earth Tech's alternative filters would provide for a 50 percent bed expansion, whereas the base design only allowed for a 30 percent expansion. This eliminated any potential permitting risk and provided additional operational flexibility.
4. The reduction in the facility footprint resulted in a significantly reduced impact at the plant site (a hillside). This produced a more environmentally friendly and aesthetically pleasing facility.

The firm proposed to implement a comprehensive O&M and capital improvement program to meet the city's objectives. Highlights of the program described in the proposal were:

- Construction of Crystal Lake Water Treatment Plant.
- Refurbishing of one primary clarifier and one intermediate clarifier at the wastewater treatment facility and rehabilitation of three sewage pumping stations.
- Installation of a security and fire detection system at the sludge landfill and Reservoir Hill booster station.
- Installation of approximately 10 miles of water main.
- Installation of 5,100 water meters.

Earth Tech proposed a systematic approach to capital preservation built upon state-of-the-art predictive, preventive and corrective maintenance programs so that, upon contract completion, the city's facility would be in excellent condition and have significant residual value.

In its proposal, the firm assured the city that the facilities would be operated and maintained to:

- Assure compliance with EPA Administrative Consent Orders;
- Assure that the health and safety of the citizens of the city and the employees of the facility would be are protected;
- Assure that the wastewater facility would fully meet all NPDES requirements;
- Assure that the water facility would fully meet the SDWA; and
- Assure that odors and noise would be reduced to the greatest possible extent.

FINANCIAL APPROACH

As described in Earth Tech's Proposal, above, the firm proposed financing the new water treatment facility and the capital improvement program with a combination of tax-exempt (private activity bonds) debt and Earth Tech's equity. However, Gardner pursued governmental loan funding and ultimately secured a zero-percent-interest state-revolving loan to fund the water plant.

CONTRACT TERMS AND CONDITIONS

Earth Tech was selected on July 22, 1998 and the service agreement was fully negotiated in approximately 90 days. The service agreement is a typical service agreement between a private sector vendor and a public sector partner for design-build-operate services. At the time of the contract signing, Gardner and Earth Tech also signed a voluntary contract termination, effectively terminating the existing contract the city had with the firm for operating the city's wastewater treatment facility, sludge landfill, and six wastewater pumping stations that had been scheduled to expire in December 1999. All warranties in effect at the time of the termination were assigned to the firm from the existing contract to the new contract.

SERVICE AGREEMENT SUMMARY

Representations, Warranties and Covenants

1. Each party made the standard representations and warranties such as, duly organized and validly existing, full power and authority to enter into agreements, no existence of any material suit or other action (except for consent orders issued by the state and EPA), etc.
2. Each party agreed that the agreement was to conform to Internal Revenue Services Procedures 97-13 and each party would not act so as to jeopardize the tax-exempt nature of the city's tax-exempt debt during the term of the agreement.

The Obligations of Earth Tech

1. Design and construct the Crystal Lake Treatment Facility in conformance with applicable law and performance criteria listed in agreement by a date certain, operate and maintain the city's water supply systems (excluding distribution system) and wastewater system (excluding collection system), design and construct additional capital improvements (water main replacement, water meter replacement, wastewater facility clarifier rehabilitation, sewage pump stations rehabilitation) at a guaranteed price and by a date certain.
2. Obtain and maintain all necessary permits and approvals.
3. Maintain all the physical assets contracted.

4. Prepare O&M manuals.
5. Implement computerized process control and maintenance programs.
6. Provide annual and other periodic reports as to the operation and maintenance of the contracted assets.
7. Provision for an independent engineer is retained by the city, acceptable to Earth Tech, to provide monitoring and oversight to ensure that the assets to be designed and built, and the operation and maintenance of the systems, are in accordance with the terms of the agreement.
8. Full responsibility for the administering and monitoring of the Industrial Pretreatment Program.
9. Collect, treat, haul and dispose of residuals from operations at the city landfill.
10. Staff the facilities with refusal by the state regulatory agency to approve any staffing changes not considered a change in law.
11. The city has access rights to the systems and data and information (operation, maintenance and financial) maintained by Earth Tech in carrying out its obligations.

The Obligations of the City

1. Ensure and provide access to systems.
2. Continue to implement infiltration and inflow abatement program.
3. Provide raw water in accordance with the quality requirements specified in the contract, provided Earth Tech meets its obligations related to watershed maintenance.
4. Provide review and approvals for reports and other submittals by Earth Tech.
5. Responsible for additional costs due to a change in law.

Design and Construction of Capital Improvements and Capital Modifications

1. For capital improvements, the city has the right to review and approve or require changes to the design and construction. The city will pay for the changes it requests.
2. Within 30 days of the execution of the contract, Earth Tech will provide the schedule for the dates for commencing and completing construction, start-up, acceptance testing, and commencement of operations.
3. For capital modifications, which is defined as all capital improvements not identified initially and required during the term of the contract, Earth Tech is required to submit a request to the city with certain key and detailed information upon which the city may base an approval. The city can reject and request a modified resubmission based upon its comments. The city also has the right for any modification exceeding, in the aggregate, 25 percent of the remaining service fee, to make the modification on its own.
4. The city will pay for such modification (capital and operation and maintenance), if required by a change in law, or at the request of the city or if the city determines that such modification will benefit the city. Earth Tech will pay for modification if it is not within the three cases mentioned above.

5. For modifications over $100,000, the city has the right to retain an engineering consultant to assist in its review. Earth Tech will be required to competitively bid the construction.
6. If modification is due to force majeure or change in law and the parties cannot agree on necessity, scope, or cost, it will be submitted to mediation.
7. Earth Tech shall cooperate and assist the city in obtaining financing for the modification.
8. If the city constructs the modification and desires that Earth Tech operate it, then Earth Tech will have the right to review and comment on design, start-up, and acceptance procedures. Earth Tech can object to the construction contractor if it believes the contractor is not qualified to perform the work; it can inspect work and receive assignment of any warranties.
9. For capital improvements and capital modifications, Earth Tech will be responsible to pay for all costs for design, construction, start-up and testing, with city reimbursement as appropriate under the contract. Earth Tech will be responsible for the work including the work of any engineer, contractor, or subcontractor retained by Earth Tech.
10. Earth Tech is required to prepare and submit design plans and specifications for city review. The city will issue a notice to proceed. If Earth Tech proceeds with construction prior to regulatory approvals, it does so at its own risk.
11. Earth Tech is required to submit periodic reports and allow for city inspection.
12. Earth Tech will be liable for any additional costs in excess of the amount the city agreed to except, if an allowable force majeure event occurred. Earth Tech will reimburse the city for any penalties or fines pursuant to the Administrative Consent Order, except for those caused by city negligence.
13. Earth Tech guarantees that the performance standards contained in the contract will be met and, if not, it will pay additional costs to attain such performance standards.
14. Earth Tech shall require warranties from all vendors and contractors, which can be transferable to the city. Earth Tech warrants that materials and equipment will be new and work free from defects and that the work conforms to contract requirements. Earth Tech shall pay for repairs or replacements from defects or workmanship.
15. Earth Tech will provide notice of acceptance tests and allow the city to witness same. With passage of acceptance tests, the city will issue a final acceptance notice to Earth Tech. Thereafter, Earth Tech will provide record drawings and documents to the city as well as appropriate O&M manuals.

Insurance and Bonds

1. Typical insurance is required, including: general liability, workers compensation, auto liability, environmental impairment, excess liability, and professional liability. The general liability has provisions, if needed, for standard blasting, explosion, collapse, and underground coverage; broad form property damage; contractual liability; protective liability against claims arising out of operations; and builders' all risk. The city is

additionally insured. Earth Tech is obligated to show proof of coverage annually. Deductible amounts have a ceiling of $100,000.
2. Performance, labor, and material bonds are required annually for an annual service fee amount. For construction, 100 percent performance and payments bonds are required.
3. The city required a corporate guarantee from Earth Tech's parent, Tyco International, guaranteeing the contractual obligations of Earth Tech.

Payments

1. The city will pay an annual service fee in 12 equal installments, subject to adjustment from an escalation factor. The city will pay within 30 days. Service fee includes OM&M fee and any fee associated with a capital repair or replacement, capital improvement or capital modification for which the city is liable and not covered by the separate application for payment that covers payment for costs incurred for design, permitting, engineering, installation, and construction related to capital improvement or modification or repair or replacement. Earth Tech and the city agreed to have Earth Tech fund minor capital improvements to the water and wastewater systems (approximately $450,000). The annual service fee included repayment for this work as well as a $400,000 one-time payment to the city for its procurement costs. Repayment was included as fixed monthly amounts. Under any termination, the city would be obligated to repay the outstanding indebtedness of these amounts financed by Earth Tech.
2. Earth Tech paid the city a one-time payment of $400,000 to cover the city's implementation costs. Earth Tech is also obligated to pay annual amounts to the city ranging from $20,000 to $40,000 to cover city oversight and monitoring activities. Earth Tech also will pay the city $10,000 per year to defray costs to the city for work due to disputes. These payments will accrue until $100,000 is reached. If the fund is reduced by the city to $80,000, Earth Tech is obligated to make annual payments until the balance reaches $100,000.
3. There is provision for interest on overdue payments.
4. Earth Tech agreed to comply with Internal Revenue Service Procedure 97-13 as to compensation for services.

Default and Termination

1. The city can terminate for the failure of Earth Tech to perform material obligations; the persistent, repeated, and unexcused failure to perform; failure to make payments as required; false representations; or bankruptcy. The city will give notice and a cure period is allowed, with extensions, if Earth Tech is diligently pursuing a cure. If there is an emergency to public health and safety, the city can terminate with only a notice. If Earth Tech is not at fault for said emergency, then termination will be treated as termination for convenience and, if Earth Tech is at fault, it will be termination for cause. In the case of bankruptcy, termination is immediate, without notice.

2. Upon termination, each party will pay the other what it is entitled to up to the point of termination, including outstanding payments for capital improvements or modifications. The city will retain monies paid to the city, as per the agreement, as liquidated damages. Earth Tech will also pay for costs to correct the default. Earth Tech is also liable for reasonable incremental costs for securing alternate services and a replacement operator, except when there is a Force Majeure or Earth Tech is not at fault. The city also has the right to pursue action for actual damages suffered as a result of default, whether or not it elects to terminate.
3. Earth Tech can terminate for city fault due to unexcused failure to perform material obligations; persistent and repeated failure to perform; unexcused failure to make payments; and bankruptcy. Earth Tech will give notice and allow for a cure period and extension to the cure period, if the city is diligently pursuing a cure.
4. Upon termination by Earth Tech, each party will pay to the other what it is entitled to the point of termination including payments for capital improvements or modifications.
5. Any dispute related to a default will be subject to non-binding mediation.
6. There is a provision to govern termination in the event of unenforceability of the contract.
7. There is a limitation on damages and there will be no damages for lost profits, loss of business good will or reputation, and, absent of willful misconduct or gross negligence, there will be no punitive damages.
8. Force Majeure events allow for cost and performance relief and include any act or event or condition beyond the reasonable control of Earth Tech or the city but not including change in law or negligent or willful action or inaction.
9. The city was granted a termination for convenience clause and has to provide a 60-day notice. Each party will pay to the other what it is obligated to, up to point of termination.
10. After a termination, Earth Tech is obligated to provide continuing services up to 180 days at the city's option and Earth Tech will be paid for such services.

Term and Indemnification

1. The term is for 20 years with a 5-year extension permissible.
2. Earth Tech indemnifies the city for willful misconduct and negligent performance.

Miscellaneous

1. The typical and standard miscellaneous provisions are included: Enforcement and Waiver; Assignment; Entire Contract; Notices; Severability; Standard of Care; Relationship; No Third-Party Rights; etc.

2. Non-binding mediation for disputes related to the contract is provided for. Mediation is intended to assist in resolution and no mediator is empowered to render a binding decision.
3. The change in law definition is typical and standard, and provides for Earth Tech to pursue actions to address change in law and be given time to pursue and be entitled to additional costs, both capital and O&M, as appropriate. If there are resultant cost reductions, then the city will benefit by service fee adjustments. If service fee increases are above certain percentages of annual service fee amount or aggregate service fee amount of remaining term, the contract could be terminated by city and be treated as a termination for convenience.

SUMMARY

The City of Gardner formed a public-private partnership with Earth Tech, a global engineering and environmental technology company, for a $33 million design-build-operate contract, while retaining ownership of its water and wastewater systems. Under the terms of the 20-year public-private partnership, Earth Tech's Total Water Management Division will provide Gardner with reliable, high quality water and wastewater services and perform much-needed capital improvements at significant savings to the city's ratepayers and taxpayers. Ultimately, Gardner elected to pursue a zero-percent-interest state revolving loan to fund the necessary capital improvements.

Earth Tech has operated, maintained, and managed Gardner's 5-mgd wastewater treatment plant since 1986. In addition to these services, the company has assumed responsibility for the operation, maintenance, and management of the city's existing drinking water facility and will provide design and build services for a series of capital improvements, including the design and construction of a new 3-mgd water filtration facility.

Other near-term capital improvements being carried out by Earth Tech under the new DBFO agreement include the total rehabilitation of certain existing processes at Gardner's wastewater treatment system and pump stations, the installation of approximately 10 miles of lined water main, and the installation and changeout of more than 5,000 water meters. Although day-to-day operation and maintenance of the city's wastewater collection and water distribution systems was initially retained by the city, Gardner officials have since approved an amendment to the partnership agreement, scheduled to be signed February 22, 1999, transferring these responsibilities to the private firm.

"We have effectively cut the costs of water and wastewater services and capital improvements by $19 million through the introduction of private sector participation," said Gardner's Mayor Charles J. Manca, who led efforts to expand and lengthen the public-private partnership approach. "These improvements will ensure that our residents continue to receive high-quality water and wastewater services at the lowest practical cost."

REFERENCES

City of Gardner, Massachusetts (1997). "Request For Proposals To Operate And Maintain Water And/Or Wastewater Facilities and To Construct A Water Treatment Facility." December: pp. 2–3.

Earth Tech (1998). "Technical Proposal to Operate and Maintain Water and/or Wastewater Facilities and to Construct a Water Treatment Facility." Submitted to Gardner, MA, April 17.

Herbst, Douglas (1997). "The Impact of Recent IRS Revisions for Management Contacts on Long-Term Public/Private Partnerships." Privately published, p. 5.

Manca, Charles J, Mayor of Gardner, Massachusetts (1998). Telephone interview, 6 November.

CHAPTER 8

Privatization of Wilmington, Delaware's Wastewater System

Kash Srinivasan

The City of Wilmington owns and operates a 105-million-gallon-per-day (mgd) advanced secondary wastewater treatment plant. The plant serves about 500,000 people, including 72,000 city residents, much of northern New Castle County and portions of southern Delaware County in Pennsylvania.* Operating costs of the plant are covered entirely by sewer charges assessed to Wilmington residents and by a service agreement between the city and New Castle County.

In December 1997, the city turned over management, operating, and maintenance responsibility for the plant to US Filter Operating Services, a wholly owned subsidiary of US Filter Corp. The 20-year professional services agreement provides for operations and maintenance services and for significant capital improvements to be made at the plant. The city paid US Filter a fee of $8,180,000 in fiscal year 1998. The fee is subject to annual adjustment according to the producer price index (PPI).

HISTORY OF PRIVATIZATION IN WILMINGTON

Privatization of wastewater services dates back to 1985 when the City of Wilmington, through its Department of Public Works, entered into an agreement with Envirotech Operating Services (EOS) for the operation and maintenance of its new sludge dewatering facility. The dewatering system replaced a sludge air-drying operation, also

Kash Srinivasan is commissioner of public works for the city of Wilmington, Delaware.

*See Figure A–4 in the Appendix.

provided through contract, which took liquid sludge and spread it out over open land to dry. The department recognized that the dewatering facility represented a significant change in technical sophistication and operational complexity over the "wastewater" side of the treatment plant. The complex facility, consisting of 14 large dewatering presses and associated sludge and polymer feed systems, had to meet strict performance standards and was anticipated to be far less forgiving of operator error than the predominantly gravity-flow wastewater processes that its labor force was accustomed to operating. Hiring and retaining skilled staff to manage, operate, and maintain this complex system represented a huge challenge.

EOS was selected to operate the dewatering system after the city evaluated proposals submitted by several firms. EOS staffed and started up the new system. The city labor union, AFSCME (the American Federation of State, County and Municipal Employees), filed for arbitration against the city's decision, alleging that the privatization was an erosion of the bargaining unit. The matter was resolved when EOS agreed to the unionization of its employees under a new AFSCME local.

The arrangement with EOS was renewed after a subsequent round of contractor selection. In this round, the city issued a Request for Proposals (RFP) featuring the requirement that firms submit separate technical and cost proposals. Initially, the proposals were ranked on technical merit alone. The cost proposals were later used in negotiating a multiyear operating agreement that provided for the preservation of infrastructure, performance incentives, and cost containment. The privatization of the dewatering system proved to be a very positive experience and contributed greatly to the city's interest in seeking to privatize all of its wastewater treatment operation.

RATIONALE FOR THE CURRENT ROUND OF PRIVATIZATION

Like many other municipalities, Wilmington faced challenging economic circumstances in the early 1990s. When Mayor James H. Sills, Jr. took office in January 1993, he was saddled with a budget deficit of more than $6 million from the previous year. A deficit was anticipated in the then-current fiscal year as well (and, in fact, it amounted to $11 million). These were significant deficit amounts for a city the size of Wilmington.

The promulgation of Executive Order (EO) 12,803 in the waning months of the Bush administration and the initiation of a pilot project in Franklin, Ohio formed the backdrop against which the city began its own evaluation of the feasibility of wastewater privatization as a means of meeting its short- and long-term financial needs. In October 1994, Raftelis Environmental Consulting Group (RECG) was chosen to conduct the study. RECG evaluated the likely cash infusion that could result from either leasing or selling the facility to a private operator and the likely impact on rates to customers. In its feasibility analysis, RECG also considered factors such as impact on employees, future capital needs, regulatory compliance, transfer of operating risk, public acceptance, and perceived loss of city control. The study concluded that the sale of the facility, coupled with a long-term professional services contract with the privatizer, would allow the city to meet its financial objectives.

THE SELECTION PROCESS

In January 1995, the decision was made to proceed with the privatization initiative. In formulating its procurement process, the city identified a number of management objectives:

1. The wastewater assets were to be sold at a predetermined net book value (NBV) of $52 million. The proceeds from the sale would be distributed in accordance with EO 12,803 to the city, state and federal governments. The city's share (estimated at $28 million) would be used to defease outstanding debt ($10 million) associated with the assets and for other financial purposes allowed by EO 12,803. The city would retain ownership of a $20 million expansion then underway to preserve the tax-exempt status of a State Revolving Fund bond.
2. The privatizer would assume responsibility for regulatory compliance and be required to maintain high standards in the operation and maintenance of the wastewater assets. The service fee paid by the city would cover these operation and maintenance (O&M) responsibilities.
3. The privatizer would assume responsibility for the future capital needs of the facility, including capacity expansion, upgrades, and replacement of worn out assets. A separate budget and reimbursement process would be established to cover these capital costs.
4. Impacts on rates and the affordability of wastewater services in the region were to be kept to acceptable levels.
5. All current employees of the facility would be offered employment with compensation, training, and career opportunities comparable to or better than those available through the city.
6. At the end of the term of the professional services agreement, the city would have the right to repurchase the assets at the then-NBV, estimated at $1.5 million.

The Request for Proposals

The RFP issued in May 1995 included draft agreements for professional operating services, the sale and repurchase by the city of the assets, and a ground lease (in recognition of the city's continued ownership of the land underlying the assets to be purchased). Participants were required to submit separate technical and cost proposals. The formats for these proposals were defined in the RFP. The cost proposal would specify the service fee for the first year of operation, subject to escalation by the consumer price index (CPI) in the Philadelphia Region. The cost proposal was not required to account for capital replacements and repairs. These costs were to be handled through an escrow account. Up to two alternate cost proposals and privatization arrangements could be submitted as well. The prescribed alternatives were an "alternative asset valuation" method and associated service fee, and a "professional services only" option with an up-front payment of $16 million.

The RFP also described the evaluation process that the city intended to use—a modified "Professional Services Procurement" model in which the technical proposals alone would be evaluated and ranked on the basis of nine predefined criteria. These criteria (listed below) were assigned explicit weights and described in the RFP document.

Evaluation Criteria

- Corporate profile
- Corporate experience and expertise
- Regulatory experience
- Key management and operational personnel
- Financial strength
- Utilization of disadvantaged business enterprises and EEO compliance
- Employee considerations
- References and reputation
- Completeness and responsiveness of proposal

A large number of documents relevant to the treatment system and a copier were made available to the participants in a so-called "war room" set up at the facility. These documents included operating records; contracts and agreements; drawings; cost information; inventories of parts, equipment and vehicles; financial reports; user charge rates; industrial discharges; and employee information. Participants were allowed complete access to the facility in scheduled blocks of time.

Evaluation of Participants

Four firms or joint ventures participated in the privatization bidding process. These firms, in the order of ranking of their technical proposals, were:

- EOS, then a subsidiary of Wheelabrator Technologies,
- American-Anglian, a joint venture of American Water and Anglian Water,
- Northern Delaware Clean Water Corp, a non-profit entity proposed to be teamed with Professional Services Group as operations service provider, and
- US Water.

Perhaps reflecting the relative seriousness of intent of these firms, the cost proposals from these firms mirrored the technical ranking.

The schedule for the project was necessarily quite aggressive, calling for completion of negotiations approximately one month after the rankings were announced. It is instructive to examine some of the reasons why the actual process of getting to a deal took more than two years.

LAWSUITS AND POLITICS

Soon after the announcement of the rankings, the third-ranked firm filed suit seeking to overturn the process. This firm believed itself to be the lowest-cost service provider (the cost proposals had not yet been made public). It alleged that the "professional services" process violated state law, and that, in addition, the city owed it to the taxpayers to select the lowest cost "bid." The Chancery Court heard the case on a expedited basis. It ruled that the selection process was well within state law and perfectly logical and reasonable in view of the technical and financial complexities of the service sought.

The Intrusion of Politics

New Castle County is the largest user of the city's wastewater treatment system. Until July 1997, the county had had a cost-sharing agreement with the city, whereby the county's cost was determined annually by multiplying the actual treatment costs by the county's percentage use of the system. In Wall Street parlance, the RFP had put the plant "in play": it had become an opportunity for the county to acquire control of an important regional asset. At various forums the county asserted that, by virtue of its payments for service, it was 70 percent owner of the facility and therefore had a right to a proportionate share of the sale proceeds. Other county positions included threats to build its own treatment system and divert all of its flow away from the plant, and offers to buy the plant from the city outside the EO 12,803 process. For a variety of political reasons, it was unacceptable to the city to become merely another municipal "user" in a county-owned regional operation.

The strained relations between the city and the county introduced a great deal of uncertainty into an already complex situation. The firms already engaged in the process stayed the course in the face of these uncertainties, a good omen for the future of wastewater privatization in this country. Negotiations with EOS were begun and continued in this strained atmosphere. The city continued to hold parallel discussions with the county to try to reach a political accommodation in the matter.

A change in the county administration brought new thinking to the table, culminating in a new city-county service agreement that became effective in July 1997. This agreement mirrors the privatization deal, featuring a fixed-cost arrangement for the county that shields it from any unforeseen costs.

A Change in Thinking

The limited experience of members of the city team with complex financial issues also affected the pace of negotiations with Wheelabrator EOS. Accounting treatments and "bottom line" impacts of various contractual scenarios had to be developed, explained, and exhaustively debated. In the process, some serious financial concerns with the original "asset sale" approach began to emerge.

The transfer of a service asset to a buyer, and the purchase of services from this buyer, is equivalent to a real estate "sale-leaseback" arrangement wherein the sale proceeds are offset by a stream of annual lease payments. Under the EO 12,803 requirements, however, the "sale" proceeds are required to be shared with the state and federal governments while the city and its customers bear the full burden of the annual "leaseback" payments. The EO 12,803 process represented an expensive way for the city to borrow money. The State of Delaware and EPA stood to recoup grant proceeds even though the now-private treatment facility would be exclusively used for a public purpose. These financial drawbacks, coupled with the political difficulties with the county, caused the city to abandon the "sale" model and turn its attention to a simpler long-term service agreement. The gradual turn-around in the city's financial situation in the intervening fiscal years (1994, 1995 and 1996) helped facilitate this decision.

NATURE OF THE AGREEMENT

The 20-year service agreement with US Filter Operating Services (the successor to Wheelabrator EOS) provides for the management, operation, and maintenance of the wastewater treatment facilities and three pumping stations. The operator is responsible for regulatory compliance, provided the characteristics of the wastewater entering the facility (the influent) are within a specified range. Capital replacements and repairs, including a package of scheduled improvements that were identified as being needed, are folded into the negotiated service fee. The inclusion of capital improvements and repairs in the service fee represents a significant and innovative departure from the original RFP. The operator also has the opportunity to propose "Value-added Projects" that, by their nature, result in increased revenues or decreased operating expenses. The city would share in the net proceeds from such projects but would not be obligated for any of the costs or risks.

The city will pay an initial annual service fee of $8,180,000. The fee has two components—a Base Component of $6,600,000 and an Electricity Component of $1,580,000. The Base Component is subject to escalation with PPI. The cost escalation is capped at 4 percent after the first four years. The Electricity Component is adjusted, up or down, in accordance with actual changes in the unit cost of electricity through an electricity index. Increased electricity costs caused by increases in consumption are not compensable.

The fee is also subject to adjustment, up or down, for changes in flow and biosolids production. These latter adjustments have a relatively minor impact on the fee. A fifth component of the service fee, initially zero, compensates the operator for costs incurred due to uncontrollable circumstances. The agreement includes sample calculations illustrating the use of the adjustment factors.

There are no significant "up-front" payments in the deal. The city does get partially reimbursed for expenses it incurred in developing and negotiating the privatization project. The operator also pays an annual site license fee of $465,000. The city retains full control over user rates and remains directly accountable to the taxpayers for the service.

A key feature of the agreement is that all plant employees are guaranteed employment with US Filter and are guaranteed not to be laid off or transferred against their will over the term of the agreement. This provision is stronger than the one originally contemplated in the RFP, which protected employees from layoffs for two years. Pay and benefits are comparable to or better than the levels under city employment. Employees who chose not to join US Filter bid into existing city vacancies or exercised "bumping rights" provided in their union contracts.

TERMINATION PROVISIONS

Several sections in the agreement deal with mechanisms for terminating the agreement. Either party may terminate for breach of contract. Either party may seek termination if a city "termination right," embodied in the city charter, is not waived by the Delaware General Assembly. This charter provision allows the city to terminate any contracts after four years without liability for damages or lost profits. US Filter may terminate if the treatment facility, by virtue of its private operation, loses its status as a Publicly Owned Treatment Works (POTW). Such a designation is critical to certain industrial dischargers to the facility because of the protection afforded by the Domestic Sewage Exclusion in the hazardous waste regulations (see 42 U.S.C Section 6903(27), 40 CFR Section 261(4)(a)(1), and Delaware Administrative Code Section A). In general, termination of the agreement triggers repayment by the city of the NBV of any capital expenditures made by US Filter in connection with the scheduled capital program, its responses to uncontrollable circumstances, or as part of any value-added projects.

IRS Rules

In 1997, the Internal Revenue Service promulgated new regulations regarding privatization of public assets. These rules make it easier for municipalities to enter into long-term service agreements with private firms. Under its previous rules, privatization agreements were limited to five years based on the premise that assets funded with tax-exempt financing could not be used to generate private profits. Under the old IRS rules, a 20-year agreement would potentially negate the tax-exempt status of municipal bonds, requiring them to be defeased. The new rules ease the defeasance requirements. Municipalities can now conduct long-term privatization transactions without fear of affecting bond holders.

Financial Benefits

The important financial components of the deal are the service fee, the scheduled capital improvements program valued at $15 million, the site access fee, and the $1 million reimbursement to the city. From the operator's perspective, the service fee constitutes a guaranteed revenue source from which it must recover its operating costs (employee salaries and benefits, materials, overhead, contracted services), amortize the scheduled capital and reimbursement payments, pay the site access fee, and make a return on its investment. The scheduled capital costs are assumed to be fully amortized within the 20-year period regardless of when the improvements actually are made. From the city's perspective, the service fee replaces a portion of its annual operating costs (direct employee salaries and benefits, materials, some contracted services). It would continue to carry other costs discussed below.

The value of the costs replaced by the service fee represent annual savings to the city of $1 million to $1.5 million. In computing these savings, the city made relatively conservative assumptions including:

- Continued cooperation;
- Annual increases in labor costs of 2.5 percent;
- A 2 percent growth rate for other costs;
- $18 million in additional capital improvements spread evenly over the 20-year period;

- 3 percent inflationary growth in the service fee;
- $15 million of privatizer funded improvements in the early years.

The PPI has hovered around 1 percent to 2 percent recently. The inflation factor used in the savings analysis is consistent with historical trends in this index. The negotiated cap on inflationary adjustments to the service fee (4 percent after the first four years) supports the assumed "inflationary growth" number used in the "Privatization" scenario. For the "Continued City Operation" scenario, the analysis assumes that the city could keep its costs below the inflation rate. Over and above the projected savings, less tangible financial benefits also accrue from the transfer of a variety of insurable risks from the city to US Filter. These risks include workman's compensation, general liability, and risk of physical loss of plant and equipment. A sample savings evaluation (Table 8–1) for FY 98 is shown below.

Other costs will continue to be incurred (and booked) directly by the city, independent of the service agreement. These include depreciation, interest on debt, city-wide and departmental overhead costs, and contractual costs associated with sludge disposal. These costs total a little over $7 million.

TABLE 8–1. ANALYSIS OF CITY OPERATION VERSUS PRIVATIZATION (USING FISCAL YEAR 1998, IN DOLLARS)

Account Title	City Operation	Private Operation
Salaries and wages	1,910,609	347,819
Employee benefits	549,930	106,255
Printing and advertising	2,786	2,786
Communications and utilities	2,080,701	0
Transportation	2,068	2,068
Rentals	14,088	2,000
Contracted maintenance	473,817	0
Professional fees	376,627	376,627
Other fees	412,417	45,900
Memberships and registrations	3,878	3,878
Miscellaneous services	5,887,040	2,000,000
Office and general supplies	5,470	5,470
Wearing apparel and safety supplies	8,608	1,530
Misc. materials, supplies, parts	268,720	20,400
Petroleum and chemicals	351,592	0
Construction and repairs	0	0
Indirect costs	1,310,952	1,310,952
Equipment and livestock	7,000	7,000
Fixed assets	1,383,413	0
Internal service fund	211,843	161,160
Amortization	15,920	15,920
Debt service	1,049,989	957,553
Depreciation expense	2,191,536	2,191,536
TOTAL COSTS	**18,518,103**	**7,558,892**
Capitalization of fixed assets	1,383,413	
Privatizer's fee		8,180,000
NET COSTS	**17,134,690**	**15,738,892**

Water and sewer services are provided and funded entirely from user rates. These rates are adjusted periodically to keep costs and revenues in rough balance. In surplus years, some revenues may be transferred to support the General Fund, which is almost entirely funded through property and wage taxes. The wastewater treatment operation represents over one-half of the Water/Sewer Fund costs. Privatization stabilizes these costs and provides the city with the flexibility of postponing rate increases and funding needed improvements in other water and sewer infrastructure. The site access fee also provides a small but assured flow of revenues to the General Fund, reducing the pressure for future fund transfers from the Water/Sewer Fund.

Operating Benefits

US Filter guarantees the performance of its obligations under the service agreement and will be a joint permittee on the NPDES permit. Provided that the city meets certain conditions regarding the quality and quantity of plant influent, its management burden for assuring compliance with environmental and other laws is considerably eased with this arrangement. The chronic problem of underinvestment in infrastructure maintenance is overcome and the annual competition with other city departments for scarce capital dollars is avoided. For example, the scheduled capital improvements will occur over three to five years. Under the normal city spending regime, such a program could take 15 years to accomplish. New technologies will also be quickly introduced if they provide an economic advantage to the operator. Improvement plans include the rapid conversion of the existing coarse bubble air diffusers to more efficient fine bubble systems, and the investigation of sludge digestion alternatives to reduce sludge volumes.

The "adjustment for flows" component of the agreement provides a strong incentive to the operator to maximize flows to the plant, thereby minimizing overflows from the city's combined sewer system during rain events. Under city operation, such an incentive would be difficult to create. High plant flows, sometimes double the norm, are challenging events for plant personnel, and the instinctive reaction is to try to cut back rather than maximize these flows. The rainy spring of 1998 has already demonstrated the efficacy of this incentive.

CONCLUSION

The protracted negotiations created considerable anxiety for the employees at the plant. A few employees voluntarily transferred to other jobs in the city. The vacancies created mostly went unfilled, with the slack being picked up through contracted services for maintenance and overtime. When finally consummated, the transfer to private operation resulted in bonus payments of $2,000 to each employee who signed on with US Filter. The transition to private operation in December 1997 and the subsequent months have been surprisingly smooth and positive. The plant employees appear to have settled in well to their new circumstances. These employees remain under the AFSCME union umbrella and are negotiating a new contract with US Filter. A single Public Works employee handles liaison, contract compliance, and transitional issues and oversees the completion of ongoing construction at the plant.

EPA approval of the transaction (under EO 12,803) was received on June 24, 1998. A key step in the EO 12,803 approval process is the issuance of "grant deviations" under the EPA Construction Grants program. The approval is conditioned on the issuance by the city of a public notice of the deal to its sewer customers by the end of the calendar year. In general, the notice is required to address the nature and disposition of up front payments or sale proceeds and the impact of the deal on rates. The notice that the city intends to issue is shown below.

The City of Wilmington recently entered into a contract with US Filter Operating Services for the operation and maintenance of its wastewater treatment plant at Cherry Island. The city will pay a fixed fee, subject to inflationary adjustment, to US Filter for its services. The contract provides that US Filter will reimburse the city $1 million for expenses incurred in conducting feasibility studies, financial analyses, compliance with federal and state regulations, and negotiating the contract. US Filter will also pay the city $465,000 annually as a site access fee. This fee will accrue to the city's General Fund and will help support the General Fund services of the city, including police and fire protection and other public services.

The service fee to US Filter is paid out of the sewer charges assessed to our customers and is intended by US Filter to compensate them for their costs in operating the plant, their payment of the site access fee, and the one-time city expense reimbursement.

As a consequence of this arrangement, we anticipate cost savings of over $1 million annually. These savings enable water and sewer rates for Wilmington residents to remain unchanged in FY 1999 (beginning July 1998). Stable rates are anticipated in FY 2000 as well.

City Council recently passed two key measures to utilize these savings in a manner directly beneficial to our customers. The first measure (Ordinance 98-001) specifies that the savings be used to improve the water and sewer infrastructure in the city. We expect to increase our investment in water mains and sewer lines in the next several years. The second measure (Ordinance 98-023) assists homeowners with the financial burden of replacing old and corroded water service lines when they fail unexpectedly. The typical cost of replacing a private water service line (which runs from the curb into the house) is about $1,000. Council's "service line initiative" will be administered by the Department of Public Works.

Also in June 1998, the Delaware General Assembly passed legislation allowing municipalities the flexibility of waiving the four-year "termination right" provisions in their charters. Recall that this right makes contracts longer than four years in duration subject to cancellation without liability for damages or lost profits. Wilmington's City Council will need to pass a suitable ordinance to make the four-year right inapplicable to the US Filter agreement.

These actions at the federal and state levels remove key impediments to the 20-year relationship. Capital improvements contemplated in the agreement can move forward because the financial uncertainties are now greatly reduced. The likely first steps include conversion of secondary aeration systems from "coarse bubble" to the more efficient "fine bubble" systems, introduction of digital controls and data acquisition, and refurbishing of key plant components.

A lot can happen in 20 years. A carefully crafted agreement that is fair to both parties is a precondition to ensuring a successful journey. So far, our experience has been positive. US Filter has reached agreement on new labor contracts with two of the three union locals that represent its employees at the Wilmington plant. New working relationships are being forged between city and US Filter staff on issues such as industrial pretreatment, combined sewer overflow (CSO) minimization, and the prioritization of capital improvements. Our confidence in the relationship is reflected in the increased management focus on basic infrastructure issues such as sewers and CSOs.

Figure 8–1. Public Notice

CHAPTER 9

Implementing Seattle's Tolt Treatment Facilities' Design-Build-Operate Project

Elizabeth S. Kelly, Paul D. Reiter, and Scott Haskins

Seattle has utilized an innovative public-private partnership for development of its first water filtration plant. Filtration was sought for the Tolt River source to increase system reliability and flexibility by permitting continuous operation of the source through periods of high turbidity. In addition, it allowed long-range conformity with anticipated regulations and increased system yield.

As an alternative to the conventional public works procurement process where the design, construction, and operational phases were distinct, Seattle sought to create a situation where integrated teams of designers, constructors, and operators could innovate to provide the most reliable and cost-effective proposal for development of the project, while maintaining public ownership of the facility. This process would achieve $70 million in savings over the $171 million estimated cost of developing the same facility utilizing a conventional design-bid-build procurement with city operations for 25 years.

The contract, executed in May 1997, was valued at $101 million. The facility is a 120-million-gallon-per-day (mgd) or 454,000-cubic-meters-per-day (CMD) filtration and ozonation plant for treatment of Seattle's Tolt River source of supply. The Tolt River provides about one-third of the water for Seattle and the city's 26 regional wholesale customers. The contract provides for online status by the end of 2000.

Elizabeth S. Kelly is project manager, Paul D. Reiter is strategic policy director, and Scott Haskins is resource management branch director at Seattle Public Utilities.

This chapter discusses issues, options, and processes associated with the use of a design-build-operate (DBO) model for Seattle's Tolt Treatment Facilities project.

BACKGROUND

The Need for Filtration on the Tolt

Seattle is supplied by two large unfiltered surface water sources, the South Fork Tolt River and the Cedar River, with a service population of approximately 1.3 million people.* Filtration was sought for the Tolt source to improve the quality of the treated water in order to meet Seattle's public health objectives as well as to satisfy forthcoming stricter surface water treatment and disinfection by-product regulations. Filtration increases reliability and system flexibility by allowing the Tolt supply to be operated over a much wider range in weather and reservoir level. Finally, it provides additional yield of approximately 9-mgd (34,000 CMD) from the Tolt source.

Project History

Working within the conventional framework of a design-bid-build procurement process, Seattle initiated predesign for the Tolt Filtration Plant in 1991. This effort included water quality studies, environmental studies and preparation of an Environmental Impact Statement, preliminary permitting, and project predesign. Predesign studies recommended development of a 120-mgd (454,000-CMD) filtration plant on a site located approximately 1.5 miles (2.8 kilometers) downstream of the existing Tolt Regulating Basin.

Based on the predesign work completed in 1995, the city anticipated total costs of $156 million (net present value, 1998 basis), construction costs of approximately $100 million, and estimated operations and renewal costs of $56 million over a 25-year period. This estimate was for a filtration-only facility (no ozonation or sedimentation), and was based on awarding a competitively bid construction contract with an independent design engineer's specifications, and city operation.

THE PROCESS

Seattle's Challenges: The Project in a Broader Context

The consideration of an alternative approach to developing and operating the Tolt Filtration Plant is best understood in the context of Seattle's broader set of challenges. Like many other large municipal utilities in the United States, Seattle Public Utilities (SPU) was confronting a number of pressing infrastructure challenges in need of simultaneous action over the next decade. Seattle's imperative in addressing

*See Figure A–5 in the Appendix.

these challenges was to provide more cost-effective and competitive services while meeting the needs of its customers and remaining sensitive to the need for environmentally sound practices of facility operations and project development.

The challenges faced by SPU included:

- responding to emerging water quality regulations, including the need to add ozone and/or filtration capacity for its Tolt and Cedar River water sources, which historically have not been filtered;
- completing significant replacement and/or rehabilitation of aging and outdated infrastructure, including combined sewer overflows, storage reservoirs, transmission pipelines, water and sewer mains, and solid waste transfer stations, and addressing the impacts of these capital requirements, including higher utility rates and associated operating costs;
- developing additional water supply sources to support the growth in demand of suburban wholesale customers;
- enhancing wildlife habitat and water quality in the city's watersheds and urban creeks; and
- encouraging citizens to conserve resources through an integrated set of recycling, pollution prevention, water conservation, and education programs.

Completion of these activities was expected to require nearly $1 billion of capital investment over a 10-year period. In view of these requirements, SPU's challenge was to address its imperative for efficiency through strategic planning and delivery systems, performance measurement and benchmarking, strategic technology investments, and organizational streamlining. The first of these challenges drove the search for more strategic approaches to project delivery and gave rise to the consideration of alternative procurement processes for developing major projects. Since Seattle was at the beginning point in the development of the Tolt Treatment Facilities project, the stage was set for such an alternative approach to implementation.

Consideration of an Alternative Approach for the Tolt Project

Seattle's initial willingness to consider an innovative approach to developing the Tolt Filtration Plant derived from its overall effort to find more efficient ways of doing business in the context described above. A variety of specific factors led to the viability of using an alternative approach over the traditional approach to design, construction and operations of the Tolt facility:

- Seattle's lack of experience in filtration plant design, construction, and operation;
- a growing body of demonstrated, private experience in plant operation;
- the emergence of capable DBO consortia;
- potential cost savings associated with an alternative approach; and
- a statute allowing design-build procurement within the state of Washington.

In considering alternative approaches, Seattle sought to ensure that the Tolt plant would perform to a standard at or above conventional public ownership and operation, but at minimum cost to the city within a preferred risk allocation framework.

The Conventional Approach in a Changing Environment

The traditional model for developing and operating water filtration plants in the U.S. has been the design-bid-build-operate (DBBO) model. This model involves three primary actors—the designer, the constructor, and the operator/agency responsible for supplying water from the plant to its customers—and a serial approach to the design, construction, and operation steps.

In the first step in this process, a private design firm is hired under contract. Their responsibilities include determining plant requirements for the municipality, including (implicitly) many of the risk elements of the problem. They then design the plant and develop bid specifications. In the second step, bids are tendered to meet the specifications and the low bidder is given the task of constructing the plant in accordance with the specifications. In the third step, the municipal operator is responsible for operating the plant in accordance with performance standards.

Advocates of the conventional approach emphasize the independence of the design function and the advantages of that independent responsibility. In addition, the segmentation of various stages in project development provides for greater involvement and control by the utility, and a larger field of potential consultants and contractors can lead to more competition.

While the conventional DBBO approach to public works contracting is well understood, and most appropriate for many applications, its shortcomings are numerous and well documented. They include:

- segmenting the design and construction process often leads to excessive design costs due to undesirable risk allocation and mixed incentives, and a failure of the designer and builder to collaborate in ways that would lead to reductions in construction costs;
- implementation budgets expand unpredictably and the low-bid method of selecting constructors has high potential for costly change orders;
- schedules become extended due to difficulty in managing and staffing large, technically complex development and project delivery systems;
- responsibility for cost overruns associated with design or construction are typically borne by the owner, and risks associated with the failure of the plant to operate in accordance with the municipality's intent are often ultimately borne by the municipality/operator, both because of the ambiguity in causation for a problem among the designer, constructor and the municipality;
- the low-bid, fixed price method of selecting constructors heightens the risk of performance failures, particularly in a situation of ambiguous responsibilities for a failure;
- municipalities are often not equipped to efficiently operate complex plants, lacking economies of scale in management and chemical procurement, and being subject to complex labor agreements.

In today's dynamic environment, many if not most of the basic roles and responsibilities implicit in the traditional approach are beginning to be rethought. A changing point of view about municipal versus private sector capabilities, the increasing prominence of integrating design and construction processes, and the emergence of capable private operators have all contributed to a broader set of opportunities for formulating a new and better approach to the conventional model.

In formulating a new approach to the development of a large new project like a water filtration plant, municipalities need to consider:

- the desired degree of integration of design, construction, and operations;
- options for operational responsibility—municipal versus vendor;
- plant ownership and financing—municipal versus private versus mixed model;
- the desired risk and control position—should it be primarily the municipality's or primarily the vendor's? and
- the method for specifying deliverables—a performance-based approach versus a prescriptive approach.

Considering the challenges, as well as its opportunities, Seattle settled into a serious evaluation of alternative contracting procedures from the available options. Several basic approaches and their distinguishing features emerged. They are shown in Table 9–1, assuming progressive departures from municipal ownership and operations.

After an initial consideration of options, which ranged from purely public ownership, financing, and operation to private ownership, financing, and operation, Seattle developed a preference for the DBO model, as it would most completely meet the city's key objectives: efficiency in project delivery, optimum allocation of risk, performance-based specifications, and city ownership and financing.

Consideration of these choices and the basic approaches outlined above leads to a set of related observations:

- The distinguishing features of the conventional approach, the DBO model, and other alternative models (*e.g.*, DB, BOOT) lie in variations in the roles and responsibilities of the different actors.
- In developing a plant using a DBO or other alternative approach, one is required to make a series of fundamental decisions about needs, control,

TABLE 9–1. DISTINGUISHING FEATURES OF ALTERNATIVE APPROACHES

Distinguishing Features	Conventional	Design-Build (DB)	Design-Build-Operate (DBO)	Build-Own-Operate-Transfer (BOOT)
Level of Integration	segmented	partial	complete	complete
Operational Responsibility	municipal	municipal	vendor	private owner
Ownership and Financing	municipal	municipal	varies	private owner

and risk related to these factors that go beyond the distinguishing elements of the approach.
- The features of any project that uses an alternative model, such as Seattle's Tolt DBO project, therefore represent a combination of decisions—some of which are unique in nature to the approach, and many of which reflect preferences about needs, control, and risks independent of the approach.

In combination, these decisions represent a *project philosophy*.

Selecting and Further Refining the DBO Approach

A DBO approach has appeal to municipalities that want to take advantage of an integrated approach to design, construction, and operation, but still wish to retain ownership and financing control. These circumstances were consistent with Seattle's needs and thus drove Seattle to use a DBO model for development of the Tolt Filtration Plant.

However, as the points above suggest, developing a project philosophy and an overall project approach involves more than simply choosing to pursue a DBO model. Major issues to be addressed in the context of a DBO approach for any municipality include:

- whether requirements are to be specified in terms of a *prescription* for a plant/process (*e.g.*, we want a filter plant built to the following specifications) or instead, in terms of *required outcomes*, with the process for getting there left open to the vendor to propose, and the municipality to review (including the physical processes and equipment proposed to accomplish the outcome);
- whether innovation in proposed technical approaches is to be encouraged in favor of potential efficiencies and costs savings, or discouraged in favor of more familiar and predictable approaches that result in known technical and cost performance;
- whether the municipality wants its role in plant operation to be an arms-length relationship, or instead, a close partnership, with participation in plant operation and a role in decision-making.

These choices are interrelated and reflect trade-offs between perceptions of control and costs. The municipality's project philosophy drives these decisions in one direction or another—that is, in a direction of more prescriptive solutions, based on the belief that this control will lower their risk and costs, or in a direction of more performance-oriented solutions, based on the belief that this path will reduce risk and costs.

Seattle's DBO Philosophy

Seattle's needs and objectives-setting process resulted in a "project philosophy" built around the following concepts:

- that the city describes its needs in terms of facility performance requirements and other key outcomes, and then lets vendors propose solutions for how to achieve these requirements;

- that technological innovation be encouraged within the range of proven technology, and competition be used to achieve *both* technical innovation and lower costs;
- that risk be allocated between the city and the vendor in a manner that minimizes overall project costs (*i.e.*, assign the risk to the party best able to manage it); and
- that as the entity ultimately responsible for providing a safe and reliable water supply, the city should maintain ownership of the facility (an economic analysis and market feedback also concluded this to be the least-cost approach over the long term).

This philosophy formed the foundation for the myriad of decisions required to describe the project and construct the Request for Proposal (RFP), Request for Qualifications (RFQ), and service agreements.

Pursuant with this philosophy, Seattle developed its overall approach to a DBO process. There are five core features of Seattle's DBO approach: integrated contractual responsibility for design, construction and operation; a long-term service agreement for operations; performance-based project specifications; public ownership and financing, and optimal allocation of risk between parties.

MAJOR STEPS IN SEATTLE'S DBO PROCUREMENT PROCESS

The Seattle DBO procurement process, from project definition through contract negotiation, was comprised of five major steps. Each step in turn had a number of elements and related milestones.

Step One—Defining the Project

In late 1995, the city began to solicit for a Support Services Consultant for assistance in the DBO solicitation, evaluation, and negotiation processes. The team of RW Beck and Malcolm Pirnie was selected for this role, bringing technical expertise relating to the design of major treatment facilities, and local aspects of project permitting, as well as expertise in the area of alternate contracting processes. This augmented a high-level city project team, comprised of water quality, engineering, and policy and administrative leadership in SPU. Additional specialty skills were brought in with representatives of the city council staff, the mayor's office, and other independent technical, legal, and financial advisors. From the beginning, this high-level involvement and process buy-off throughout the city's bureaucracy were major factors in the success of the project.

Creating a Decision Process for Approving the DBO Approach Having made a preliminary choice to pursue a DBO approach for development of the Tolt Filtration Plant, it was clear to the project team and elected officials that there was not a sufficient base of experience from which to firmly project cost-savings or likely market interest in the project. The city was interested in proceeding with the DBO alternative only if sufficient savings could be obtained from capable firms with proposals that could meet Seattle's requirements. At the same time, the vendor community was concerned about the potential for an arbitrary and prolonged decision process following a substantial investment in the development of proposals on their part.

This uncertainty led the city to develop a process by which final adoption of the DBO approach would be subject to a threshold determination of its viability. This process was intended to protect both the interests of the city and those of prospective proposers.

Under this threshold determination process, the city agreed to proceed with the DBO approach if one or more proposers could guarantee:

- cost savings of at least 15 percent, compared to the life-cycle cost of a benchmark plant that Seattle would have built under a conventionally sequenced design, construction, and city operation;
- water quality, quantity, and cost performance through the application of proven technology; and
- agreement with the city's published risk allocation requirements.

This threshold assessment was to be made in the context of proposal evaluation.

Strategy Session In early 1996, the city, the support services consultants, other advisors, and major stakeholders conducted a strategy session with the objective of establishing the overall framework and direction for the DBO solicitation. Products resulting from this session included the detailed workplan, project schedule, approach for facility and performance specifications, identification of key issues and stakeholders, and development of the preliminary risk allocation matrix; an abbreviated version of the latter is shown in Table 9–2.

TABLE 9–2. RISK ALLOCATION MATRIX

RISK/RESPONSIBILITY	CITY	COMPANY
Financing	X	
Site acquisition	X	
Permitting		X
Design		X
Construction		X
Operation		X
Raw water quality	X	
Treated water quality		X
Change in law	X	

Step Two—Soliciting Vendor Qualifications and Developing the Short List

The city issued an RFQ in April 1996. The RFQ set forth the city's objectives in using the DBO process, as follows:

Facility design and construction objectives:

- Optimization of present and future water treatment processes.
- Minimization of design and construction costs.
- Sound design and quality construction for long-term operational reliability.
- Minimization and mitigation of environmental impacts.
- An acceptable project schedule.

Plant operation and maintenance objectives:

- Integrated operation with the city's water system.
- Reliable, efficient water treatment services to the public.
- Ongoing compliance with all applicable regulations.
- Effective response to both standard and unforeseeable or emergency operating situations.
- The lowest reasonable operational costs.
- Prudent management and protection of public resources.

The city requested that those submitting Statements of Qualifications (SOQ) comment on specific concepts presented in the RFQ, including treatment processes, financing options, risk allocation, guarantor requirements, and confidentiality restrictions. The city's preliminary risk allocation matrix was included as an appendix to the RFQ.

In addition, the RFQ discussed the concepts of the threshold decision to proceed and the use of benchmark comparisons listed the evaluation criteria and specified that the city would pay an honorarium, or stipend, in the amount of $100,000 to firms submitting proposals responsive to the Request for Proposal (RFP) that were not awarded the DBO contract. The RFQ also stated the city's intent to require a single, financially strong guarantor for all phases of the project. The primary reasoning behind this was that the city did not want to find itself in the center of disputes among the members of the DBO consortium. Continuous operations were imperative and resolution of problems would need to occur in a manner that insured this. In May 1996, 10 SOQs were received. Based on the recommendation of an evaluation committee, the city selected four teams to be asked to submit proposals.

Step Three—Developing Performance/ Risk Specifications and the RFP

Starting with the discussion of the overall objectives for the Tolt facility, the development of the technology and technical performance specifications of the plant formed the core of the discussions, which occurred in the course of developing the RFP. Highlights of these discussions and the resolution of key issues follow.

Level of public health protection

Issue: Should control of the infectious protozoa, *cryptosporidium*, be achieved in addition to recently required control of giardia cysts?

Resolution: The Tolt specifications provided for alternative bids for these two levels of control.

Treatment unit process efficiencies

Issue: Should planned filter rates of eight gallons per minute per square foot (gpm/sq ft) (30 m/h) from the preliminary design of the plant be used, or should higher rates be allowed in combination with flocculation alternatives, and should direct filtration with flocculation be required or should in-line rapid mix be allowed?

Resolution: The Tolt specifications provided for flexibility on both of these issues.

Instrumentation and control

Issue: Should the plant be attended full time and what should be the level of automation?

Resolution: The Tolt specifications provided for the possibility of unattended operation with suitable remote controls and alarms.

New technology

Issue: Should technologies that are new to the U.S. and that have recently been proven effective be allowed as acceptable alternatives?

Resolution: The Tolt specifications allowed for proposals based on membrane filtration for particulate removal and ozone as an alternative disinfectant, but disallowed some technologies including diatomaceous earth filtration.

A wide range of issues, including the technical issues described above, were addressed in the course of developing the performance specifications, contract elements, and risk allocation matrix. Overall, the DBO specification process allows for the resolution of these issues by obtaining proposals that balance cost with performance proven processes and the appropriate sharing of risk between the city and the potential contractor. During this phase, Seattle also approached the regulatory agencies—King County Department of Development and Environmental Services and the Washington State Department of Health—to discuss alternative facility permitting approaches. It was concluded that there was sufficient flexibility in permitting processes and a desire to work cooperatively with the city for successful project implementation within both these agencies.

The city issued the RFP to the four short-listed proposal teams in August 1996. The RFP presented proposal requirements and provided proposers with information to facilitate preparation and submission of proposals, and included 25 schedules that set forth baseline conditions and requirements for the treatment process, facilities, and operations. The city issued five addenda to the RFP and provided written responses to questions submitted by proposers. Responses to questions were issued in writing to all proposal teams. The Draft Service Agreement, which detailed the city's preferred risk posture, was issued in September 1996.

Proposal A versus Proposal B The RFP required proposers to submit two separate proposals: Proposal A and Proposal B. Proposal A's facility specifications were intended to result in a facility that met both current regulations and those that could be reasonably anticipated as of the projected facility online date. Proposal B's specifications were intended to result in a facility that met current regulations, took a more expansive outlook on potential Enhanced Surface Water Treatment Rule requirements, and addressed other enhancements that could be desired in a long-term operational period for the facility. It was expected that Proposal B would require additional treatment processes, such as ozonation.

In addition to the requirements for the water treatment process, the RFP set forth the following general facility objectives:

- The facility must provide the necessary unit processes, process control, monitoring and control, and system redundancy in order to continuously meet the treated water quality requirements;
- The facility, project features and operations must be environmentally compatible with the project and facility site(s);
- Project features must be designed and functions must be grouped together to provide for easy and efficient operation, monitoring, and collection and consolidation of long-term operating data;
- The facility layout on the site must provide space and flexibility for future process and capacity additions, taking into consideration impacts of future expansion on wetlands and other sensitive areas;
- The facility layout must provide for efficient and safe access into and throughout the site;
- The facility must be designed and constructed using materials and equipment consistent with a 50-year life for buildings and structures;
- Materials and equipment must be of sufficient quality to assure long life expectancy, a low incidence of failure, manufacturer support and service, and compatibility with Seattle's system.
- Structures, piping, and equipment must be designed so that they can be maintained while continuing to meet the supply and quality performance criteria; and
- Equipment, buildings, landscape, and other facilities must be designed to require infrequent maintenance while presenting a well-kept and pleasing appearance.

The RFP also set forth facility performance objectives with respect to construction; process design; hydraulics; civil work; architecture; structural work; building service; power supply; process instrumentation and controls; and remote monitoring and control. Objectives for plant operations, including plant classification; facility management and staffing; and coordination of plant operations with the city were also set forth in the schedules.

Step Four—Proposal Development, Evaluation, and Contractor Selection

The RFP set forth specific criteria to be used by the city in evaluating proposals. These criteria included both team/technical criteria and economic criteria. Team/technical criteria were given a weight of 60 percent and included: project implementability; technical reliability; technical viability; environmental, proposer, and staff past performance; and women and minority business enterprise (WMBE) utilization. Financial criteria were weighted 40 percent and included cost-effectiveness and financial qualifications.

Consistent with the desire to obtain a broad base of input from individuals with a variety of expertise and interest, an expert committee was appointed to evaluate proposals and recommend the best-qualified proposal for initiation of negotiations. The evaluation committee was charged with conducting an evaluation process that was fair and consistent across proposals, thorough and well-informed, and comprehensive in scope.

The evaluation committee included individuals with expertise or responsibility for utility and project management, water quality facilities development, DBO contracting, financial analysis, permitting, engineering, and construction. Technical teams consisting of evaluation committee members and advisors were created for detailed investigation into the various proposal components.

The evaluation committee conducted a multistep process to understand and evaluate proposals. This process included individual evaluations to obtain input from technical work teams, consultants, and advisors in order to better understand proposals and proposal teams, and group discussions on the merits of each proposal. The evaluation committee asked the design/build and operations technical teams to integrate their findings and emphasize the necessity of a robust proposal package wherein the treatment process and various operational features (staffing, instrumentation, process control, and reliability) would be evaluated concurrently.

In addition to reviewing technical information contained in proposals, the evaluation committee conducted a process of reference checking and site visits. This activity was carried out in four major phases.

Phase 1—Interviews Regarding Proposed Personnel. Telephone interviews were made to gather information regarding the performance of individuals given key responsibilities by each proposal team.

Phase 2—Interviews Regarding Related Project Experience. Each of the SOQs contained a listing and description of relevant project experience. Telephone interviews were held with an individual or individuals at several referenced projects for each of the major organizations involved with each proposal team. In total, 71 Phase 1 and Phase 2 interviews were conducted.

Phase 3—Site Visits. A representative group from the evaluation committee visited one project involving contract operations for each of the proposal teams.

Phase 4—Team Interviews. A panel including all voting members of the evaluation committee interviewed representatives of each of the proposal teams in February 1997.

Following receipt of proposals, the city conducted a series of internal work sessions to better understand various aspects of the proposals and to help determine whether issuing requests for Best and Final Offers would serve the city's interests. Based on these work sessions, the city issued Request for Clarification and Best and Final Offer letters to each proposal team in mid-January 1997. These letters clarified certain information contained in the RFP, requested additional information on issues that were unclear in the proposals, and solicited pricing and approaches for the addition of certain process enhancements.

The evaluation committee reached a unanimous recommendation to select the CDM Philip team for initiation of negotiations. The committee also determined that their recommendation would not be materially affected by the choice of whether to develop the Proposal A or Proposal B facility.

In March 1997, the Seattle City Council formally authorized SPU to proceed with the DBO process based on their findings that all Best and Final Proposals met the city's technical requirements and substantially exceeded the required 15 percent savings over the benchmark.

Proposal Cost Range and Savings Calculation The cost range for the four Best and Final A Proposals was $83 million to $108 million and for the B Proposals was $103 million to $124 million (net present value, 1998). Although the Proposal B facility was not used to create the benchmark cost estimate of $156 million, Seattle estimates a Proposal B facility would cost $171 million using a conventional design-bid-build process. (This estimate was produced by addition of the Proposal A benchmark of $156 million with the average cost differential between Proposal A and Proposal B facilities received from the proposers.) Thus, with the final negotiated contract valued at $101 million, the city estimates savings of 40 percent ($70 million) over a conventional approach to design, construction, and operation of a similar facility.

Step Five—Contract Negotiations

Negotiations between Seattle and CDM Philip commenced in March and continued until late May 1997. Early in the negotiations, the city decided to pursue the proposal B facility. This decision was based on the higher public health protection provided by the B facility over the A facility, and was aided by the appealing price for the CDM Philip B facility, even when compared with the cost benchmark, which was based on the proposal A facility.

The objective of the city at the initiation of negotiations was for each party to develop a strong understanding of the others' interests and to develop a basis of trust that would sustain the long-term partnership. The negotiations were conducted in three phases.

The first phase consisted of the city's identification of major issues and concerns, and the resolution of these issues through the initial collaborative work session. The second phase involved several intensive joint work sessions for facility refinement, and exploration of various facility enhancements. This phase proved to be very beneficial to the project, resulting in an improvement over the proposed facility and a reduction in cost. Facility improvements included enhanced facility redundancy, increased water storage capabilities, greater flow commitments, more incentives for efficient chemical use, and a more comprehensive definition of raw water quality parameters. The third phase included finalization of all service agreement terms and conditions, and other related documents.

As contract negotiations were completed, CDM Philip and SPU had developed a solid basis of trust, cooperation, and partnership from which to launch project implementation. In May 1997, the City Council Utilities and Environmental Management Committee advanced an ordinance authorizing SPU to enter into a DBO Service Agreement with CDM Philip for the Tolt Treatment Facilities.

CDM Philip is a Washington-based joint venture company, originally established to undertake the project. CDM Philip is built on the technical foundation of Camp Dresser & McKee's (CDM's) experience in water treatment facilities design and in CDM and Philip Utilities Management Corporation's (PUMC's) experience in treatment facility operations. Financial guarantees are provided by Philip Services Corp. In its capacity as prime contractor, CDM Philip entered into a DB Agreement with Dillingham Construction N.A., Inc. for the engineering, procurement, construction, and startup of the Tolt Treatment Facilities. CDM serves as the design engineer for the project. Once the completed facility has passed the acceptance test and is deemed suitable for delivery of treated water, the operations and maintenance of the facilities will be provided by CDM Philip.

THE DEAL

Contract Provisions

Seattle described its water quality and supply needs in the RFP and developed performance requirements detailing expected outcomes, then asked proposers to provide solutions for meeting these standards. Thus, the Service Agreement is primarily a performance-based contract. The agreement stipulates output requirements for water quality and quantity, but does not provide detailed specifications for facility design and construction. Water quality standards are higher than those required by current and anticipated state and federal regulations and supply obligations are conservative enough to allow the city ample flexibility for conjunctive use with its other supply sources.

Performance Requirements

Some of the performance requirements for the facility are listed below:

1. The facility must be capable of providing 120-mgd (454,000-CMD) of treated water (meeting all performance requirements) for raw water

turbidities up to 15 NTU (nephelometric turbity units). Plant production is reduced at turbidities above 15 NTU. For example, 55-mgd (208,000-CMD) of treated water will be produced at a raw water turbidity of 40 NTU.
2. The facility must provide up to 5 logs or one one-hundred-thousandth (99.999 percent) of cryptosporidium removal through filtration and inactivation (ozonation). A minimum of 2.5 logs must be achieved with ozonation based on the best available information at the time the service agreement was executed (Peeters, 1989).
3. It must address upcoming Disinfection By-Product (DBP) Phase I regulations by providing water with less than 64 µg/L (micrograms per liter) Total Trihalomethanes (TTHMs) and less than 48 µg/L Haloacetic Acids (HAA5) based on a Simulated Distribution System (SDS) test.
4. Residual disinfection will be provided using gas chlorination with flexibility to add ammonia for chloramination in the future.
5. Fluoride will be added at the new facility, as it has been at the existing facility.
6. pH and alkalinity will be adjusted (using calcium oxide and carbon dioxide) to allow for corrosion control optimization.
7. The facility includes emergency generator capacity for the entire plant. No unozonated water will enter the distribution system from the Tolt River Treatment Facility, even if the power supply is interrupted.
8. CDM Philip is responsible for disposition of backwash solids. They are planning to utilize on-site disposal of the solids in drying beds.
9. There will be no pumping of finished water; the Tolt will remain a gravity supply.
10. The facility will have redundancy in all process units, which ensures that at least 60-mgd plant capacity is available at all times, including periods of scheduled plant maintenance.
11. Comprehensive, real-time, plant operating information will be provided to the city at its Operations Control Center.

Other contract provisions include payment of prevailing (in the case of construction workers) or comparable (in the case of operators for whom prevailing wages are not available) wages to workers, and subcontracting to women- and minority-owned businesses at 29 percent of the total contract amount for design and construction.

Proposer-Determined Facility Specifications

The facility will include the following treatment processes: ozonation and flocculation (30 minutes detention time in a combined ozone contactor and hydraulic flocculation basin); and filtration with a maximum rate of 12 gpm/sq ft (30 m/h) through 72 inches (1.8 m) of anthracite monomedia.

The total value of the contract is approximately $101 million (net present value, 1998 basis). This includes $65 million for design and construction, and $36 million for 25 years of facility operations. The term of the contract is 25 years (15 years with options for two five-year renewals with the same contract terms, at the city's discretion). Seattle estimates the one-and-a-half-year procurement process to have cost $1.5 million to $2.0 million.

Structure of Capital Payments

The capital payments to be made during the development and construction phases of the project are structured in two different methods. Payments to be made during the development phase—the time during which CDM Philip completes design and permitting for the facility—are to total no more than 2 percent of the capital component of the contract. These payments are divided equally into monthly payments, assuming the development period will be complete one year following contract initiation. The remainder of the capital payments are to be paid out based on a predetermined milestone payment schedule, with 5 percent retainage held out until final acceptance testing is complete.

Structure of O&M Service Fee

The structure of the service fee was very important in meeting the city's objective to allocate incentives appropriately. The city's interest in development of the service fee was to assure incentives that mirror the city's foremost objective of reliable delivery of high-quality water to customers. Other objectives, including minimization of chemical use and environmental impacts, and aggressive maintenance practices for the long-term health of the facility, were conveyed in the service fee structure as well.

The O&M service fee is divided into two primary components, a fixed base fee and a variable component. The fixed base fee includes the base operations and maintenance service charge (for labor, facility maintenance, including costs for equipment repairs under $10,000, and all other non-chemical and non-pass-through costs of operating the facility) and the base chemical charge (the negotiated fixed cost of chemicals used in the treatment process). The variable component consists of pass-through costs primarily for insurance and electricity, which are "passed through" for city reimbursement without markup.

Major Equipment Renewal and Replacement

An important city objective for the facility included the requirement that it be designed and constructed using materials and equipment consistent with a 50-year life for buildings and structures, and that materials and equipment be of sufficient quality to assure long life expectancy, and a low incidence of failure. Another city objective was to assure that the facility be maintained over time to a consistent high standard of appearance, longevity, and functionality. Accordingly, the city wanted to ensure 1) that preventive maintenance be performed to specified standards, and 2) that appropriate investments would be made in minor and major equipment replacement, both on an ongoing basis and especially toward the end of the contract term.

The city's approach in meeting the first of these objectives was to require the contractor to include in their operations and maintenance plan schedules listing all major items with replacement frequency and type of each, and to require that the facility be maintained to that schedule unless the city authorizes otherwise. In addition, a requirement of the Service Agreement was that third-party facility audits be performed semiannually throughout the term of the agreement in order to assure proper maintenance, repair, and replacement.

The approach for meeting the second interest was to require proposers to submit, as part of their proposal, a plan for major equipment renovation throughout the term of the agreement, including preventive maintenance plans, related operator training, and a Renewal and Replacement Plan for replacement of major items (over $10,000) over the term of the Service Agreement. The city will make annual payments as designated in the Service Agreement to a Renewal and Replacement fund to cover major equipment cost replacements that have been identified in the plan. Higher or unforeseen maintenance or replacement costs are the responsibility of the contractor, and if the contractor maintains equipment to lives longer than those specified in the plan, cost savings will be shared between the city and the contractor. Repair and replacement of items of $10,000 or less are considered ordinary maintenance activities and are covered within the fixed base fee.

Liquidated Damages and Incentive Payments

Seattle's fundamental obligation is to provide safe drinking water to its direct service customers and wholesale purveyors. In this context, a major challenge in crafting the RFP and draft Service Agreement was to definitively convey the essence of this public trust to the contractor. The city used the Service Agreement as a primary method of meeting this objective. The contract includes modest financial incentives for exceptional facility performance and substantial liquidated damages for nonperformance (addressing regulated and enhanced water quality and water supply). Eighteen treated water quality parameters are to be monitored for which noncompliance liquidated damages vary depending on the water quality or public health consequences and severity of the infraction. All of the more serious infractions include grounds for termination after a specified period of noncompliance. In addition, the contract includes significant liquidated damages for supply of unacceptable treated water which may or would result in drinking water regulatory actions up to and including a "boil water" notice.

Incentive payments are to be made to CDM Philip by the city in the amount of $5,000 per month for any calendar month in which the total number of noncompliance events is zero. Lesser amounts of monthly incentive payments will be made if certain less serious noncompliance events are less than two or three in a calendar month.

Schedule

CDM Philip fulfilled their development period obligations (site permitting and substantial completion of design) by May 1998. Notice to proceed was issued in May 1998, and construction began soon thereafter. The contract provides for 926 calendar days for completion of construction. The contract provides for completion of construction and online status by the end of 2000.

Project Financing

From the outset, the city desired to own the new Tolt Treatment Facility and believed that utility tax-exempt financing would achieve a lower cost of capital for the project. The city did invite proposers, as part of the RFP process, to submit alternatives for private financing or suggest other alternatives. No workable responses were received and the market supported our basic assumptions.

For a time, the city planned to use private activity bonds as a preferred financing approach because they would accommodate a long-term operations contract with a private vendor while still allowing the use of tax-exempt debt. It was important not only for the city to secure a long-term operations contract but also to include certain performance features. However, the IRS released new private activity regulations that became effective in May 1997, allowing for more flexibility in the structure of long-term operations contracts between private operators and governmental entities. These new regulations and the final negotiated terms of our Service Agreement make it possible in the opinion of our bond counsel to issue governmental purpose tax-exempt bonds.

Labor Issues

No significant labor issues arose during the procurement process or execution of the contract. This is primarily due to the newness of the function of design, construction, and operation of a major water treatment facility to the city; there was no displacement of existing labor as a consequence of the DBO approach. The contract includes a provision for joint hiring of operators prior to initiation of CDM Philip operations. In the interim between contract execution and facility operations, this provision allows Seattle to fill open operator positions with personnel selected jointly with CDM Philip and with the intent that these operators will transition to the new facility once it is operational.

City Oversight of Contract

The Service Agreement is primarily a performance-based contract. The agreement stipulates output water quality and quantity requirements, but does not provide detailed specifications for facility design and construction. In certain instances, however, the city finds it necessary to maintain close involvement and oversight of various project issues:

Project Staffing Project team staffing was an important consideration during proposal evaluation, and corporate commitments of key personnel continues to be a major component in project success. For this reason, Seattle desired to specifically obligate individuals to their project role and time commitment. The Service Agreement lists names of individuals and their required project involvement level during each phase of the project. In addition, the city and CDM Philip will jointly select the facility manager, who will begin full-time project responsibilities at least one year prior to the scheduled acceptance date in order to participate in preparations for operations start-up.

Project Permitting For three primary reasons, Seattle maintained oversight and involvement in the facility permitting process. First, the city had initiated discussions with the regulatory agencies during the conceptual development of the DBO process, and it was important to provide continuity with them as the primary contact for the project transitioned to CDM Philip. Second, since permitting was a critical path item on the project schedule, Seattle felt it beneficial to be present in order to provide assistance in the way of project background, city interests, plans, or priorities, should it become warranted at meetings. Finally, and most importantly, the city enjoys ongoing relations with both major regulatory agencies and maintains participation and oversight in order to ensure the preservation of positive interactions between city representatives and regulators.

Design Seattle's approach to design and construction oversight was to ensure compliance with the Service Agreement with primary emphasis on the areas that 1) have plant longevity considerations, 2) could cause major maintenance problems, 3) would be extremely difficult to repair, replace, or rebuild, and 4) have major facility reliability considerations. Seattle participated in the review of four major design packages throughout the development phase of the project. These city reviews were conducted by city staff and the city's support services consultant with the primary objective of ensuring conformance with the Service Agreement. Although many comments and questions were transmitted from the city to CDM Philip, only a limited number of them were identified as significant issues for which a response, change, or clarification was requested.

Construction The city's primary internal staffing objective going into the construction phase of the project was to maintain an owner's presence on the job, to ensure implementation by CDM Philip of their QA/QC program (primarily in the four areas identified above for Design oversight), and to be available to assist in interface issues if necessary. The city expected that its role on site would evolve as experience dictated, and was mostly interested in observing the execution of quality construction practices.

Start-up The transition and acceptance testing procedures are defined in the Service Agreement. The city staff participated in the review of the transition plan and acceptance test plan, close monitoring of the transition and start-up activities, and training of facility operators.

Facility Operations Water quality standards are strictly defined in the Service Agreement and higher than those required by state and federal regulations at the time of contract execution. The city can maintain a "window" into operations of the facility through monitoring of instrumentation and control at Seattle's Operations Control Center. City staff can participate in facility operations training at the discretion of the city, and can monitor water quality and quantity control parameters within the facility at its discretion. CDM Philip's facility water supply obligations are flexible enough to allow the city ample control over the conjunctive use of the Tolt supply with its other supply sources.

CONCLUSION

Developing an overall project approach and process is crucial to the success of any project, as well as thoughtful development of project procurement documents, legal agreements, and technical requirements. Our objectives determined the specific approach, project staffing, decision-making and communication process, evaluation criteria, and implementation strategy. However, in the use of an alternative approach to project development, like the DBO model for the Tolt project, it is critical that there be a very well designed and executed process of implementation. Since the alternative approach is new, utility staff and management, elected officials, and the public at large are not entirely sure of all of the steps. The process provides a roadmap of goals, activities, expectations, decisions, and points of involvement.

In reflecting on the Tolt DBO procurement process, Seattle can turn to the underlying objectives and core features of the project in evaluating its success. The city maintains ownership of the facility and the contract includes strong financial guarantees for the three phases of the project. The Service Agreement appropriately allocates risk, is fair to both parties, and sets forth a clear roadmap for project implementation. The length of the agreement is important in ensuring the viability and long-term reliability of the process design and facility construction. The contract is performance-based, but is clear about those water quality, supply, maintenance, and engineering standards that must be incorporated into the facility. The performance-based nature of the contract allowed proposers the rare (for public works projects) opportunity to do a project in the most efficient and practical manner—cooperatively among site permitting specialists, designers, constructors, and operators, and allowed them to push the design envelope and be innovative in their proposals. A single point of accountability has been established within the CDM Philip team rather than segmenting responsibility and liability. This same consolidated accountability is driving the team to coordinate design, construction, and operations phases to achieve better quality, a faster schedule, and lower cost.

We observe a market that is ready to develop projects using a DBO process and similar approaches. Since many consider Seattle's Tolt project to be a test case (or a controlled experiment) for implementation of projects using alternative contracting approaches, it is important to recognize that there are several distinguishing attributes of Seattle's situation that drove the decision to use a DBO approach and aided in the project's success.

Finally, the keys to Seattle's success with the development of the Tolt DBO project has been multifaceted:

1. The project's diverse and multidisciplinary team was critical in bringing in diverse perspectives and better solutions to complex issues.
2. The project philosophy, developed early in the process, was key to balancing performance versus prescriptive interests, choices between innovation and traditional approaches, and tradeoffs between control and costs.
3. The project partnership between the utility and Seattle's elected officials ensured that stakeholders were on board throughout the process, and clear points of involvement for elected officials were identified early in the process.
4. A relentless and continual emphasis on maintaining process integrity provided the assurances needed by the private sector in participating in the competitive process.
5. Thoughtfully developed performance specifications, contract incentives, technical requirements, and other contract provisions have provided the basis for an approach of *trust*, *cooperation*, and *partnership* with CDM Philip for contract implementation.

REFERENCES

Kelly, E., Haskins, S., and Reiter, P.D. (1998). "Implementing a DBO Project—The Process of Implementing Seattle's Tolt Design-Build-Operate Project Provides a Road Map for Other Utilities Interested in Alternative Contracting Approaches." *Journal AWWA*, June 1998, pp. 34–46

Peeters, J.E. et al. (1989). "Effect of Disinfection of Drinking Water With Ozone or Chlorine Dioxide on the Survival of *Cryptosporidium parvum* Oocysts." *Appl. & Envir. Microbiol.*, 66:6:1519

Westerhoff, G.P., Gale, D., Reiter, P.D., Haskins, S.A., Gilbert, J.B. (1998). *The Changing Water Utility—Creative Approaches to Effectiveness and Efficiency*. American Water Works Association.

Chapter 10

Cost Savings for Atlanta's Water System Through Contract Operations and Management

Remedios del Rosario, David L. Seader, and John E. Salo

As with many metropolitan areas of the United States, Atlanta's antiquated water, wastewater, and sewer systems must comply with increasingly stringent environmental requirements while providing system users with high-quality service. With the need to immediately implement its major capital improvement program, in order to accomplish these twin objectives, Atlanta was faced with two choices in paying those costs: either continually raise user rates to meet operations and debt service requirements, or find savings in existing operations to offset rising costs.

After having recently raised water and wastewater rates considerably, the city decided in June of 1997 to assess the potential to reduce operations costs. To help evaluate its options for cost savings and embark on a course of action, in July 1997, the city retained a team of consultants to perform a phased operations assessment and implementation plan.

The first phase analyzed water, wastewater, and sewer operations, and identified possible alternatives for cost savings and technical improvements. The assessment, delivered in October 1997, provided a database of improvement measures, arranged into different alternatives for action. It also presented a menu of strategies that Atlanta could use to develop a comprehensive cost savings program.

Remedios del Rosario is commissioner for Atlanta's department of water, David L. Seader is director of the water industry practice of Pricewaterhouse Coopers, and John E. Salo is senior vice president of Brown and Caldwell.

After Mayor Bill Campbell selected outside contract management of the city's water system and its largest wastewater treatment plant, and re-engineering of other operations, the consultant team began the second phase, delivering an implementation plan supportive of and consistent with the city's preferred course of action.

While the re-engineering program was getting off the ground in the spring of 1998, a procurement was conducted to select and contract with a private sector team to take over the complete operation of the water system. The successful team was chosen and it assumed the operational duties of the system as of January 1, 1999, less than one year from the issuance for the Request for Proposals for the contract. This contract represents the most comprehensive and complex privatization of a major metropolitan water system in the U.S. to date.

SYSTEM BACKGROUND

Water and Wastewater Facilities

Atlanta provides water and wastewater service to more than 1.5 million people.* The city owns and operates three wastewater treatment plants, which together currently handle 150 million gallons per day (mgd). The sewer system consists of 12 pumping stations and 2,200 miles of sewers, including a 305-square-mile area served by combined sewers. The city owns and operates two water treatment plants, with a capacity of 201.4-mgd; three major pumping stations; 90.5 million gallons of storage; and 2,400 miles of water mains.

The Water Department, Sewer Division, and Wastewater Services 1996 operating budget (including indirect expenses) was $116.4 million. The city's CIP will require more than $850 million over the next five years to replace and upgrade antiquated facilities, provide for future capacity, and to meet more stringent regulatory requirements.

Functions and Responsibilities

The responsibility for the treatment and distribution of potable water and the conveyance, treatment, and disposal of wastewater resides with the Department of Water and two divisions in the Department of Public Works: the Division of Wastewater Services and the Sewer Division. Functions of the Department of Water include obtaining and treating raw water from the Chattahoochee River and distributing it to about 1.5 million people in metropolitan Atlanta. The Sewer Division is responsible for maintaining Atlanta's sanitary and combined sewers and conveying wastewater from a service population of about 1.2 million to the city's wastewater treatment plants. The Wastewater Services Division is responsible for operating and maintaining the city's three wastewater treatment plants and related combined sewer overflow (CSO) control facilities.

---

*See Figure A–6 in the Appendix.

Staffing Levels and Costs

In 1996, the baseline year of the assessment, the Water Department, Sewer Division, and Wastewater Services Division had 763, 286, and 419 authorized positions, respectively, for a total of 1,468. In addition to these direct staff, support was provided from several other city departments to maintain water and wastewater services. The 1996 operating budgets are shown in Table 10–1.

PHASE I: THE OPERATIONS ASSESSMENT

Methodology

Analysis of Existing Systems The consultant team first embarked on an investigation of the existing systems' cost-effectiveness, conducting numerous interviews and reviewing dozens of documents. The analysis was organized around 40 or so functions, facilities, and support service categories that comprise the three systems. Using the system review, together with comparisons to other similar systems (benchmarking) and its extensive experience and expertise, the consultant team identified numerous operations areas in which cost savings and improvements could be made.

Competitive Issues Workshop The city held a day-long competitive issues workshop in August 1997. The purposes of the workshop were:

- to provide a forum for the city's managers and employee representatives to learn about practical techniques that other major U.S. cities had found to lower the cost of operations, raise productivity and efficiency, and improve the quality of service;
- to provide the consultant team with an efficient means of confirming its view of the state-of-the-art of alternative service delivery techniques (what is currently "doable");
- to provide everyone with any updating of benchmarking-type information on current cost savings available for the various functions and facilities being analyzed.

TABLE 10–1. ATLANTA WATER AND WASTEWATER SERVICES' OPERATING BUDGETS (1996)

Water Department	$49.5 million
Sewer Division	15.1 million
Wastewater Services Division	33.9 million
Indirect Expenses*	17.9 million
TOTAL	$116.4 million

*Indirect expenses include overhead, and charges by other city departments, including Personnel, Finance, Purchasing, Law, Contract Compliance, and Architectural Services.

The workshop brought together invited city professionals and union representatives with their colleagues from other cities, both in the public sector and with the private firms that provide water and wastewater operations, maintenance, and management. Each presentation was designed to provide a better understanding of particular cost savings and improvement techniques, including internal re-engineering, managed competition, outsourcing, contract operations, and other types of public-private partnerships.

Selection of Alternatives The consultants outlined several different methods by which those savings might be obtained: internal re-engineering (including bid-to-goal), outsourcing, managed competition, contract operations of facilities, and private system management. With this array of tools and the operational areas in which to apply them, the consultant team then estimated the amount of savings that could be derived from each of the measures under each of the options, and assembled them into savings packages or alternatives.

The consultant team calculated the potential savings for each alternative over the next four years, estimating the implementation time, investment cost, and timing of gains for each piece of the package. Variations also exist within the alternatives, like using managed competition rather than outsourcing/contracting, or trying bid-to-goal rather than straight re-engineering; and between alternatives, like mixing and blending pieces of the packages. Internal re-engineering savings are assumed to be realized gradually over a three-year period.

Eight alternatives were presented to Atlanta in the operations assessment phase:

"Light" Re-Engineering/Outsourcing (Alternative 1): Making the straightforward changes that can be accomplished within a single system/department, and those that will result from the implementation of the CIP. Also included in this alternative is the outsourcing of non-core functions to the private sector.

"Heavy" Re-Engineering/Outsourcing (Alternative 2): Adding to the light re-engineering package those more difficult internal actions that involve the cooperation of two or more departments and/or a basic change in the way that the city has conducted its business. Also included in this alternative is the outsourcing of non-core functions to the private sector.

Contract Operations (Alternatives 3, 4, and 5): Selecting through competition a qualified private firm or firms to operate one or more of the five treatment facilities in the system, while implementing the "heavy" re-engineering package on the non-contracted facilities and functions that remain under the city's management. Three specific alternatives were addressed under this approach:

- Contracting out the two water plants—Chattahoochee and Hemphill (Alternative 3);
- Contracting out two co-located plants—Chattahoochee Water Plant and R. M. Clayton Water Reclamation Center (Alternative 4); and
- Contracting out all five water and wastewater plants (Alternative 5).

System Management (Alternatives 6, 7, and 8): Selecting through competition a qualified private firm to manage the water system, the wastewater/sewer system, or both, while implementing the heavy re-engineering scenario on the remaining system:

- System management of water (Alternative 6).
- System management of wastewater/Sewer (Alternative 7).
- System management of both (Alternative 8).

Note that for each case other than light re-engineering, each and every function and facility is "renovated," either internally or through privatization. This makes every choice a comprehensive, top-to-bottom choice.

Evaluation of Alternatives

To assist the city in evaluating the impact of these alternatives on the customers of the systems, the consultant team constructed a computer spreadsheet-based financial model. The model reproduced the flows of revenues and costs through the systems, including operations, indirect costs, capital spending, debt service on bonds, outside revenue sources, and so forth, to show what future rates would look like under the various scenarios. One goal of the city was to see if enough savings could be generated under any alternative to avoid rate increases other than for increases in the cost of living. The model projected the extent to which rates would increase if there were no additional savings generated in the next four years. The resulting annual average rate increase would be about 8.5 percent. Next, estimated savings were included in each of the alternatives for each of the next four years, and then rates were projected for each case. The results of the modeling are contained in Table 10–2.

TABLE 10–2. ESTIMATED ANNUAL SAVINGS AND USER RATE IMPACTS

Alternative Description[a]	Annual savings (in $ millions) 1998	1999	2000	2001	annual rate increase[b]
No Savings—Baseline	0.0	0.0	0.0	0.0	8.5%
1. Light Re-Engineering/Outsourcing	6.4	13.9	15.0	14.9	6.4%
2. Heavy Re-engineering/Outsourcing	9.1	20.8	25.1	25.4	4.8%
3. Contract Operations—Chattahoochee & Hemphill	12.4	26.4	30.3	30.6	4.0%
4. Contract Operations—Chattahoochee & R.M. Clayton	11.0	25.0	29.1	29.5	4.2%
5. Contract Operations—All Five	14.0	29.1	33.0	33.5	3.6%
6. System Mgmt—Water	17.0	30.5	31.3	32.1	3.8%
7. System Mgmt—Wastewater/Sewer	17.1	29.9	30.7	31.5	3.9%
8. System Mgmt—All Systems	21.6	37.6	38.6	39.6	2.7%
Inflation Only	9.3	35.8	40.4	37.3	3.0%

[a]Alternatives 2 through 8 each include comprehensive re-engineering of all facilities and functions either internally or through privatization.

[b]Required compound annual growth rate in both water and sewer rates.

As the table demonstrates, projected rate increases can be cut in half by implementing practically all of the alternatives except light re-engineering. Only the most dramatic alternative—placing all of the systems under private management—has the potential to limit rate increases to less than inflation (assumed in the analysis to be at 3 percent per year).

For comparison purposes, a "typical" residential customer billed for 7,000 gallons per month currently pays approximately $36 per month for water and sewer service. An 8.5 percent increase in water and sewer rates would result in an average increase of $3 to $4 per month in each successive year.

While user rate impact has been the most important consideration in the definition, analysis, and evaluation of the results, other criteria were also important to the city. In conjunction with city managers, the consultant team derived a set of criteria to use in defining the differences among alternatives. In brief they were:

Other Financial Criteria: immediacy and guarantee of savings, access to adequate capital, financial control by city, contract control by city, impact on the city's General Fund.

Operational Considerations: maintenance of service quality, degree of risk, impact on construction activities.

Legal Considerations: complexity of required legislation, maintenance of permits and public status, maintenance of operational powers, compatibility with existing legal requirements.

Policy/Institutional Criteria: impact on current employees, opportunities for disadvantaged groups.

A multi-criteria evaluation matrix was used to weigh and display the relative merits of each of the eight alternatives. Armed with the financial results and the evaluation matrix, Atlanta was prepared to make a decision on its course of action.

RESULTS

The mayor and his staff reviewed and analyzed the consultant team's Operations Assessment, keeping five broad policy goals in mind: being fair to city employees, making operations more efficient, keeping water and sewer rates competitive, delivering quality customer service, and reducing the need for rate increases.

Timing was a critical factor to consider as well, since the city had to issue revenue bonds in mid-1998 in order to keep construction of its regulatory-mandated phosphorus reduction program on track. Thus, a combination of "guaranteed" savings and rate increases had to be in place before the city could issue its bonds.

Recognizing that re-engineering is a long-term effort, the city concluded that private management of water and wastewater facilities needs to be a major part of its cost-savings program. On that basis, the mayor recommended that the city engage a private operator or operators to manage the existing water system and 100-mgd R.M. Clayton Wastewater Reclamation Center (WRC) under multiyear contracts—a hybrid of what the operations assessment had identified as Alternatives 4 and 6.

CONCLUSION

Three major conclusions can be reached by reviewing the performance of alternatives against the criteria:

1. The more the savings, the more the overall risk. This is not surprising given that reward generally correlates with risk in any investment/change environment. The alternative with the most savings potential (Alternative 8) was one that has not been used before in the U.S., though the private management of both water and wastewater services has been the norm in several countries, notably the United Kingdom. The private management of water utilities of Atlanta's size is less of an anomaly, as there are many investor-owned water utilities throughout the world. Proceeding to less intensive alternatives, there is more experience with contract operations of individual facilities, so unknown risks are less. If only light re-engineering is done, as in Alternative 1, then there would be little risk accompanying more modest savings.
2. Every alternative demands a major commitment from the city. Even the alternatives that place one of the systems in private hands still require that the city's management demonstrate leadership and commitment to re-engineer thoroughly the other systems left under its direct operation. All of the options will require some degree of training, increased employee empowerment, incentive and bonus plans, cooperation among several departments, and perhaps citywide changes to support water/wastewater/sewer system changes. All alternatives would reduce the level of direct and indirect payments from water, wastewater, and sewer operations to the other city departments or the General Fund itself.
3. The success of the public-private alternatives depends on good contract management. The response to many of the concerns embodied in the criteria is to write requirements, prohibitions, penalties, and incentives properly into the contract with the private firm (or the agreement with the employees in bid-to-goal and managed competition), and to manage the contract tightly over its lifetime. This will take some preparation and dedication of city management, and a change in skills needed from day-to-day operations management to contract management and policy planning and development.

All alternatives can accommodate the city's key nonfinancial concerns. By specifying the city's policies clearly and directly, most all of the city's requirements would be satisfied through the relationship developed with any private firm involved in the systems. This is most important with respect to the treatment of the city's workforce. The city has stated its goal that no employee will be displaced by any alternative for up to the first three years of implementation. The consultant team believes that the projected staffing levels contemplated in all eight alternatives can be met by limiting reductions in force to attrition and for cause. Similarly, the same can be said about working with the city's affirmative action goals, maintenance of city control over the systems, maintenance of access to tax-exempt financing, and so forth.

PHASE II: PROCUREMENT OF A LONG-TERM CONTRACT OPERATOR FOR THE WATER SYSTEM

The Overall Program

On December 8, 1997, Mayor Bill Campbell announced his decision to bring private management to Atlanta's drinking water system and to the R.M. Clayton WRC, as well as to overhaul the remaining functions of the wastewater and sewer systems. Six staff subcommittees were established to develop detailed plans for various parts of the program: legal and financial issues, employee-related issues, contract management and organization, facilities, re-engineering, and executive coordination. The objectives of the program were to:

1. Prepare a Request for Proposal (RFP) for services that will be contracted out, evaluate proposals and negotiate with the most responsive proponent so that contracts could be put in place during 1998 for water, and in 1999 for the Clayton plant.
2. Develop a reengineering plan for the wastewater and sewer functions that will be retained by the city.
3. Provide training and technical assistance to city employees and management so that they will be able to design and operate the remaining city functions in an efficient and competitive manner.
4. Reorganize the retained city functions as needed to reflect the new operational environment.
5. Conduct a comprehensive communications effort to inform and educate all stakeholders who will be affected by the city's privatization/reengineering initiatives.

Privatization of the Water System

Functions and Facilities A key element of the plan was to continue the procurement process for the private operation, maintenance, and management of the city's entire water system using a 20-year contract term. The elements to be placed under private management included raw water supply, treatment and distribution, billing, collection and other customer services, as well as the provision of certain capital repairs, replacements and some of the system's capital improvements. The functions to be taken over were comprehensive, so that the contractor would essentially become the city's water department and take on full responsibility for operating and maintaining the system in accordance with all policies, rules and regulations, and provide the required services to all water system customers. Table 10–3 lists the specific functions the contractor was asked to perform.

TABLE 10–3. WATER DEPARTMENT FUNCTIONS ANTICIPATED TO BE PROVIDED BY A PRIVATE CONTRACTOR

Operations and Maintenance	**Customer Service**
Chattahoochee water production complex	Field operations (meter reading and repair, turn-on/off, inspections)
Hemphill water production complex	One-call customer service
Distribution system	Billing (does not include collections/includes sewer)
Grounds maintenance	Meter applications
Laboratory	**Support Services**
Sample collection and analysis	**Personnel**
Cross connection control	Employment
Disinfection of mains	Labor relations
Response to water quality complaints	Employee records
Quality assurance/quality control (QA/QC)	Training
River water withdrawals	Employee assistance program
Technical Services	Classification and compensation
Administration (operations, etc.)	**Finance**
Day-to-day engineering	Collections
Construction and maintenance	Employee benefits
Construction design and inspection	Accounting
Substations and repair crews	Financial analysis—general
Heavy construction	Financing—general
Craft shop (14th Street)	Capital improvement program input
Pipe, meter, valve repair and maintenance	Contract services—(fire hydrant maintenance, etc.)
Administration	Purchasing
Information systems	**Public Works**
Maintenance management system (MMS)	Repave streets/sidewalks
Local area network (LAN)	Supply materials (*e.g.*, cement mix)
Customer information system (CIS)	PW dumpster pickup
Computer-aided design – Computer-aided mapping (CAD-CAM)	Motor vehicle services
Supervisory control and data acquisition (SCADA)	Howell Mill Road garage
Commissioner Support	Repair (vehicles and other motorized equipment)
Public Relations—Routine	**Procurement**
Conservation	**Fuel**
Safety	**Legal**
Equipment Pool	**Fire Department Coordination**
Materials Management	

The contractor was also required to provide its own support services, including personnel and human resources management, purchasing, motor equipment maintenance, legal and financial services, building services, management and customer information management, and general administration and management. All of the city's facilities, assets and systems were to be made available to the contractor, including the department's offices, shops, garage and facilities.

City-Retained Functions The city retained certain functions and responsibilities, including rate making authority; some functions of the Commissioner's office, including public relations, regulatory/legislative matters, and wholesale customer pricing contract relations; the Geographic Information System; and sufficient planning and engineering capability to plan and update the CIP and administer CIP engineering contracts handled by other contractors. In addition, Atlanta retained sufficient staff for contract management, quality control, and audit capability to oversee the water system contractor.

At the city's sole option, certain CIP projects, including those that relate directly to the performance of the water system, were to be handled by the water system contractor on a design-build basis. To help price those projects, they were included in the solicitation and were bid on by the proponents submitting proposals. Other portions of the CIP could be handled exclusive of the water system contractor and procured through other design-build contractors or traditional design-bid methods at the city's option. Thus, the city retained both control over its CIP and leverage in selecting cost-effective contractors, but also retained flexibility in CIP implementation by having the water system contractor available to complete projects as the city desired if it were the most effective mode of implementation.

Objectives In transferring the performance of the system to a private operator, the city hoped to achieve a balanced set of goals and objectives in both the short and long-term. The key objectives that the city expected the contractor to accomplish were:

Economic and Institutional Objectives

- Create a more efficient, less costly system while maintaining or enhancing service quality;
- Become a good corporate citizen of the Atlanta metropolitan community;
- Participate with and advise the city on water resource-related policy issues;
- Improve the public image of the system;
- Maintain a professional, courteous, and equitable relationship with the city;
- Implement the Equal Business Opportunity and Equal Employment Opportunity Program goals established by the city for this initiative.

Employee Objectives

- Maintain a safe, secure and productive work environment for employees;
- Provide a smooth transition for employees from the city workforce to the private workforce;
- Protect the employees' salaries and provide comparable benefits during and after the transition;
- Provide enhanced opportunities for career development and professional growth.

Operations and Maintenance Objectives

- Produce water that maintains quality standards in conformance with all applicable laws, rules, regulations and, to the extent consistent with the city's cost-savings objectives, enhance water quality;
- Provide high-quality, uninterrupted service, meeting the daily demands of all users of the system;
- Operate the system in conformance with good industry practice;
- Preserve, protect and modernize the system;

- Promote the reliability and efficiency of the system through state-of-the-art predictive, preventive and corrective maintenance practices.

Capital Planning and Implementation Objectives

- Assist the city in developing and implementing the city's capital improvement plans for the system;
- Implement designated capital improvements and repairs to the system in a timely and cost-effective manner.

Billing, Collection, and Other Customer Service Objectives

- Provide courteous, professional and helpful service to all customers and users of the system;
- Provide timely and efficient billing and collection of revenues and maintain current and accurate customer records;
- Improve customer billing, collection and reporting services, by installing a new, computerized customer information, billing and collection system;
- Efficiently and effectively communicate with and inform customers and the general public regarding appropriate aspects of the operations, regulations, rules, policies and requirements concerning the system.

Schedule Atlanta set an ambitious schedule for the procurement process, looking to have a contractor on board less than one year from the issuance of its RFP. One reason for the tight scheduling was that the savings anticipated through privatizing the water system had to be "bankable" when the city offered its next issue of tax-exempt bonds in order to be able to use the savings to offset any rate increases that would otherwise need to be in place at that time. Because the major idea of privatization was to offset rate increases and pay for new capital spending, the city needed to have a contractor identified with a guaranteed level of savings when it went into the bond market in late 1998 or early 1999.

One of the tools used to compress the schedule was to issue a combined Request for Qualifications (RFQ) and RFP. This was intended to give proponents the maximum opportunity to become familiar with the system and the required services, including the Draft Operating Agreement, which was prepared by the city's legal team. All proponents were warned that any work they put into the preparation of a technical and cost proposal prior to being found qualified from their RFQ would be at their own risk and expense. Meanwhile, those proponents that were confident enough of being found qualified could begin their due diligence and work on their proposals as soon as possible.

The original schedule gave teams about five weeks to prepare their RFQs and, simultaneously, 14 weeks to complete their technical and cost proposals. After that, there was to be a third round of submittals—best and final proposals—that were to be specified and scheduled following the evaluation of the technical and cost proposals. Given the complex, momentous, and highly political nature of the solicitation, it is remarkable that the city kept fairly closely to its original schedule for the entire procurement.

Table 10–4 contains the original and actual schedules for the major milestones of the process. Atlanta was able to conduct a three-step procurement process with substantial competition from established teams, and obtain all the sign-offs, approvals, and signatures it needed in less than one year from the time of issuance of its RFQ/RFP.

STEP ONE: THE QUALIFICATIONS STATEMENT

The first submittal for each proponent was a Statement of Qualifications (SOQ) describing the overall team capabilities. The purpose of this step was to allow the city to select a pool of proponents that were capable of adequately and dependably serving the customers of the system over the long term. Proponents were given strict guidelines about the format, organization, and contents of their SOQ, including page limitations and required forms covering conflicts of interest, certification of the accuracy of the qualifications and a directory of contacts for the procurement. The city was looking for teams with the specific experience and financial strength to undertake this project.

Contents of Qualifications Statements

Proponents were asked to provide information on the composition of their team, the team's relevant experience, and their financial capability. With regard to the proponent team composition, the city asked for identification of and details about the legal structure of the team and each of the team's members. Included in the team were to be minority- and woman-owned firms to be responsible for 35 percent of the team's work effort. For each team member, the city asked for a short business history, detailed information about the ownership and organization, an ownership chart, a discussion of the team member's roles and responsibilities, and a letter of intent certifying the team member's participation.

With respect to the relevant experience of the team and its members, the RFQ asked for information about the track record of team members with contracts similar to the one being offered by Atlanta. It was stressed that any work listed as a qualification for the team relate to the services being sought, and that the proponent show

TABLE 10–4. ORIGINAL AND ACTUAL SCHEDULES FOR WATER SYSTEM CONTRACT PROCUREMENT

Events/Milestones	Original Schedule (1998)	Actual Schedule (1998)
RFQ/RFP issued	March 25	March 25
Qualifications Statements Submitted	April 29	April 29
Pre-Qualified Proponents Notified	May 14	May 14
Technical and Cost Proposals Submitted	July 2	July 2
Best and Final Proposals Submitted	August 5	August 21
Evaluation Completed and Top-Rated Proponents Notified	August 13	August 27
City Council Consideration	August	September-October
Execution and Delivery of Operating Contract	August 27	November 10

how any work done indirectly by affiliates, parents, or subsidiaries would be brought to bear on the Atlanta contract. Teams were asked to identify each water system for which they have had complete or significant responsibility—through ownership or contract—for operations, maintenance, and management. The one specific qualification required was at least one citation for a water supply system serving at least 250,000 persons for which the proponent had complete operations, maintenance and management responsibility. Lastly, teams needed to demonstrate their capability to undertake design-build projects like those in the Atlanta RFQ/RFP by citing at least three examples of similar projects completed by any team member.

Given the magnitude of the project and the nature of the work, Atlanta required that each proponent have a minimum net worth of $100 million according to their current financial statements. If the proponent didn't meet this test, then it was required to have a guarantor of its obligations that did meet the minimum required net worth test. In addition to this minimum, teams were required to submit documentation about their financial condition, three years of financial statements, credit rating information, and information about their securities and outstanding debt. A final requirement was the submittal of a $50,000 payment, accompanying the SOQ as an evaluation fee. The purpose of this fee was to help defray the city's costs of undertaking the procurement, and secondarily to assure that only "serious" proponents submitted qualifications packages.

Evaluation of Qualifications Statements

The intent of the review process for the Qualifications Statements was to make a determination on a pass/fail basis as to the suitability of a proponent to perform the work. The general review criteria categories for the Qualifications Statements were:

1. **Corporate Profile**—Do proponent firms have the necessary expertise, resources, and experienced managers available? What kind of corporate infrastructure exists to support this project? Can the proponent firms identify ways and means by which this project fits into their overall operations?
2. **Corporate Experience**—What is the quality and quantity of the experience of the proponent team members? What is the comparability of the experience?
3. **Unique Strengths**—In what ways has the proponent distinguished itself? What specific innovations, cost savings, awards and other achievements can it claim?
4. **Financial Strength**—Does the proponent or its guarantor meet the minimum standards? What is the quality of the proponent's assets; has it ever been bankrupt; has it failed to meet any financial obligations?
5. **Contractual, Regulatory and Legal**—Does the team have the requisite participation of minority- and woman-owned firms? What is its record of environmental compliance, legal claims, criminal offenses, non-performance, etc.?

Details of all of the review criteria were included in the RFQ/RFP for the guidance of the proponents submitting Qualifications Statements. Atlanta received five Qualifications Statements. Each proponent was led by a major firm in the industry:

- Atlanta Water Alliance (Compagnie General des Eaux/Vivendi)
- Atlanta Water Corp. (SAUR)
- OMI Atlanta (OMI/Thames Water)
- United Water Services Atlanta (United Water/Suez Lyonnaise des Eaux)
- U.S. Filter Atlanta Water

Review teams were established for each of the five major criteria categories to analyze each of the submittals and to report whether any failed to meet the requirements for qualification. Each team used specific guidelines and worksheets to report on their findings, including written reports from checking any references given by proponents. All five teams were found qualified to continue in the process and submit their technical and cost proposals.

STEP TWO: TECHNICAL AND COST PROPOSALS

At the completion of the qualifications phase, all five bidders were invited to submit separate technical and cost proposals. The RFQ/RFP for these proposals provided such information as historical and current water utility operating statistics, staffing, budgetary and other related financial data, proposed capital improvement programs and various policies and procedures. In addition, the RFQ/RFP included various required levels of service and performance measures to which the winning proponent would be held during the contract term. In general, the proponents' proposals were to describe: 1) the detailed technical project approach, 2) proposed key personnel and staffing plans for the project team, 3) the proposed annual cost for the term of the contract, and 4) the mechanism for the escalation of costs over time.

A significant part of the RFP was the inclusion of an entire draft of the proposed operating agreement between the city and the successful proponent—the contractor. This document was a definitive statement of the proposed relationship between the parties over the course of the contract with specific and measurable performance standards, detailed scope of activities, mechanisms for the participation of the contractor in the CIP, incentives and penalties for performance, and all the procedural requirements for the implementation and conduct of the relationship between the city and contractor.

Access to Information

In large part, the success of such a comprehensive procurement as this one depends upon the dissemination of information about the system, and the proponents' understanding of system operations, status, conditions, plans, procedures, and so forth. The city provided several opportunities for the proponents to learn about the details of the system and to question city managers about their operations. Information was made widely available, but also carefully controlled for fairness, so that no proponent gained an advantage over others by having special access to data about the system. Contact with city employees was strictly limited, and communications with the city were channeled through a single representative. Several means of access and communication were employed over the course of the procurement.

Pre-proposal Conferences Three opportunities were provided to proponents to seek clarification of the RFQ/RFP. The first two conferences were held before the due date of the SOQ. One of them was devoted solely to explaining to prospective bidders the Equal Business Opportunity and Equal Employment Opportunity requirements of the procurement, emphasizing the importance that Atlanta places on these programs. The third conference was open only to pre-qualified proponents and was mandatory. During the course of the procurement, additional conferences were scheduled with all five proponents together to clarify any major issues that arose. Any questions that were answered in the conferences were also confirmed in writing afterward.

Facility Visits Nothing really substitutes for direct observation of the facilities and operations of the system. Before the deadline for the SOQ, an overview facility tour was arranged for all prospective bidders. Great care was taken to make sure that all participants received the same information about the facilities visited. After the proponents had been qualified, the city conducted more extensive tours of the system's facilities, including the one-call center, craft shops and executive offices. These facility tours were optional on the part of the proponents. Following the formal tours, each proponent was given 48 hours allotted in two-hour increments over the course of 11 days to schedule any follow-up visits that they desired with whatever experts they wanted to bring to any of 30 or more sites in the system. All but one proponent took advantage of the entire time allotted for these visits.

Data Room Because of the volume of information needed and available for the understanding of the operations of the system, it was impractical to distribute all of the documents to all prospective bidders. Instead, the city gathered all available relevant documents and placed them in a data room that was accessible to proponents. Items like facility drawings, plans and specifications, operating logs, compliance reports, and other detailed data were generally available in the data room during all business hours up to one week prior to the submittal date of the technical and cost proposals. Visits to the data room were carefully logged and monitored to assure fairness among the bidders. Photocopying services were available at bidder expense for any data requested; some important documents were copied beforehand and were sold at cost to any proponent desiring them.

Public Information Sessions A priority of the city was to keep the public informed about the plans to contract out the water system. Beginning at the issuance date of the RFQ/RFP, the city held over a dozen public information meetings throughout the city to explain the goals, benefits and procedures of the program to select a contractor to run the system. Held in local schools, churches and other meeting venues, the sessions were run by the chief operating officer of the city, and were attended by senior city managers and consultants connected with the procurement.

Contents of the Proposals

The contents and formats of the technical and cost proposals were carefully specified to maximize the clarity and consistency of information received and to assist the city in its review process. Each section was page limited for city convenience. For maximum flexibility, proponents were asked to provide proposals for both 10- and 20-year contracts. The outline of the technical proposal was as follows:

Cover letter—signed by an authorized representative of the proponent and stating at a minimum that the proposal represented a binding offer.
Title page—with all necessary contact information.
Table of contents
Executive summary—including a statement of the difference between the bidder's 10- and 20-year offers.
Water system technical approach—details of the specific measures to be taken by the proponent to ensure that the city's objectives would be successfully achieved and that all required services would be provided. In particular, the city wanted assurances that the bidder had plans for operating the facilities at or above the desired levels of service with significant efficiencies and cost savings for the city. The section was subdivided as follows:

- **Operations and Maintenance Plan**
- **Staffing Plan**
- **Maintenance Management**
- **Innovative Technologies**
- **Customer Information System**
- **Interaction with Regulatory Agencies, the Public and the City**

Management Team—The city was especially concerned about the quality and experience of the key managers, especially the general manager or lead professional that would be responsible for running the system.
Employee Relations and Transition Plan—A requirement of the city was that all water system employees desiring so be hired by the contractor at equivalent pay and benefits and not be laid off for a period of three years following the initiation of the contract. Proponents were required to describe their plan for accommodating this requirement.
Capital Improvements—A plan for implementing the CIP of the city as represented in the RFP.
EBO Compliance Plan—including the requirements not just for a joint venture with minority/woman owned businesses, but also a detailed plan for subcontracting services or supplies to minority/woman owned businesses.

The cost proposal was to be submitted separately from the technical proposal. In it were cost forms to be filled out detailing the cost to the city of the services to be provided, the means proposed to escalate those costs over time, and the amounts of utilities to be used by the system. The quantities of utilities would be used to determine the amount of utility costs to the city, which were to be pass-through costs because of the inability of the contractor to control utility prices as opposed to usage amounts. The cost categories used were: personnel, equipment, vehicles and related supplies, chemicals, other materials and supplies, contractual services, other, overhead and profit.

Proponents were also asked to provide their bids for the design-build of the capital projects listed in the RFP. Although not used in the cost evaluation process, these bids would establish a basis of negotiating with the winning proponent on the implementation of the projects after the contract was signed. Finally, the proponent was asked to make any comments or take any exceptions to the draft-operating contract included with the RFP. Each proponent was required to post a proposal bond of

$1 million, which would be forfeited if the proponent were chosen and didn't negotiate in good faith.

After the submittal of proposals, the city interviewed each team. Each interview lasted two hours and consisted primarily of questions posed by the city management team to the bidders. Proponents were allowed to bring a maximum of five people to the interview, including the most senior commercial representative of the team, the proposed general manager, and the key representative of its minority/woman owned partner. No general presentations, slide shows or other audio-visual materials were allowed, except to be used in answer to questions. Questions raised by either side that could not be answered during the interviews were submitted and answered in writing as a follow-up. The interviews were videotaped by the city, and in a gesture of openness, the press was invited to attend interview sessions. After the conclusion of the interviews, the city was able to begin the evaluation process.

Evaluation of Technical and Cost Proposals

Because of the close public scrutiny of the procurement, the city wanted to be certain that the evaluation process was clear, comprehensive and fair. The following principles for the evaluation system were established:

1. Ranking of proposals should be based solely on their content, with as much objectivity as possible.
2. Each aspect of the proposals should be rated as much as possible by those most able to judge the quality of the content.
3. Scoring and ranking should be fair to all proponents.
4. Scoring and ranking should be clear.
5. Rankings should accurately reflect the relative quality of the proposals.
6. No individual rater should have undue influence on the outcome.

In determining the most responsible and responsive proponents to invite for providing best and final proposals, the city used a 100 point evaluation system divided into six categories (see Table 10–5).

Within those categories, evaluation criteria were established consistent with the city's procurement code. The point system itself was a requirement of that code. Twelve teams, encompassing 46 city professionals, were used to make the evaluations. Each team was responsible for evaluating various specific aspects of the proposals, rather than entire proposals, to further decentralize the evaluation process.

TABLE 10–5. POINT EVALUATION SYSTEM

Cost-Effectiveness	30 points
Quality of Technical Approach	20 points
Quality of Management Team	20 points
Equal Business Opportunity Plans	15 points
Employee Relations and Transition Plan	10 points
Performance Capabilities	5 points
Total	100 points

Teams were selected by using the point assessment categories subdivided as necessary to reflect the weight of any particular category. For example, Quality of Technical Approach had three teams evaluating separately: 1) Operations, Management and Staffing; 2) Maintenance and Innovative Technologies; and 3) Customer Information and Relations. The other categories had two teams each, except for Performance Capabilities, which had only one. Total scores were compiled with the help of the consultant team. Scores and year-one base costs were announced publicly. Because the five teams' scores were grouped closely, all were invited for the next step of the process. Table 10–6 summarizes the results.

STEP THREE: BEST AND FINAL PROPOSALS

The final step in the procurement process was the submittal of best-and-final proposals (BAFPs) from each of the proponents. In this phase, the city refined its requirements and standardized the draft operating agreement. The goal was to make the proposals as comparable as possible so that the maximum advantages to the city could be obtained. From the technical and cost proposals, the city took any of the ideas and offers that it wanted and incorporated them into the final specification for BAFP submittal. For example, in the previous step, each bidder was invited to propose an escalation factor for increasing its fee year to year.

In the BAFP stage, the city standardized the escalation factor to be used by bidders as the Atlanta Regional Consumer Price Index (CPI). The city also added a capital project to the capital improvement program (CIP), reflecting a suggestion from one of the proponents. Each aspect of the procurement was thus refined and finalized for the proponents to use in submitting their BAFP. Proponents were asked to submit full technical and cost proposals again, reflecting any changes they wanted to make to accommodate the changes in the contract specifications. Thus, another complete set of technical and cost proposals was submitted by the bidders.

Key Contract Provisions

The final draft operating agreement that accompanied the call for BAFPs contained several key provisions. Among them were:

1. All employees were to be taken on by the contractor for a period of three years, and terminated only by attrition or for cause; employees were to get equal pay and reasonably comparable benefits.

TABLE 10–6. TECHNICAL SCORES AND FIRST-YEAR COSTS FOR PROPONENTS FOR ATLANTA LONG-TERM WATER CONTRACT

Team (Leader)	Technical Score	1st Year Service Fee Bid
Generale des Eaux	81.221	$29.8 million
OMI	85.649	$27.6 million
SAUR	66.502	$31.5 million
United Water	79.797	$27.0 million
U.S. Filter	71.446	$25.8 million

2. The contractor would get a fixed annual fee for its services; the fee would escalate each year by 90 percent of the increase in the CIP.
3. Certain utility costs, such as electricity, gas, sewer service and waste disposal, would be passed directly through to the city, subject to maximum usage limits imposed in the contract.
4. Payments to the contractor could be increased from several sources, including significant increases in water use, costs due to new capital projects or uncontrollable circumstances, savings in utility usage below targets, increases in water quality above target levels, or part of other specific cost savings achieved by the contractor.
5. The contractor was responsible for any costs above the target amounts and could be penalized for inadequate performance.
6. The contractor was responsible for the first $10,000 of the cost of any capital repair or replacement in the system; above that amount the city would pay if it chose to undertake the repair or replacement. All normal maintenance expenditures were the responsibility of the contractor.
7. The contractor would share in any operating cost savings resulting from any investments it made in the system (such as automation).
8. The city had the right to use the contractor on a design-build basis for any capital improvement implemented; alternatively, the city could bid out the project or engage another design-build contractor for capital improvements.
9. The contractor was required to install new computerized customer information and maintenance management systems.
10. The city had the right to terminate the contract at any time for its own convenience, subject to reimbursing the contractor for its investments outstanding at the time of termination.

Evaluation of BAFPs

The goal of the evaluation process for the BAFPs was to select the proposal that offered the best overall value to the city, based substantially, but not exclusively, on pricing components. It was important that the process be, and be perceived as, objective and transparent, with all proponents treated fairly. To be as objective as possible, the consultant team developed the Value Index to be used to select the top-ranked proponent. The value index is defined as follows:

$$\text{Value Index} = \text{BAFP Technical Score} / (\text{BAFP First Year Service Fee})^2$$

The Value Index is the ratio of the score received by a proponent in the evaluation of its technical BAFP proposal, divided by the first year service fee proposed by the proponent in its cost BAFP proposal. It represents a balance between quality and cost-effectiveness. The technical score in the numerator was derived by using the same process as in the previous step for the evaluation of proposals, this time using the same score categories without the 30-point Cost-Effectiveness category. Thus, a perfect score would be a 70. The cost in the denominator was the dollar amount proposed on the cost form in the BAFP cost submittal.

The technical proposals were scored by the same teams that performed the previous evaluation, in order to maintain consistency of approach. The cost proposals were collected unopened and were kept under lock and key until August 27, 1998, the day chosen for calculating the Value Index. Under the guidance of only the consultant team, the technical scores were collected, and the cost proposals were opened in secret. The consultant team calculated the Value Index score of each proposal and forwarded the results in a sealed envelope to the mayor, who opened and read the results at a press conference.

The winner of the competition was United Water Services of Atlanta, the team headed by United Water Services, a company owned by United Water Resources and Compagnie Suez Lyonnaise des Eaux. The Value Index scores are shown in Table 10–7.

On November 10, 1998, the city and United Water entered into a 20-year contract for the operation, maintenance and management of the water system. The total anticipated first-year cost of operating the system under United Water management was $29.5 million, consisting of $21.4 million in basic service fees, and $8.1 million in utility pass-through costs, indirect costs, and city oversight costs. This amounts to a savings of $20 million per year, or $400 million over the life of the contract. United took over the system as of January 1, 1999.

LESSONS LEARNED

In a speech delivered to the 12th Annual Conference of the National Council for Public Private Partnerships in Atlanta on October 15, 1998, Mayor Bill Campbell identified seven lessons that he learned from successfully completing the process of privatizing the city's water system through the 20-year contract with United Water Services. His remarks are edited and summarized below.

> This unprecedented 20-year contract is both visionary and the broadest in scope anywhere in the country. It is my belief that the two most compelling issues for every city in this country over the next 10 years will be crime—always number one—and infrastructure repair. Atlanta's infrastructure is, in terms of our water and sewer system, about 100 years old, actually in its infancy compared with some of our northeastern cities in terms of the length of time that the infrastructure has been in place and not had a major overhaul. This proposal that we have today will save us $400 million over the life of the contract and will allow us to use that $400 million to rebuild our water system.
>
> Let me tell you why we felt privatization was our best option for our water system here. No matter how much you see growth in a city, there's one dynamic: that you

TABLE 10–7. THE VALUE INDEX SCORES

Category	United Water	Other Proponents
20-Year Contract Value Index	0.130	0.111–0.094
1st Year Service Fee	$21.4 million	$22.8–25.9 million
10-Year Contract Value Index	0.110	0.094–0.087
1st Year Service Fee	$23.3 million	$24.0–28.8 million

really cannot grow without water. You can grow without land because you can simply build up rather than out. You can grow with bad zoning, no matter what it is, but water is the one indispensable factor that you must have in order to grow. That's why our water system is so vitally important. And that's where this privatization will be most helpful.

A very little noticed change in the federal tax regulations in February of 1997 allowed cities to have long-term privatization of their water and sewer systems, going from five years to 20 years. And it is that change which really allowed us to begin this process of saving $400 million. The reason that this is important is because we, in essence, get guaranteed funding of our savings, at least that's one of the options that we have, and that will be one of the options that many cities will consider. Every city is going to have to fix its water system, not just because they are old and crumbling but also because there are more stringent requirements from the EPA or from the state environmental agencies across the country or simply because we serve more customers.

So, let's look at where we are and what this public-private partnership taught us. There are seven lessons we believe we learned from this privatization, seven key lessons.

Lesson 1: Be bold, think big. You've got to be innovative and prepared to face challenges and the criticism that will inevitably come when you propose something as massive as this. While I do not believe that privatization is the answer for all of the ills of government, what I do understand is that when you have a separate enterprise fund for a water and sewer system, as most cities have, there is only one other way of getting the money to fix the system—raise the rates. And, in fact, even with this water privatization, we will still have to raise the rates. But as opposed to raising the rates 82 percent we would only have to raise the rates on a blended rate of about eight percent over the next four years. So there are massive savings involved here.

What we were also able to do here is to look at the approach and find ways of muting the criticism from those that inevitably are commentators on every aspect of human life. So the first lesson is to be bold because the little incremental changes are going to get you just as much criticism as the broad bold changes so why not go for everything and take the criticism as it comes.

Lesson 2: Make the process open. We've held 13 public hearings. We've researched, produced and made public literally thousands of pages of documents. I appointed a nine-member independent privatization task force in August of 1997. We held a privatization showcase for city employees, union members and officials. I personally went to all of the workforce that will be affected and talked to them about this privatization effort. We brought in representatives from other cities like Indianapolis, Houston, and Charlotte, and vendors to share information about what they had done—the event, by the way, was completely financed by the vendors. We also allowed, for the first time, reporters to actually sit in on the presentation process. I will tell you that that is a mistake, and to whatever degree that we assess what we did and we tell you "do not do this" as you are going forward, that is a mistake. The mistake that we made was that while we made the process as open as possible, you can never, ever satisfy the press!

For example, after we completed our process the newspapers consulted what they believed to be the preeminent expert on the privatization process, and you know what his main objection was to the way Atlanta had conducted our bid process? That we let the newspaper in on the actual evaluation panel and that we released the first ranking of scores! That was their biggest criticism of what we did. Now what does that tell you? That the newspaper, which actually threatened to sue us over this matter, had its own expert identify it as the main problem in our process. And the reason it was the main problem was because, of course, the company that was ranked No.1 in the first round then became the target of all the criticism from both the newspaper and, of course, all the other companies that did not win. Things like this simply obscured what we were seeking to do, and so I would urge you, no matter how open the process is, and it should be very open, there is nothing that you can do that will ultimately satisfy the press. So, you might as well do it the way you think its the best and just take the criticism as it comes because you cannot, *cannot* avoid the criticism no matter how much you open the process up.

Lesson 3: Make the process fair and competitive. The evaluation of the bidders has to be comprehensive and it must stand the test of scrutiny. All the proponents must be given equal access to the information; the process must be designed to minimize anyone having undue influence on the outcome. This is vitally important. Our evaluation panel had 47 people. The reason that's important is because even assuming that you could influence one or two, you can't influence 47 different people, not in Atlanta or not in most cities. Forty-seven people is really a lot of people to have as an evaluation team, but I think that at least tells you that we tried to broaden it so that no one part of it would be subject to any undo influence. Based on the city statutory evaluation criteria and the required scoring systems, individual teams were assembled to review each aspect of the proposal and the entire process was monitored independently by the consulting team; so the process was fair and competitive.

Lesson 4: Ensure absolute integrity and professionalism in the process. Our process was conducted, we believe, with the highest degree of integrity. Since the beginning we were advised by a team of respected consultants. We hired the team of Brown and Caldwell, PricewaterhouseCoopers and Harrington, George and Dunn, the engineers and accountants and consultants. They oversaw the process from beginning to end. So we knew early on what our benchmarks and our boundaries really were. In addition we hired another team to look over the first team to provide oversight and advice. Now you may ask: why so many layers of oversight? Well again, it's to respond to the criticism about the process. We hired independent experts and lawyers and consultants and then we hired consultants to look over what the consultants were doing and then we appointed the outside independent team to review what all of them were doing and even then we were criticized. But that's why you have to go back to Lesson 1, be bold, because you're not going to do this without criticism. We paid over $2 million to make certain that our citizens had the best technical, legal and financial advice available in the world and it worked.

Lesson 5: Get the best value for the city and the ratepayers. With regard to this, timing is everything. It's one of my favorite sayings, "timing is everything." We selected United Water not only on the cost considerations but a value index which looked at factors such as experience of the company, track record on customer service, the commitment and level of minority participation, which we believe in fervently in the City of Atlanta, and the company's willingness to invest in our inner city by taking advantage of our Empowerment Zone designations. It's not just that we selected a great company, although all the companies that bid were terrific companies, and we could not lose no matter who we selected.

But, there were two or three factors that really made this proposal the winning one. First and foremost was cost. The reason that we believe we got enormous savings is because Atlanta is the first one to embark upon a privatization of its water system with this magnitude and, as I indicated, every city is going to do this—so timing is everything. Whether you like privatization or hate it you cannot ignore the inevitable saving that will come as a result of the change in the Treasury regulations, especially when faced with the huge dollars that are necessary to pay for infrastructure improvements.

There also were other things that were important as well. Remember we have a two-step bidding process—we were criticized for that also. The company that ultimately won changed their minority participation to 35 percent—from 15 to 35 percent—from the first step to the second. That's important for us in Atlanta. I don't know where you live or what you feel about it or how you believe it's important, but in Atlanta we believe it's vitally important and we live by it. The company more than doubled their minority participation from the first round to the second round. That's just one benefit we got from the two-step process.

They also had a very innovative commitment on public policy. They made a commitment to put 20 percent of their employees into the Empowerment Zone, to live in the Empowerment Zone. They will have their national headquarters in the Empowerment Zone. They are giving a $1 million research grant to one of our African-American colleges, Clark Atlanta University, for the establishment of a water institute. They are providing $1,000 stipends for those who have purchased homes in the Empowerment Zone and $500 rental subsidies for those who are going into the Empowerment Zone, $100,000 a year in entrepreneurial funds for the development of small businesses in the Empowerment Zone, as well as the adoption of a school and the mentoring of kids in the Empowerment Zone. This is a pretty remarkable public policy initiative that comes about when you're privatizing your water system and getting a $400 million savings as well—so get the best value for your city and your rate payers.

I happen to believe you can combine the efficiencies and the cost savings with other good social, good public policy objectives when companies are bidding. It's harder to do if you don't having something like the Empowerment Zone, but you certainly can be very creative with the companies that are bidding and I think that's important as well.

Lesson 6: Make certain the contract protects the city's interest. Our contract provides for management that specifies clear outcomes and expectations. It provides for 24 hour, seven day a week access and oversight by the City of Atlanta. Should the contractor fail to meet the deadlines, the company, not the city, will pay the fines. The contract requires prior city approval of all improvements the company proposes and we can terminate at any time without cause on fair terms to both parties.

Lesson 7: Trust the process. If you followed Lessons 1–6 then you have a fair, open honest process—a good system is in place—but the most important lesson I can share with you is that you must trust the process and stay with it because there will be bumps and bruises along the way. Ahead of us lies the privatization of our sewer system as well. We expect that process to be a little easier only because we've done it the first time.

So, we have learned that rebuilding our city is much like two wings of a bird. It's the public and the private sector working together in harmony to lift the bird and allow it to fly. And of course, metaphorically, our symbol is the Phoenix bird. So the rebuilding of Atlanta, which started from the ashes of little more than a century ago, really has been fairly new and we believe that this new partnership with United Water and the privatization of our water system really will allow Atlanta to move progressively into the twenty-first century.

CHAPTER 11

Wastewater: A Public-Private Partnership in Oklahoma City

Kirk Humphreys

Public officials have the responsibility of monitoring the resources of the municipality and ensuring the operation, management, and delivery of a wide variety of essential services. In striving to meet budgetary demands and to operate under increased fiscal scrutiny, maintaining the highest standards of service has become more challenging. In recent years, the trend toward privatizing governmental services has eased the burden on several communities. In many cases, privatization offers an alternative means of operation without a reduction of services to the citizens.

In an effort to increase efficiency and control rising costs, Oklahoma City began investigating the option of contracting out a portion of the wastewater treatment plant operation in 1984.* At that time, the contract operations partnership approach was a relatively new concept, making its mark primarily in small communities. Since then, several other large cities have discovered that partnering with a private enterprise can be an effective means of streamlining operations and meeting the constraints of decreasing budgets.

BACKGROUND

Oklahoma City envisioned that private service providers could assist in reducing costs and meeting growing service needs without the addition of city personnel. By 1987, city personnel were operating three wastewater treatment plants with the

---

Kirk Humphreys is mayor of Oklahoma City, Oklahoma.

*See Figure A–7 in the Appendix.

assistance of four separate companies providing contracted services. Realizing that it would be easier and less expensive to oversee a single source contractor, the city began the task of reorganizing the wastewater treatment operation.

In December 1987, Oklahoma City and the Oklahoma City Water Utilities Trust entered into a five-year contract with Professional Services Group, Inc. (PSG) for Wastewater Treatment Plant Operation and Maintenance and Sludge Disposal Services. Under this contract, PSG became responsible for plant operation, maintenance, and management (OM&M) in compliance with state and federal Environmental Protection Agency (EPA) permit requirements. It also included repair and replacement of equipment, the implementation and management of an odor control plan, and an area-wide sludge management program.

This agreement required PSG to offer employment, with no loss in compensation and benefits, to all city personnel employed at these facilities. Upon approval of the contract, approximately one-half of the eligible employees chose to work for PSG.

The initial contract was so successful that the scope of services was expanded several times within the first four years. By teaming with PSG, Oklahoma City saw some very impressive results. Today, PSG and the city in partnership operate four wastewater treatment facilities with a combined capacity of 101 million gallons a day (mgd), an increase from 81-mgd in 1987. The city's contract includes the operation of a 75-mgd pump station and biosolids management of 18,000 dry tons per year.

A TWO-WAY STREET OF BENEFITS

This partnership with a private enterprise allows the city to increase efficiency and realize a substantial reduction in operating cost. The plants now operate closer to EPA permit limits due to PSG's ability to offer financial incentives to their employees. These incentives are tied to plant performance and therefore motivate employees to increase efficiency. Similar incentives are generally unavailable within municipal operation.

The automation of facilities, including remote monitoring, has increased operation efficiencies and further reduced personnel costs. Many of these system improvements were financed by PSG and, along with staff restructuring, have reduced man-hour requirements at the plants by 60 percent and by almost 100 percent at the pump station. The treatment labs at the three plants were consolidated and moved to one centralized location, improving process control and reporting.

Operations were streamlined by idling select plant basins and redirecting flows to maximize operational efficiency. Treatment operations remained within EPA permit limits and a substantial savings on electrical services was realized. Investments in technology and improvement to existing facilities continue to occur. The implementation of a Sludge Disposal Program has decreased transportation and disposal costs. As a private, for-profit organization, PSG has more flexibility in dealing with manpower issues.

PSG operates similar facilities around the country and has a national network of industry experts available for consultation. This reduces the need for the city to hire outside consultants to assist in resolving operational concerns.

Achievements include OM&M cost savings of more than $30 million and capital cost savings of more than $25 million due to PSG recommendations on the city's capital plan.

PSG has also realized numerous benefits from their association with Oklahoma City. This partnership was an important business opportunity that has become a model for successful public-private partnerships. As an established operational base, it enabled efficient expansion to other Oklahoma municipalities. It opened up the opportunity to develop new or different technologies and processes that are applied at other PSG projects and the opportunity to develop new operational and management approaches to meeting ever-increasing financial and regulatory demands under stringent cost controls. By combining resources, this partnership has allowed both parties to achieve their goals.

PARTNERSHIP HIGHLIGHTS

PSG was instrumental in recommending the expansion of existing facilities to meet an EPA and state mandate to upgrade Oklahoma City's North Canadian Wastewater Treatment Plant from 60-mgd to 80-mgd. Computerized process control data demonstrated that the plant was capable of greatly exceeding its design capacity. By upgrading the existing facility, the city saved $23 million previously estimated for a new expansion.

As a leader in biosolids dewatering technology, PSG developed a post-dewatered lime process for Oklahoma City. This process reduces polymer-conditioning aids by 50 percent and has drastically reduced hauling costs. Formerly, 420 to 450 truckloads of liquid sludge left the project each week, now that average has dropped to 90 to 100 truckloads per week. Local farmers receive the treated sludge for fertilizer, increasing their bushel-per-acre yield and helping them produce more drought-resistant crops. This land application program earned the EPA's 1991 "Environmental Excellence Award" for the beneficial use of municipal wastewater biosolids.

LESSONS LEARNED

Pursuing a partnership with a private enterprise is a decision that must be made locally on an individual basis. In addition, several tasks should be completed before entering into a public-private partnership. A comprehensive evaluation of the operation is necessary to identify the specific goals to be achieved through privatization and to target services for outsourcing. The screening and pre-qualification process, negotiation and drafting of the contract, and establishing monitoring and oversight practices are all essential components of the process.

In order to minimize risks, a screening and pre-qualification process should take place to ensure that only the most qualified private partners are considered. There are many individuals and firms interested in performing such services for a city. Some firms, although well intentioned, can cost the city more money rather than reduce costs. It is crucial that a city develop a clear definition of the services the city is already performing with city personnel and evaluate which services would be good candidates for privatization. The contract must ensure accountability with a clear assignment of responsibility and definition of consequences.

Consideration must also be given to how the remaining city personnel should be organized and how oversight of privatized services will be accomplished. Since there will still be a need for in-house administration of the wastewater treatment program, retaining a knowledgeable staff to supervise and monitor treatment operations is key to a successful partnership. Oklahoma City's agreement is reviewed annually, for changes in the scope of service, and pricing is adjusted according to the Consumer Price Index.

A private operator's recommendations must be scrutinized to ensure that the benefits to the city are worthy of consideration, or if modifications are necessary to retain the cost savings initially gained by privatization. The municipality remains ultimately responsible for permit reporting but has a lack of control in the day-to-day operation of facilities; therefore, a monitoring agreement is essential to the long-term success of the project. Monitoring ensures contract compliance and accountability.

SUMMARY

The true measure of success is often the test of time. Frequently, new associations gain notoriety from the flurry of changes and activity that accompany new ventures. In Oklahoma City, it is plain to see that the union of public and private agencies can achieve a lasting, positive result. Since 1988, the five-year contract has been renewed twice; the current contract extends through December 2002. Currently, the annual operating costs are approximately $4 million lower than they were before Oklahoma City contracted with PSG. These savings have created more funds for wastewater capital improvements and the maintenance of lower sewer charges for the city's customers.

Part III

Managed Competition

CHAPTER 12

Philadelphia's Biosolids Recycling Center: Transforming a Public Operation

Guru P. Bose, William E. Toffey, and James E. Golembeski

The Philadelphia Water Department (PWD) is a city-owned and operated utility providing water and wastewater treatment services to 2.3 million residents of Philadelphia and adjoining townships.* PWD operates the Biosolids Recycling Center (BRC), a 72-acre centralized facility for dewatering and composting of solids collected during wastewater treatment (formerly termed "sewage sludge," but now called "biosolids"). This facility produces 4,500 tons each week of a semisolid material called "cake" that is suited for recycling or further processing.

In 1993, the PWD was confronted with high costs, low productivity, and other adverse issues at the BRC. PWD undertook a study of options for the privatization or contract operation of the BRC. The study concluded that contract operations could shave between $1 million and $6 million off a $21 million baseline budget (the 1992 expenditures were $25.6 million). The contract operations study served as a challenge to municipal employees and management at BRC alike to accomplish improvements in operations and to reduce costs.

Three years later, through actions taken by municipal managers and with the cooperation and support of labor leadership, the expenditures of this unit were reduced

---

Guru P. Bose is manager of southwest operations, William E. Toffey is a biosolids water manager, and James E. Golembeski is plant manager of the Biosolids Recycling Center at the Philadelphia Water Department.

*See Figure A–8 in the Appendix.

by an amount equivalent to the high end of the contract operation study's estimate. Expenditures in fiscal year 1995 for biosolids management were $14.9 million. The staff of the Biosolids Management Unit at the BRC had been trimmed from 235 employees in 1992 to 135 employees in 1995, with employees transferred on a mostly volunteer basis elsewhere in PWD or to other city agencies.

A new management team at the BRC employed several strategies to meet the challenge posed by the contract operations study. The strategies included:

1. Institute a cost accounting system for budgeting and productivity purposes;
2. Adjust production goals and targets to achieve the cost target, including substantial reduction in compost processing;
3. Change work rules to improve worker productivity and supervisor flexibility;
4. Make reductions in staff to meet the requirements of the revised production plan;
5. Repair the relationship with labor and its leadership in line with principles of Total Quality Management;
6. Selectively use competitively priced contracted services in place of city-employee services; and
7. Procure new equipment or make modifications to improve production, particularly with centrifuges for biosolids dewatering.

The PWD did not ultimately resort to contract operation or management of the BRC. The managers of this facility did, however, employ strategies common to private business, notably those of cost accounting, work performance standards, and contracting to outside service providers. The experience of this utility demonstrates that public sector organizations can successfully employ business tools to achieve cost-efficient operations. It also shows that facilities operated by municipal employees may deliver services at costs comparable to, or lower than, services provided by private firms.

Flexibility in dealing with unforeseeable events proved to be an additional advantage to the city's operation of its biosolids facility. During the past several years, contamination of biosolids, changes to state regulation, and the increased amounts of solids were factors that would have been troublesome in a contract for private operation. Flexibility in disposal options also enabled the city to take advantage of increased competition among service companies, which in turn led to lower unit costs for handling biosolids in recycling and disposal programs.

FACILITY DESCRIPTION

The Philadelphia Water Department operates one of the largest facilities in the United States for the processing of biosolids, formerly called sewage sludge. The Biosolids Recycling Center consists of a centralized biosolids dewatering station and a biosolids composting plant that handles all of the solids generated by three water pollution control plants. These plants serve a regional watershed of 2.3 million, including the City of Philadelphia and adjoining portions of Delaware,

Montgomery, and Bucks counties. The BRC is equipped with 10 large-capacity centrifuges that handle 15 million gallons weekly of anaerobically digested liquid biosolids. The dewatering equipment yields approximately 4,500 tons weekly of a semi-solid product called cake. On an annual basis, the BRC produces 240,000 tons, or 300,000 cubic yards, of biosolids cake.

The PWD maintains a diversified program of biosolids processing and distribution. One-third of the cake is put into a composting process referred to as "aerated static pile composting," involving a four-step process of aeration, curing, drying and screening. The product of composting is an excellent soil amendment distributed to homeowners and landscapers for turf and horticultural applications. Half of the remaining portion of biosolids is recycled to farmlands in the rich agricultural district in and around Lancaster County (about 60 miles west of Philadelphia) and to mine lands in the bituminous coal district of northcentral Pennsylvania (about 250 miles west). The balance of solids is co-disposed in privately owned municipal solid waste landfills located in southeastern Pennsylvania.

BACKGROUND ON BRC CHALLENGE

In the late 1980s, the PWD's centralized dewatering and composting facility, now called the BRC, had many problems. It suffered from high costs of operation, low productivity, community distrust, extremely high overtime expenditures, labor unrest, improper equipment and, worst of all, a consent decree imposed by the Pennsylvania Department of Environmental Protection for removal of stockpiled products from unpermitted areas. The BRC was a facility "ripe" for contract operation, or "privatization," as it was called among the employees.

The BRC was the target of unfavorable union action and media attention during protracted municipal union negotiations in the summer of 1992. This occurred after management made public the "past practices" that kept productivity low. After settlement of the contract, the city administration set for itself a goal to reduce the operating costs at the BRC. Contract operation of the BRC was generally presumed to be the only realistic avenue available to the city to accomplish large cost savings.

In 1993, PWD undertook a study of options for the privatization or contract operation of the biosolids facility. While the primary focus of this study was reducing costs of operation and maintenance, the study also investigated such issues as controlling excessive inventories of biosolids products, regulatory compliance issues (including a consent decree with state regulators), high overtime expenses, unfavorable media attention, and labor-management conflict. The study concluded that between $1 million and $6 million could be shaved off a baseline budget of $21 million, depending on the scope of the private operation. The actual fiscal year 1992 expenditure was $26.2 million, excluding additional costs to dispose of stockpiled materials.

A new management team for the Biosolids Management Unit at the BRC held the belief that its operation was capable of significant change and improvement. The contract operations study had served as a challenge to municipal employees and management alike to accomplish improvements in operations and in costs. In

August 1993, at the close of the study, the city administration granted the affected employees in this unit a one-year time period within which to demonstrate substantial progress toward implementing cost reductions. The city thereby held off issuance of a Request for Qualification to initiate contract operation services.

A vigorous time of change ensued for all workers, supervisors, and managers at the BRC. The urgency of schedules to meet reduced budget targets was critical. Managers developed a long-range plan to direct processing and utilization programs, and proceeded to guide their personnel and equipment needs in accordance to these plans. The primary focus was in accepting the financial challenge and consistently meeting aggressive "expense goals." Starting from a benchmark point in December 1993, the operating budget was decreased sharply, in advance of making changes to personnel and process plans necessary to meet those targets. A comprehensive cost reporting system evolved for several purposes: identify costs by product and program, give direction on cost savings, hold supervisors responsible for specific budget tasks, and demonstrate success in cost containment to individuals outside the unit. Cooperative relationships had to be forged with other divisions within the PWD, with other departments of city government, and, most importantly, with labor's union leadership. The collaborative nature of these relationships proved essential for success in making the fundamental changes to operations that were necessary to meet the targets set by the BRC managers.

STRATEGIES FOR EFFECTIVE OPERATION

Institute Cost Accounting

Cost accounting was a key tool used by the managers, supervisors, engineers, and Total Quality teams to focus attention on where money was being spent in the process. Cost accounting also became a powerful tool in negotiating issues with the union. The BRC's engineering group and the plant's administration unit designed a cost accounting system. The system captured costs associated with all unit processes at BRC, including personnel costs, maintenance costs, and any and all parts and supplies. When combined with the utilization program costs, this information provided to managers the basis for a cost-effective allocation of biosolids quantities to different programs.

This information also helped convince others within PWD that changes to these programs were saving money. Senior managers were shown that changes weren't "smoke and mirrors." Workers at the facility were also interested in the cost information, as they had never had this data shared with them before. Because they were in danger of losing their jobs because of high cost levels, the workers were receptive to getting this information and being part of the solution.

Cost accounting is typically prepared at a macro level by accountants and administrators. At the BRC, the plant's managers, engineers, and supervisors undertook cost accounting at the micro level. This was an "eye opener" for the plant management team because previously, their focus had been on completing the job, not on how it was costing them to do it.

Cost accounting compelled the BRC managers to make the hard choice of closing out long-standing programs for recycling that had high total costs. The unit reduced composting, increased landfill disposal, and set as its mission the goal of recycling biosolids in its less expensive cake form through farm and mine reclamation programs. Also, the unit suspended a popular program in which its crews used compost to improve neighborhood playing fields. The choice of programs also was influenced by external factors, most significantly a sharp decline in the prices for landfill disposal in the Philadelphia area. The shift of programs is shown in Table 12–1, which summarizes the allocation of biosolids products from 1992 to 1997.

Once the Biosolids Management Unit seriously demonstrated its intention to select programs on the basis of cost, a strong signal was sent to the service industry. Prices offered by bidders on biosolids utilization contracts responded to the clear evidence that the unit was willing to forego programs and direct biosolids to those with lower prices. Bid prices for the programs declined on a dollar-per-ton basis, and differences in costs between programs became narrower than they previously had been.

Adjust Production Targets

A serious focus at BRC was in closely matching biosolids production plans with the needs of biosolids "customers." In its past, the BRC did only a fair job of assuring that production and disposition programs were "in sync." Adopting the Total Quality focus, the managers' new focus was "customer satisfaction." Matching production with product demand was part of this focus.

The idea that BRC had "customers" had not been an obvious one; at the outset, PWD employs contractors to transport and apply its biosolids to lands, employing "requirements contracts" that enable the city to choose how much and when a contractor receives biosolids, but pose the risk of unreliability to contractors. Managers came to recognize that competitive pricing and dependable service for biosolids product removal could be gained from contractors by delivering to them the best product on time and in planned quantities. The BRC managers also worked with financial officers to pay contractors on time. Through this kind of collaboration, BRC managers were even able to negotiate reduced prices on one of its contracts.

TABLE 12–1. BIOSOLIDS PRODUCT ALLOCATION, FISCAL YEARS 1992 TO 1997

Biosolids Product (in wet tons)	Fiscal Year 1992	Fiscal Year 1993	Fiscal Year 1994	Fiscal Year 1995	Fiscal Year 1996	Fiscal Year 1997
Biosolids Cake Produced for Direct Disposition	107,467	114,910	116,901	106,960	168,729	106,960
Unscreened Compost Produced for Recycling	35,607	57,897	3,068	0	0	0
Screened Compost Produced for Marketing and Give-Away	17,712	23,587	27,535	27,243		19,134
Mine Mix Produced for Recycling	86,070	40,625	33,805	33,030	15,949	33,030

Developing a "customer" orientation had a second advantage as well. The BRC has limited storage space, and is thereby vulnerable to costly double handling of materials. Working with the customers in a cooperative relationship gained the managers some flexibility in responding to unintentional stockpiles of biosolids.

Change Work Rules

A principal challenge in the redirection of BRC operations was for management to reassert its right to direct employee work assignments. This challenge confronted the long and tightly held "past practice" principle. A past practice is an established procedure for carrying out an assignment that could not be fairly altered without the union steward's prior consent. At the BRC, a number of serious "past practice" traditions had led to significant inefficiencies. During the course of municipal employee contract negotiations and labor-management discussions at the plant, important flexibility was gained for management. The following are examples of some of those past practices.

The Great Circle Route Composting is a labor- and equipment-intensive activity, requiring the movement of large volumes of compost. The past practice involved a step-wise movement of compost, encircling the composting facility. Management was not allowed to alter the steps or direct the route trucks were to follow due to this past practice. Site supervisors were allowed to alter the routes of truck movements within BRC only after the master contract between city management and labor had been concluded.

Loading Site Restriction Past practice restricted site supervisors from shipping materials that were not stored on specified storage areas within the BRC. This practice required equipment operators to move biosolids products from one location to a designated storage area prior to shipment. Operators were supported by union stewards in refusing to load customer trucks with products that were stored in sites other than those designated as loading areas. In reality, no valid reason existed for not shipping from any location within the facility. The right of management to direct equipment operators to load products at the most efficient location was regained by the union contract.

Shift Change Coverage A major element of BRC's work is in the dewatering of biosolids using large-capacity centrifuges. Operation of the equipment is best handled on a continuous basis, requiring machine operators to be stationed at control panels and truck operators to be in place to receive product at all times once the centrifuges are in operation. Past practice was a discontinuation of centrifuge operations at breaks and between work shifts. Compared to continuous operations, discontinuous operations produced a poor quality biosolids cake, limited the capacity of the process, and caused additional wear and tear on equipment. Union leadership had established a past practice that required all employees have simultaneous break periods and had asserted that changes to employee work hours at shift change had to be arranged by consent of the union. Under this situation, significant processing time was lost during business hours that resulted in scheduling of work for weekend overtime. Management at BRC was able to reach an agreement with union stewards for staggering of lunch and break periods to accomplish continuous operations.

Flexible Operator Assignments Employee job classifications for wastewater treatment operators are written broadly, but at the BRC union stewards had established a principle that operators could not be trained and assigned for tasks outside their principal areas of responsibility. Under this principle, dewatering, composting and screening functions would require staff levels to compensate for vacancies arising from generous vacation, holidays, and sick leave benefits offered to municipal employees. Union and management negotiated an agreement that provided supervisors the flexibility to assign treatment plant operators to any one of several locations.

Lateness Policy Enforcement Lateness policy had been poorly enforced, and progressive discipline for repeated lateness was not regularly followed. The reintroduction of a program to enforce the lateness policy had to be discussed with union leadership. Enforcing lateness rules contributed to an improved atmosphere of worker discipline at the BRC.

Reduce Staffing Levels

A principal means of reducing expenditures at BRC was cutting the number of employees at the plant. The plan for staff reduction was founded on the plan for biosolids production and utilization. As discussed above, this plan called for reduced composting and increased disposition of cake to recycling or landfill disposal. Instead of two shifts daily devoted to preparing a compost blend, only one shift was undertaken. If efficiency goals were not achieved, the possibility of eliminating compost processing entirely loomed over the workforce.

A second major component of the staff reduction effort was productivity done by a technical support group. This analysis consisted of an industrial engineering approach to evaluating the equipment, manpower, and time resources required for each step of biosolids handling within the BRC. The analysis yielded a sharply reduced staffing need in all components of processing. The results of this analysis were presented to all supervisory-level employees and to union leadership.

Combining the revised production plan with the productivity analysis findings yielded the new staffing plan. It called for an ultimate workforce for biosolids management of 135 employees, a substantial decrease from the 231 employees that comprised the staff at that time. Over the next two years, staff levels were brought down in line with the plan, in close collaboration with the PWD Human Resources Division and the city's central personnel office. Table 12–2 shows employee reductions in compost handling operations that were the result of implementing the productivity analysis.

Improve Labor–Management Relationship

The relationship between the blue-collar employees union and management was at odds in 1992. The atmosphere was extremely volatile. The attitude on both sides was more challenge than cooperation, more disruption than harmony, and more chaos than planned action. Shop stewards exerted their influence over the rank and file to shut down operations over trifling matters. Job actions were a matter of routine life. Management leadership was faced with the essential task of closing the gap between these two groups and changing the adversarial relationship to a working relationship.

Trust between labor and management had to be reestablished. The first item on the agenda was to implement trust between the business agent and the chief of the biosolids program. Then, from this trust could be established between the shop stewards and the plant manager. Finally, trust had to be established between workers and their supervisors. This was a time-consuming process, but both management and labor were able to demonstrate responsibility for keeping their respective parts of the bargain, as promises were honored, and as fair and just treatment of the employees was carried through.

A common mission was established between the business agent and the chief of the biosolids program. All workers would be treated fairly, work would be performed more efficiently, and the common goal of all employees would be to make the processing facility the best unit in the department—best in terms of work ethics, work environment, and the cost of operation.

Once a personal relationship grew between the parties, the whole picture started to change. Mutual respect for each other and trust and reliance were clearly the key fac-

TABLE 12–2. EMPLOYEE REDUCTION AS A RESULT OF PRODUCTIVITY STUDY

	Loader Operators			Truck Operators		
Processing Task	Before Study	After Study	Employee Reduction	Before Study	After Study	Employee Reduction
Dewatering	1	0	1	4	3	1
Mixing	3	2	1	4	3	1
Composting	2	1	1	0	0	0
Curing	2	1	1	4	2	2
Drying	2	1	1	4	2	2
Screening	6	4	2	3	0	3
Total	16	9	7	19	10	9
Percentage Reduction			44%			47%

tors of this relationship; these arose from the most common human elements of commitment, pride, a need for recognition, and the stability that a job provides.

As part of the team-making decisions, the union played a significant role in the reduction of employees at the BRC. The tasks undertaken by the union included: selecting the class of personnel who would be transferred, scheduling transfers of personnel, making presentations of transfer issues to the rank and file, and assisting with alleviating the tremendous anxiety and stress of the workers during this difficult period.

Use Contract Services Selectively

Outsourcing, or the selective use of contracted services, was examined for savings compared to inhouse activity. Seasonal activities and special projects are prime candidates for an economic analysis to determine whether contract services are a cost-effective option.

A fundamental step in the use of contracted services arose from issues concerning the landfill disposal contract. Prior to 1994, operations entailed city forces delivering cake for disposal to a nearby private landfill for municipal solid waste. This operation suffered from numerous problems—automatic overtime for truck drivers, difficulty in maintaining the fleet of trucks, delays from traffic, weather, and landfill problems, and others.

Fundamentally, city employees, who were paid by the hour, had no incentive to maximize the time efficiency of the work. This situation contrasts strongly with that of a private sector driver, who is compensated on the basis of tonnage of material delivered.

The city bid a contract for both transportation and disposal of biosolids cake. The price offered by the lowest bidder was less than one-half the cost of operations by city drivers. The union was hotly opposed to granting permission for the implementation of this contract—union leadership at the highest levels negotiated with city managers to prevent having this contract put into place.

During the period of negotiations, discussions occurred between the entire non-uniformed workforce in Philadelphia and city government officials. The standard that was reached was that management had the right to implement a service contract that provided services at less cost than the cost of service provided by municipal workers. Every opportunity was afforded union leadership to present the costs of municipal worker service in comparison to contractor cost.

The watershed of the transformation of the workplace at the BRC was when the union acceded to the city managers' prerogative to put the new landfill contract in place. While past experience reasonably led managers to fear acts of vandalism and work stoppage, none occurred. For a time, contractor-trucking services were performed side-by-side with city forces. Eventually, drivers were reassigned to tasks within the composting site, and solely contracted over-the-road hauling services were employed.

Improve Processing Equipment

The unit confronted serious limitations with its processing equipment. These limitations were present at the start-up of both the composting and dewatering facilities at the BRC. They exacerbated the issue of employee performance, as a commonly held suspicion among the PWD leadership was that equipment was a minor component of the overall deteriorating situation at BRC compared to poor labor discipline. In return, operators, seeing issues with equipment remaining unsolved, held senior managers responsible for poor performance.

Once the priorities for improving process equipment were sorted out, the primary focus became upgrading the capacity of the centrifuge equipment that, although newly installed, was demonstrated to be capable of being retrofitted for higher performance. The economic justification was clear for undertaking the installation of automated hydraulic backdrives in place of manual eddy current drives, a step that improved the performance of the centrifuges. The payback of the $2 million investment in upgrading essentially new machines was less than two years. The remarkable reduction in mass of biosolids resulting from improved centrifuge equipment is demonstrated in Table 12–3.

Mobile equipment also presented limitations. A majority of trucks were of a tandem axle dump type, with a limited carrying capacity. These have been largely replaced with tractor-trailers. The buckets on the front-end loaders were enlarged and their reach extended to aid in loading of customer trucks.

Working conditions were improved. As originally designed,the BRC provided inadequate space for shop facilities, lockers, stores, and offices, trailers were added, and any available space was converted to shops.

Projects under construction in 1998 include a vehicle wash facility and an improved system for distributing air under composting biosolids. Upcoming projects include replacement of centrifuges with modern, high-solid machines, as the original centrifuges began operation in 1989 and have had three cycles of refurbishment. Another project is the installation of silos for the temporary holding of cake prior to delivery to customers.

TABLE 12–3. ANNUAL PRODUCTION OF BIOSOLIDS AT THE BRC, DRY SOLIDS AND CAKE

Fiscal Year	Annual Dry Solids Production of Northeast Biosolids (in dry tons)	Average Percent Cake Solids	Total Annual Northeast Cake Production (in wet tons)	Annual Dry Solids Production of Southwest Biosolids (in dry tons)	Average Percent Cake Solids	Annual Production of Southwest Cake (in wet tons)	Total Biosolids Cake Production at BRC (in wet tons)
1992	23,009	17%	134,879	36,851	19%	196,933	331,812
1993	21,140	18%	118,887	35,994	20%	175,910	294,797
1994	19,325	24%	79,227	35,959	26%	139,942	219,169
1995	16,559	22%	75,344	36,388	24%	149,620	224,964

WINNING THE CONFIDENCE OF CITY LEADERSHIP

Competitive Contracting Committee

Philadelphia Mayor Edward Rendell established the Competitive Contracting Committee (CCC) in 1992. The committee had three purposes:

1. to identify services that, if subject to competition, could be provided at lower cost or at higher quality;
2. to ensure successful implementation of contracting-out initiatives; and
3. to ensure compliance with collective bargaining agreements.

The city's labor contract stipulated that the city perform an economic analysis showing that contracted services were more economical than services provided by municipal employees. The CCC reviewed all projects in city government involving contracting out work currently performed by municipal employees and the economic analyses supporting a decision to contract out.

The CCC developed standards for preparation of the economic analysis. For a fair apples-to-apples comparison, all costs associated with salary, benefits, supplies, equipment, overtime, leave coverage, amortized capital, and the like had to be included. The analyses were prepared by the individual departments, with advice from the city's procurement department, and reviewed by the CCC before distribution by the CCC to the union. The union had 30 days to review the analysis and to request an opportunity to meet and discuss the proposal. The CCC also reviewed progress on implementing contracting out at 90 and 180 days after initiation of the contract.

The Biosolids Management Unit submitted to the CCC an economic analysis of the proposed bid specification for the transportation and disposal of biosolids at a municipal solid waste landfill. Prior to the bid, truck drivers from the BRC were hauling the cake to the landfill in city vehicles. The proposal recommended replacing 11 existing drivers with contract trucking services. The analysis showed that the savings attributable to contracted services was $17 per ton of biosolids handled. The CCC approved the contract, and employees were reassigned within the BRC or assigned to other units.

The CCC also maintained an active interest in the overall performance of the BRC. The CCC oversaw the work by the water department's consultant on the potential use of contract operations for the entire BRC. Through the action of this committee, the BRC managers were provided the time and assistance to implement its production plans and staff reduction plans. The committee also received quarterly updates from the manager of the Biosolids Management Unit on progress toward meeting the timetable for cost reductions. It was the CCC that finally concluded that the biosolids unit had met its target and further action to seek contract operators would not be pursued.

Cost Audit

Reliable cost data in the operation of public utilities is often difficult to find. Public agencies are not accustomed to providing its middle and upper managers with the kind of expense and cost reports that address effectiveness and productivity of operations. In 1997, the first national surveys were published permitting "benchmarking" of operations and costs among utilities treating wastewater and biosolids.

In 1992, when it embarked on its efforts to reduce operational costs, the biosolids management unit had to develop its own tools for tracking expenses and devising costs for the different steps of production. Managers and supervisors assigned employees, supplies, and services to each unit process or final product. Overtime usage was monitored biweekly and compared to budgets assigned to supervisors. Expenditures under every purchase order were tracked at the facility so that quarterly estimates of expenses could be reported.

Two key products came of this cost auditing effort. First were the quarterly reports described above that reported on the progress toward meeting budget reduction goals set by the Water Commissioner and by the CCC. Second was the annual statement of unit processing costs, that is, the cost per unit of biosolids cake, for processing and disposition of biosolids. Between 1992 to 1995, these reports demonstrated the genuine and striking reductions in costs that were being achieved. The reliability of the internal cost auditing effort was verified by an independent audit of records and finances by the water department's financial office. A summary of budgeted and actual O&M costs at the BRC are presented in Table 12–4.

EXTERNAL FACTORS AFFECTING UTILITY OPERATIONS

Changes in Federal and State Regulations

Federal and state regulations governing the disposition of wastewater solids have evolved significantly over the 25-year history of the federal Clean Water Act. In 1992, federal standards governing biosolids use and disposal had not been established, and uncertainty as to the direction of regulation was a hurdle for biosolids managers. Today, the standards promulgated by the U.S. Environmental Protection Agency are serving as a basis for changes to state standards that favor biosolids recycling. The regulatory system now promises to offer a comparatively dependable and timely approval for biosolids recycling. The system that evolved helps to reduce the cost penalty that regulatory uncertainty imposes.

TABLE 12–4. SUMMARY OF BUDGETED AND ACTUAL O&M COSTS

	Fiscal Year 1993		Fiscal Year 1994		Fiscal Year 1995	
Budget Categories	Budget	Actual	Budget	Actual	Budget	Actual
Personnel Services	$7,368,517	$7,909,975	$7,648,000	$5,929,115	$5,000,000	$4,967,081
Purchase of Services	$17,068,000	$14,564,581	$10,396,000	$8,170,801	$6,950,000	$6,560,069
Materials and Supplies	$6,025,000	$4,279,906	$5,693,000	$4,171,463	$3,500,000	$3,320,188
Equipment	$100,000	$123,486	$190,000	$141,535	$230,000	$38,690
TOTAL	$30,561,517	$26,877,948	$23,927,000	$18,412,914	$15,680,000	$14,886,028

Changes in Marketplace for Biosolids Disposal Services

Over the past two decades, the biosolids service industry has both matured in its professionalism and increased in competitiveness. Both of these factors, together with a more flexible biosolids processing facility, contributed to the ability of the city to obtain better biosolids handling services at a lower unit cost than in the past.

Changes in Biosolids Quality

No aspect of biosolids programs poses more difficulty for public acceptance than the perception that biosolids are contaminated. For that reason, the quality of PWD biosolids has been a focus of managers and regulators. The very earliest component of the management program, dating to the early 1970s, was implementation of an effective industrial pretreatment program, through the PWD Industrial Waste Unit. Concentrations of pollutant metals in PWD biosolids are now typically better than one-quarter of what the concentrations were in the early 1980s, and in some cases much lower. The biosolids produced by PWD are for the most part indistinguishable from residential wastewater solids from nonindustrial communities. Yet, PWD must continually watch for unusual pollutants that might interfere with the suitability of biosolids for recycling. The BRC must always remain responsive to changing conditions if it wishes to remain economical.

CONCLUSIONS

Local government agencies do not typically employ business tools in their management and operation. Part of the reason for this is that they need not withstand the scrutiny of banks, tax consultants, and stockholders to their financial recordkeeping. In part, their "customers" are not entitled to exercise a purchase decision, or a "vote," on the quality and price of municipal service, except indirectly through public elections. But the distinction between public and private services is lessening as public managers are challenged to meet the price of private sector competition. Ed Rendell, the mayor of Philadelphia, stated on numerous occasions that it was not the idea of privatization that he liked, but the idea that city employees need to believe they have competitors. To meet this challenge, public managers need to use the kinds of tools described below.

Develop a Strategic Plan

The BRC based its production plan on a four-year strategic plan. This plan was developed in a participatory process involving supervisors and managers representing all of the different work areas within the BRC. Specific staff assignments for the first and second year of this project were included in the plan. Plant performance against the strategic plan was to be evaluated on a six-month basis.

Prepare Budgets Based on Real Expenditures

A common approach among operating units in a public agency is to prepare annual operating budgets in a manner that allows for substantial "contingencies." There are many reasons for this: the long lag between budget proposal and budget implementation; the cumbersome procedures for subsequent budget amendments; past histories of unforeseen and costly changes in operational needs; the slow and unpredictable issuance of purchase orders against budgeted funds. The managers of the biosolids unit operated against the grain of this tradition by budgeting close to anticipated actual expenditures.

The Biosolids Management Unit used the budget as a tool for addressing issues with contracting and purchasing that had interfered with plant performance. Management developed an operating budget on a "zero base," meaning that every line was justified and compared closely to expenditures in prior years. Over subsequent years, the unit was able to bring actual expenditures to nearly 95 percent of the budget.

Develop Good Vendor Relationships

Public agencies frequently encounter prices for services far higher than those procured outside of government. While this, in part, is due to the higher level of guarantee and insurance stipulated in public contracts, several aspects of the relationship between vendor and agency can be improved to reduce prices for services. One approach is to ensure quick payment to vendors for services through prompt processing of invoices and vouchers for payment. A second approach used for biosolids utilization services was contract language that provided flexibility to the contractor, imposed no unreasonable performance targets, and shared risks arising from unforeseen circumstances. A reputation for business-like relationships, free of the suspicion of graft, helps encourage fair participation in bids for contracts with the city. The unit also developed a collaborative relationship with Philadelphia's Procurement Department to help expand the list of potential vendors qualified to do the work.

Make Plans for Capital Asset Improvements

Public agencies have the same obligation as private-sector managers to protect their capital. Few agencies create the necessary inventories of capital assets, complete with information on age, condition, and expected life. Such an inventory can be used to develop plans for capital equipment replacement. For facilities such as the BRC, key pieces of equipment require special order and custom equipment, such as the centrifuges, large pumps, and screens for separating woodchips from compost. Leadtime may be two years between order and delivery. If an order is submitted after equipment failure, the capacity of the entire facility may be hampered and inefficiencies will arise.

Organize for Preventive and Predictive Maintenance

Computerized maintenance management systems (CMMS) are being installed in Philadelphia Water Department's wastewater facilities and at the BRC. The focus of this program, long established in private industry, is on increasing resources for preventive maintenance and decreasing the need to respond to emergency maintenance. When completely implemented, CMMS will also provide management reports useful for evaluating asset costs, for predicting maintenance needs, for determining the adequacy of maintenance staffing, and for gauging the productivity of maintenance programs.

Significant Cost Savings Are Possible in Municipal Operation

Science and technology change rapidly. In the biosolids recycling area, change is in the direction of improved machine performance and drier biosolids cake. Compared to private operations under a service contract, the public operator is now better able to capture for the benefit of the public the cost savings arising from new science and technology compared to private operations under a service contract have accrued to the PWD as a consequence of public operations using improved technology.

Centrifuge technology continues to improve. Even without replacement of original centrifuges, the centrifuges have been retrofitted with improved backdrives, automatic controllers, and other devices to increase the solids content of the biosolids cake. Also, the BRC has been on the forefront of experimentation with a new generation of polymers designed to improve dewatering. As the systems for the mixing and delivery of the polymer to the centrifuges are developed, the BRC personnel expect as much as 10 percent unit increases in solids content. This may result in millions of dollars in savings. Because of the eagerness of vendors to develop new polymers and pumping systems, the trials of new chemicals have been largely performed at no cost to the city.

In part, due to the momentum created by the cost savings at the BRC, the wastewater plants have been working to improve their side of the operations to help reduce biosolids management costs. Increased destruction of volatile solids in anaerobic digesters can reduce the total mass of solids delivered to the BRC. The wastewater plants are working on ways to increase the percent-solids content of the primary and secondary sludge streams within the plant. Thicker sludges result in a thicker feed to the centrifuges atthe BRC, yielding savings in workforce, equipment, and chemicals. In a contract operation, the incentive to optimize sludge handling within the wastewater plants may be lost, and the potential for savings may not be realized.

Flexibility Can Be Maintained to Respond to External Factors

Quantities and qualities of biosolids vary unpredictably, making the flexibility of biosolids operations a key attribute of the service required by the plants. Performance of biosolids operations by public personnel enabled PWD to take advantage of opportunities to reduce or avoid costs.

Several instances serve as examples. The quantity of solids delivered by the wastewater plants has decreased since 1992 (refer to Table 12–1), presumably as a result of reduced influent solids from industrial sources. This amounted to a 10 percent reduction in historical biosolids production, in addition to the reduction in biosolids mass arising from improved dewatering. This reduction was realized primarily as savings in costs for biosolids product disposition. This savings might not have been realized in a contract for private operation of BRC.

Following the reduction in solids production was a change in operation of water filtration plants. The largest of these plants began the discharge of flocculent sediment to the sewer, which increased the solids through the system. The total biosolids production swung sharply upward, to an end point three percent higher than the historical basis. Whereas this inert material was high in ferric compounds, it exerted a positive effect on dewaterability, which enabled the PWD to produce a higher solids cake. This phenomenon largely offset the cost penalty of increased total solids. Had the change in filtration operations occurred in the case of a privately operated contract, the vendor would have been presented with an opportunity to negotiate increased costs exceeding the true costs.

Degradation of biosolids quality is another example of an external factor affecting operations. A spill of a contaminant material to the sewer has barred the solids from use in composting. Because BRC employed flexibility in its operations, the use of those solids was shifted in a way that caused no additional direct cost to be borne by the unit. In the situation of a private operation, interference in biosolids use arising from deteriorated quality might have well yielded additional charges to the public.

These examples indicate that an important benefit of municipal operation of the BRC has been the ability of public employees to respond to external factors in an environmentally and fiscally responsible manner.

Public Opinion of Government Can Be Improved

Philadelphia's biosolids operation presents an opportunity to provide a "good news" story about government. A diverse and effective program for the recycling of disposing of these solids has been put in place at the BRC. Management and labor at the BRC remain flexible to changes in the biosolids marketplace so that they can be fiscally responsible.

The compost produced at the BRC has improved in quality, and it is produced in quantities well matched to the capacities of the outlets or markets. Through the collaborative efforts of BRC and union leaderships, BRC is now capable of producing in a single shift of operation as much of its saleable product, EarthMate Compost, than it produced before on two shifts with greater labor and equipment inputs. Further, EarthMate is a product with an excellent reputation among gardeners and landscapers in the Philadelphia region.

Public utilities can be run more like a private business. They need to have the genuine challenge of competition, their leaders need vision and courage, and they need to be flexible in the face of changing technology and markets. In so doing, they serve the public's best interests.

CHAPTER 13

Managed Competition in Charlotte-Mecklenburg Utilities

Kimberly S. Eagle

Managed competition has become an important device for municipalities to use in achieving the performance levels of private sector enterprises. Managed competition for the Vest Water Treatment Plant of the Charlotte-Mecklenburg (North Carolina) Utility system will be discussed at length as a practical example of a managed competition process. This chapter will address the steps in the process, from inviting proposals to contract monitoring and employee incentive programs. Allowing a municipal unit to compete against private sector vendors has become the primary vehicle for ensuring provision of the best service at the lowest cost.

A municipality may decide to privatize a service or enter into managed competition with the private sector for varying reasons. Financial pressures, budget constraints, operational problems, labor issues, political interest, private sector interest in the purchase of system components, and the desire to "get out of the business" are common reasons. Under the philosophy adopted by Charlotte, if a service can be purchased from the private sector, it is a candidate for competition. This offers the best chance that the most effective service at the lowest cost can be obtained.

COMPETITION IN THE CITY OF CHARLOTTE

Charlotte focuses on providing the most cost-effective services possible to its customers from its 13 business units. In May 1992, a Mayor's Privatization Taskforce was formed to study the issue of privatization and the potential functions within the city that should be considered for full or partial privatization. Seventeen activities ranging from wastewater treatment to solid waste collection were identified (Sizer,

---

Kimberly S. Eagle is the assistant business manager for Charlotte-Mecklenburg Utilities.

1998). Thirteen key business units were formed in 1993 to replace the 26 existing city departments. An emphasis on operating closer to the standards of private businesses was declared.

In late 1993, a privatization/competition policy was established for services contracting and asset management. The Privatization/Competition Advisory Committee, a nine-member citizen committee appointed by the mayor and city council, was established in November 1993 to provide advisory assistance to the city with regard to competition and privatization matters.

Guidelines for services contracting and asset management were developed in July 1994. This *Competition Process Handbook* functions as a generic user's guide for key businesses in completing a competition process. The guidelines, designed to ensure a level playing field, provide direction for contracting procedures, including bid development methodology, and employee impact statement and cost elimination plan development (City of Charlotte, 1997a). The goals of the competition effort in Charlotte place an emphasis on making the city open to competition, establishing a focus on priorities and resources by outsourcing noncore functions. In December 1994, with the implementation of a five-year competition plan, all 13 key businesses identified services for competition or potential outsourcing. Each key business unit maintains an individual plan, which is updated annually.

Charlotte-Mecklenburg Utilities

Charlotte-Mecklenburg Utilities (CMU) is the largest public utility providing water and sewer services in North Carolina.* Providing service to a population of more than 555,000 in the City of Charlotte and in Mecklenburg County, CMU maintains more than 165,500 service connections and 5,500 miles of pipeline with 656 employees and an annual operating budget of $93.5 million. Formed in 1972 as a consolidated business unit of the City of Charlotte and Mecklenburg County, CMU is committed to providing safe, sufficient drinking water supply while conserving natural resources and protecting the environment.

Managed competition has become a commonplace phrase in the water utility industry over the past several years. CMU's competition philosophy is centered around several fundamental concepts. CMU competes or contracts significant activities in all parts of the organization as described in the competition plan. The determination of noncore functions is made and these functions are outsourced to the private sector. Ongoing assessments of competition savings are conducted and revisions to the plan are made in order to place concentration on areas where maximum savings can be realized. Multiple approaches, including managed competition, outsourcing, and benchmarking are utilized.

CMU was not spurred on toward competition because of high rates, capital needs, or lack of financing. Instead, ensuring the best service at the lowest cost is the driver behind the use of managed competition. Focus has been directed toward making the

---

*See Figure A–9 in the Appendix.

city and the region competitive by using competition to drive costs down and using outsourcing to focus priorities and resources. The comprehensive five-year plan directs CMU through the competition process. Competition has produced cultural and organizational change as well as a breakdown of some internal bureaucracies.

CMU's FY 1996 competition plan included managed competition for the operation and maintenance contract of the Vest Water Treatment Plant. CMU successfully competed with national and international firms for the operation and maintenance of the Vest Plant. Proposals were received in April 1996, and contract operations by CMCon-Op (the internal bidding team of the city—Charlotte-Mecklenburg contract operations) began on July 1, 1996.

The Vest Water Treatment Plant: A Case Study in Managed Competition

The Vest Water Treatment Plant was selected for managed competition for two reasons. First, the scope of work for the operation and maintenance of this facility could be readily identified. Second, no major construction was planned for the facility within the contract period.

Built in 1924, the Vest plant has a water treatment capacity of 24.8 million gallons per day (mgd). Vest has undergone several reconstruction and renovation projects, including major expansions in 1937 and 1947, and further modifications in 1981 and 1996. Water treated at Vest is run through a series of five treatment processes, including coagulation, flocculation, sedimentation, filtration, and disinfection, to ensure the highest standards of water treatment. In 1990, the Vest plant received the first American Water Works Association (AWWA) Historic Landmark designation in the state of North Carolina. In addition to the Vest Plant, CMU operates two other water treatment facilities in their system.

COMPETITION PROCESS

The primary goals of the Vest competition included the determination of the most cost-effective service provider, promotion of competition from private operators, development and utilization of objective evaluation criteria, and establishment of a level playing field. The desire for all competition to take place on a "level playing field" is strong in Charlotte. An internal wall was established within CMU to separate those assigned to work with the procurement documents and proposal evaluation (procurement team) from those preparing the staff proposal. Separate consultants were assigned to each group to provide assistance throughout the process.

In most managed utilities, either a one-step or a two-step procurement process is generally used. CMU elected to implement a two-step process for the operation and maintenance competition for the Vest plant. A formal Request for Qualifications (RFQ) was issued in August 1995. Ten firms responded to the RFQ.

A draft Request for Proposals (RFP) was issued to give the proposers an opportunity to provide comment concerning content and format. A formal RFP was issued to the nine qualified firms and, following the issuance of two addenda, proposals were received and evaluated. Every company that qualified in the RFQ stage, including CMCon-Op, submitted proposals.

The Statement of Qualifications (SOQ) and the proposals submitted were reviewed and evaluated by a six-person evaluation team consisting of two citizen members of city Advisory committees (The Privatization/Competition Advisory Committee; and the Utilities Advisory Committee, a citizen committee for CMU activities), two non-CMU staff members of Charlotte, and two members of CMU's management staff. (See Figure 13–1 and 13–2.)

As noted, the CMU bid team, CMCon-Op, was established to prepare the in-house bid. The bid team had no part in the development of the specifications or in the evaluation of the proposals. Again, internal walls were established to eliminate conflicts of interest. Although some of the largest national and international firms in the world bid on the Vest plant operations contract CMCon-Op submitted the lowest bid. CMCon-Op reduced personnel, chemical, and utility costs. Increased training, provision of employee gainsharing incentives, and increased automation also resulted in significant cost savings. A Memorandum of Understanding (MOU) formalized the operational agreement between CMCon-Op and the city.

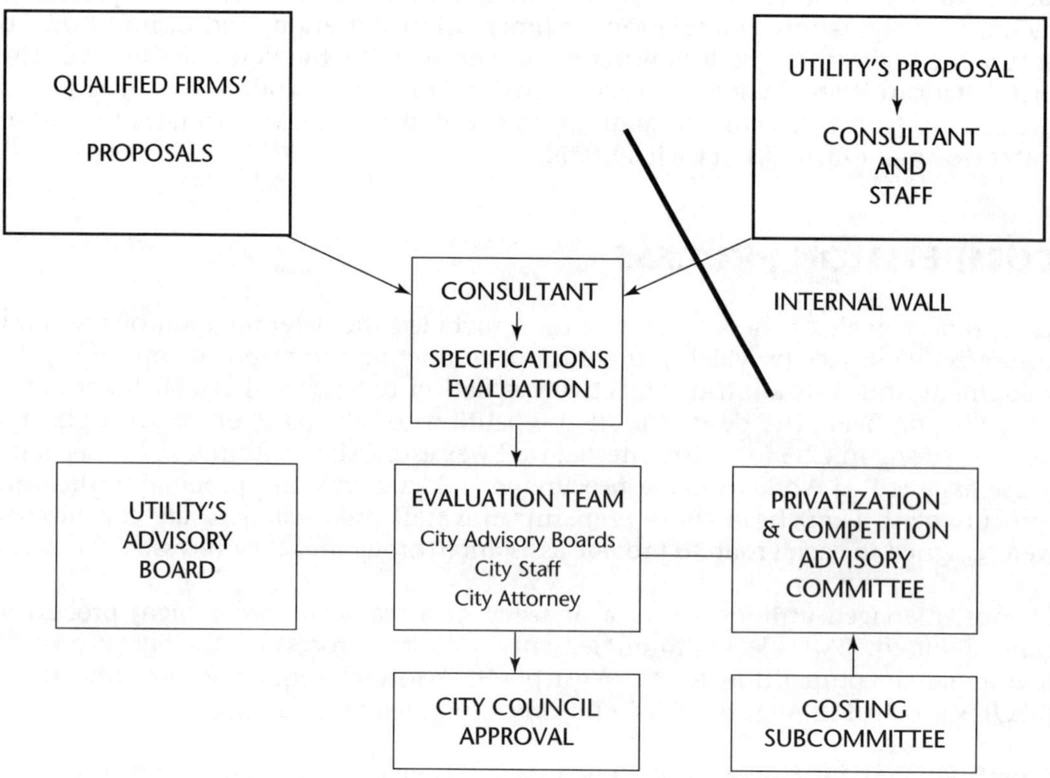

Figure 13–1. Competition Process Overview

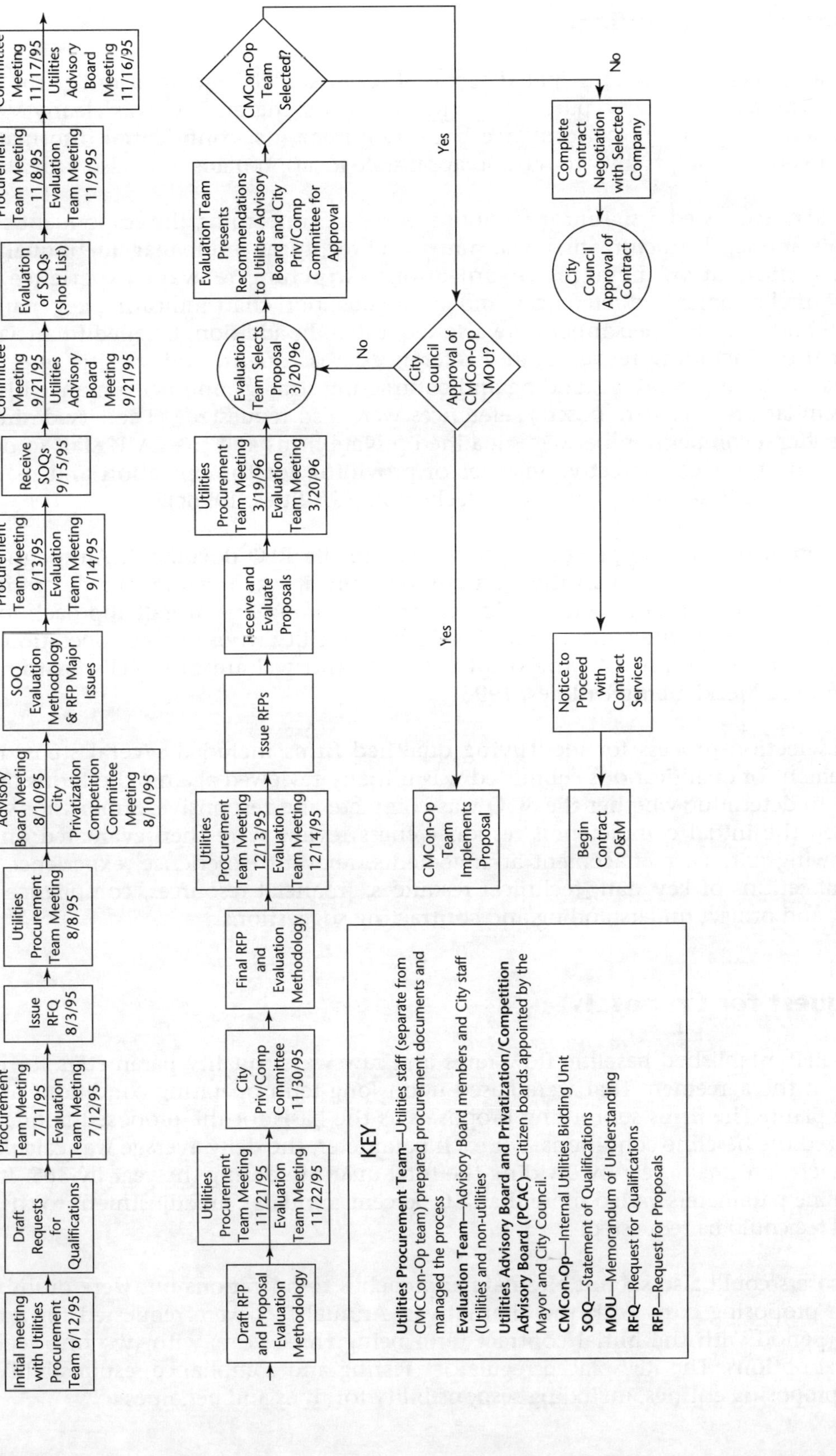

Figure 13–2. Charlotte-Mecklenburg Utilities Vest Operation and Maintenance Services Competition Process Detail Flow Diagram

Request for Qualifications

The RFQ included the anticipated scope of services to be provided under the contract. The basic objective pursued by the city in issuing an RFQ was clearly stated. Instructions regarding performance bond requirements, confidential information, submission format, facility tours, and applicable location maps were also provided.

The RFQ requested a significant amount of detail concerning the respondent's corporate and legal structure, financial status, and technical and management qualifications. Information including organizational structure, relevant experience, key personnel resumes and references, and an organization chart indicating key management and technical personnel were also requested. In addition, detailed financial information including resources and references was sought, and specific questions concerning project understanding and contracting strategy and performance history on similar projects were posed. References were also requested. "The city desires to engender a competition between qualified private firms and the CMU staff to determine the most cost-effective manner of providing for the operation and maintenance of the Vest Plant" (Charlotte-Mecklenburg Utilities, 1996c).

The intent of the RFQ process was discussed in the RFQ document. Although cost was an important factor in the Vest competition, it was not the sole determining factor. "The city is looking to obtain the most advantageous overall approach to the operation and maintenance of the Vest Plant, a facility whose proper operation and maintenance is critical to the health, safety, and welfare of CMU's customers" (Charlotte-Mecklenburg Utilities, 1996c).

The selection process for identifying qualified firms included several steps. Each statement of qualifications submitted was initially reviewed against the terms of the RFQ to determine whether the SOQ was complete and responsive. Submissions that passed the initial completeness/responsiveness review were then evaluated on the following criteria: management arrangements, relevant experience, experience and qualifications of key staff, technical resources, financial resources, compliance history, and project understanding and contracting suggestions.

Request for Proposals

The RFP established baseline flow rates and raw water quality parameters for each year of the agreement that were based upon long-term operating conditions at the Vest plant. The firms submitting proposals as the basis for the proposed annual fee utilized the baseline conditions. For each parameter, the daily average was calculated for each agreement year by dividing the total quantity during the year by 365. If the baseline parameters varied more than 10 percent annually, an adjustment to the annual fee could be requested.

Proposers could also submit alternate approaches to operations but were prohibited from proposing costs with the alternatives. Annual fees were requested for a five-year period with the initial contract term being three years with two one-year renewal options. The RFP placed regulatory testing and compliance responsibility on the proposing entities, including responsibility for fines and penalties.

Multiple levels of maintenance were developed for purposes of the competition. Routine, preventive, and predictive maintenance were included in the annual fee and were to be monitored as line-item expenditures on a monthly basis. The annual fee included costs for labor, services, and materials, including lubricants, filters, belts, and all other consumable materials. All minor corrective maintenance and its costs were included in the annual fee. Costs for equipment, materials, and specialty contract services associated with major corrective maintenance, where costs for single items exceed $2,500 but are less than $25,000, were to be paid from a separate maintenance allowance and included in the annual fee. The maintenance allowance was set at $50,000 annually. Costs for major corrective maintenance for a single item of $25,000 and above, unless caused by the successful proposer's failure to perform any of maintenance or other obligations under the contract or by the successful proposer's negligence, are the city's responsibility.

Performance specifications for operations and maintenance were included in the RFP with significant detail to ensure adequate and appropriate information for proposal preparation. The draft operations and maintenance service agreement was also included in the RFP in order to provide an opportunity for review.

Proposal Evaluation

Technical support for proposal evaluation was provided by a consultant. Each proposal was initially reviewed to determine whether the proposal was responsive to the requirements of the RFP. Proposals determined to be responsive were evaluated and ranked in order of lowest price. Technical approaches of only the proposals falling within a reasonable range of the lowest price were evaluated and ranked on the basis of nonprice considerations.

1. **Price Evaluation.** The evaluation team evaluated the proposed pricing and technical approach components of each proposal separately. The price proposals were evaluated by calculating the net present value of the proposed annual fee for each of the five contract years using a discount rate of 5.75 percent per annum. The resulting five-year, lump-sum present value was then used to compare the proposed pricing among the private firm proposals. Except for specific provisions for adjustments to the unit costs of chemicals, there were no provisions for adjusting the annual fee each year for inflation.

   The CMCon-Op proposal assumed certain levels of inflation, except for chemicals, that were incorporated into the proposed fee for each year. Potential annual fee adjustments included adjustments for changes beyond the baseline operating conditions as described earlier and adjustments for changes in the unit costs of chemicals. Within 60 days of the end of each agreement year, CMU and CMCon-Op will review the flow and parameters measured at the plant for the year, and if the flows and/or loadings differ (+/–) from baseline conditions by greater than 10 percent, the annual fee will be adjusted to reflect the additional direct costs, plus gross margin, for chemicals demonstrated by CMCon-Op to be directly incurred or saved by such variances to the extent it exceeds 10 percent.

Electrical costs were not part of the contract for the Vest plant. The cost of electricity for the plant is greatly dependent on the time of day used, and since pumping operations for the Vest plant are directed by another water plant in the Charlotte-Mecklenburg system, it would have been extremely difficult to quantify potential electrical power costs in a manner fair to all proposers.

In comparing the proposed pricing of the CMCon-Op proposal to proposals submitted by private firms, the relative impacts on various indirect costs of CMU were analyzed and the additional costs to CMU for administering of a private operations and maintenance agreement were estimated. It is important to note that the city has a policy indicating how indirect costs are to be included in the proposed pricing of all managed competition bids. The CMCon-Op proposed pricing was required to comply with the requirements of the city's "Guidelines for Services Contracting and Asset Management," as indicated earlier in this chapter. Specifically, Guideline No. Four: Costing Methodology provides a description of the categories of cost and supporting detail that were required for inclusion in the city proposal. Again, the purpose of these city guidelines was and is to provide a "level playing field" for comparing internal bids and private bids.

The CMCon-Op bid included all direct and variable costs associated with the performance of the contract, an allocation for capital costs (including depreciation and financing costs) for capital equipment and assets, and an allocation for semivariable and fixed costs that could be affected by the level of services that the city provides internally (Charlotte-Mecklenburg Utilities, 1996b).

The costing subcommittee of the Privatization/Competition Advisory Committee developed a recommendation for overhead adjustment to be applied to the CMCon-Op bid. These overhead allocations are tracked on a monthly basis as a part of the financial monitoring plan.

2. **Technical Approach Evaluation.** The proposed technical approach was evaluated by examining the quality and reliability of various aspects of the required operation and maintenance services as demonstrated in the proposal, and by conducting an assessment of specific areas of risk associated with each proposal. The evaluation team developed very specific criteria for the review of the technical approach. The level and skill mix of operations staffing, the level and skill mix of maintenance staffing, the experience of management staff, the transition plan, the operations plan, the maintenance management program, and the training program were all considered.

Memorandum of Understanding (MOU)

After careful consideration, as outlined above, CMCon-Op's proposal was selected. The Vest MOU between the city and CMCon-Op formalized the RFP and proposal terms. The MOU established the employee gainsharing program and requires that the plant be operated and maintained according to the provisions set forth in the RFP and the CMCon-Op proposal. The MOU also stated that the city accepts the program and policy changes contained in the CMCon-Op proposal (Charlotte-Mecklenburg Utilities, 1996a).

Monitoring

During the first 18 months of the contract, monitoring procedures focused on two areas: quality assurance for operational and financial performance. Contract compliance officers were established within the CMU's administration division to monitor the financial and operational performance of the Vest plant. A detailed monitoring plan was developed, which includes regular meetings, scheduled and unscheduled site visits, comprehensive review of all costs for assurance of correct invoicing and cost center coding, and complete internal records verification. Although the existing procedures seem to be effective in evaluating the status of the contract, an annual review of the monitoring system is undertaken.

On a monthly basis, comparisons are made of report totals and individual line items to cost center information in the city's general ledger. Discrepancies are investigated and resolved. A comparison of report totals and line items to the bid also is made to identify possible developing trends. Compliance with all federal, state and local laws, rules, and regulations as determined in the RFP is reviewed and the internal audit section also monitors the CMCon-Op Proposal and the MOU. The Privatization Advisory Committee and the Utilities Advisory Committee serve important roles in the monitoring process. Financial and performance monitoring is completed in conjunction with CMU contract compliance officers and is reported to the Utilities Advisory Committee and the Privatization/Competition Advisory Committee on a quarterly basis (see Table 13–1).

CMCon-Op provides comprehensive monthly, quarterly, and year-to-date reports in a format developed by the financial contract compliance officer. The reporting format closely follows the line-item detail of the city's operating budget format. The reporting is completed in detailed financial spreadsheets with accompanying narrative and a full accounting of all expenditures, overhead costs, and operations and maintenance activities. Operational problems or situations are fully documented and a detailed forecast of work for the upcoming month is included. All deliverables, including specialized reports, inventories, and contractual services information as required by the RFP, are part of the written report.

TABLE 13–1. MANAGED COMPETITION MONITORING PLAN IN CHARLOTTE-MECKLENBURG UTILITIES

Performance Criteria
- Request for Proposals
- Memorandum of Understanding

Cost Evaluation
- Internal — Financial Compliance Officer (Administrative Staff)
- External — Internal Audit
- Quarterly Reporting
- Utilities Advisory Committee
- Privatization Competition Advisory Committee

Operations and Maintenance Evaluation
- Internal — Financial Compliance Officer (Administrative Staff)
- External — Regulatory Agencies
  - Internal Audit

1. **Financial Monitoring.** The MOU states that CMCon-Op will maintain such financial records as are reasonably necessary to determine that CM-ConOp is operating and maintaining the plant in accordance with the provisions of the RFP, the MOU, and the proposal (Charlotte-Mecklenburg Utilities, 1996b). CMCon-Op was established as a separate accounting entity with new and separate cost center identification numbers. Inventories of chemicals, general supplies, spare parts, and fuel were taken on June 30,1996, the day prior to the beginning of the contract on July 1, 1996. This was performed in order to maintain the clear separation of CM-ConOp and other city cost centers from the onset of the contract. It is of importance to treat all aspects of CM-ConOp like a private operator in order for complete performance and costs to be monitored. At the conclusion of the contract, CM-ConOp will replace supplies to the level inventoried at the start of the contract or reimburse the city for the inventoried amount.

   Targeted versus actual expenditures are tracked using a spreadsheet for all line items in the contract. Personnel services, operating expenses, and other costs such as workman's compensation, division overhead, department overhead, city overhead, depreciation, finance cost, vehicle liability, and capital improvements are monitored. Performance criteria, including average mgd and permit compliance, are also tracked in this format. Primary areas of focus are use of overtime, use of contingency funds, and interdepartmental contract services. Laboratory services, maintenance and instrumentation, and electrical services are contracted with internal service providers within CMU at an actual hourly rate charged to CMCon-Op. Services are charged through a work order system and are monitored by the contract compliance officers on a monthly basis. Invoices due for the services rendered are made from the contract cost center to the cost center of the internal service provider on a quarterly basis. The hourly rate for services is a complete cost with appropriate overhead included to maintain an appropriate cost allocation structure within the key business unit.

2. **Performance Monitoring.** Performance standards for operation and maintenance of the Vest plant include the following:

Staffing	Sludge Dewatering
Finished Water Quality	Routine Maintenance
Water Delivered	Predictive/Preventive Maintenance
Fluoride Residual	Minor/Major Corrective Maintenance

   Various types of specialized required reporting are used to include predictive maintenance surveys, infrared surveys, detailed scheduling of maintenance activities, and annual cost development for anticipated capital expenditures for major corrective maintenance. Capital improvements by the city are not included in the CMCon-Op annual fee, however, the CMCon-Op bid included capital improvements to improve operation and reduce costs (amortized over the five-year period of the contract for competitive payback).

3. **Auditing.** The city's internal audit section is charged with auditing the CMCon-Op operation of the Vest plant. The internal audit process verifies the completeness and accuracy of reported costs.

   A quarterly review is performed by internal audit to confirm the data in the monthly report and to verify that all key objectives and parameters are met and documented. Verification is made on payroll records for the bid through the comparison of the city payroll register to a monthly monitoring report. Accurate monitoring confirms that only CMCon-Op employees are completing contract activities.

   All utility-controlled cost centers are reviewed to verify that Vest cost center items that should be paid under the contract are not miscoded to a cost center outside of the MOU. The number of vehicles onsite, rental equipment, contractors, and employees present are reviewed by the internal auditor to ensure compliance with the MOU.

   Compliance with all federal, state, and local laws, rules and regulations outlined in the RFP, the CM-ConOp Proposal, and the MOU is monitored by internal audit.

Gainsharing/Employee Incentive Program

A gainsharing program has been implemented to provide an employee incentive for exceeding performance and budget targets. All performance goals for permit compliance, budget expenditures, and safety (no lost time accidents) must be achieved in order for employees to participate in the program. Expenditures are evaluated monthly to determine operational savings. Savings are the bid amount minus actual costs. Any operational savings below the bid amount at the end of the quarter is shared equally with employees. One-half of the savings remains with the plant in a reserve account available for use in operations as needed or for minor repairs or improvements. Distributions of incentives to employees are made quarterly with one-half of the total amount saved being reserved for year-end distribution (see Table 13–2).

RESULTS

The first quarterly report for CMCon-Op operation of the Vest plant presented savings. Expenditures were under established targets, and the savings by CMCon-Op operations totaled approximately $72,000 for the first year of operation. (See Figure 13–3.)

TABLE 13–2. VEST WATER TREATMENT PLANT GAINSHARING RESULTS, YEAR 3

July 1, 1998–June 30, 1999

Expenditure Target		$497,543
Actual Expenditures		$435,877
Year 1 Savings		$61,666
Employee Gainsharing		
50% to the Plant	=	$30,833
50% to Employees	=	$30,833
Amount per Employee	=	$4,405
(seven employees in the Vest unit)		

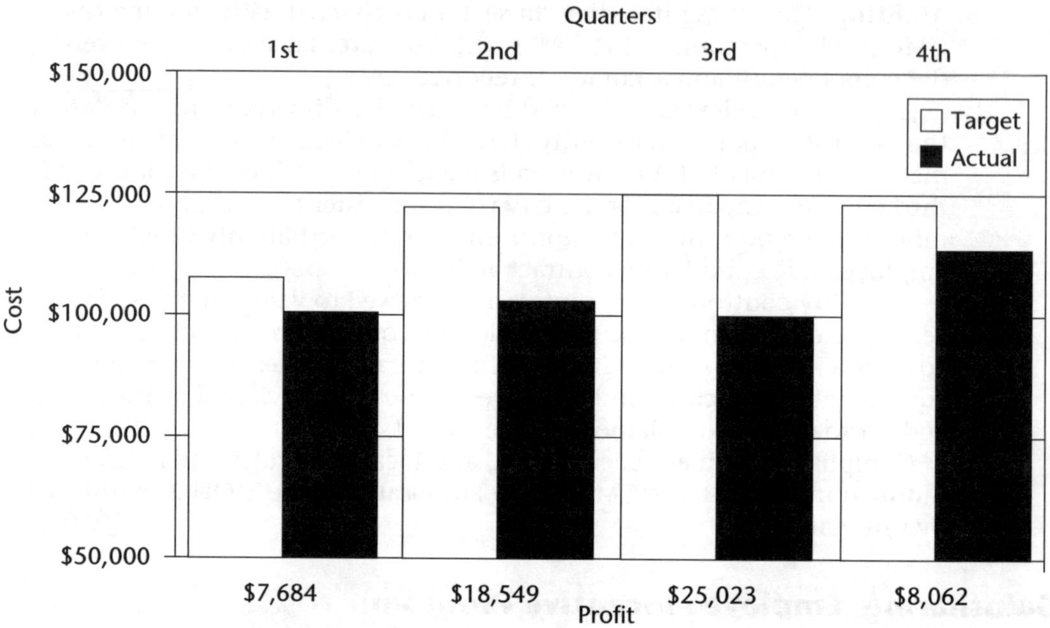

Total Profit	$59,318
Less Employee Gainsharing	($28,359)
*Unspent Contingency	$41,271
Net Profit	$72,230

*Unspent contingency represents $ in th CMCon-Op bid not included in the gainsharing targets.

Figure 13–3. Target Cost vs. Actual Cost

As described above, the CMCon-Op operation is externally monitored and validated by the city's internal audit section. As noted earlier, such monitoring includes financial testing of expenditures, performance reviews, unannounced site visits, and administrative review of operations. The contract compliance officers and the internal audit section continue to work to ensure that all contract requirements are met; established staffing plans are followed; qualified personnel are utilized in all areas of contract requirements; necessary equipment and supplies, as specified in the RFP, are available; detailed records of expenditures, deliverables, maintenance and accomplishments are maintained; and all expenditures are documented to the fullest extent.

CONCLUSION

Given the time and resources, municipal employees can compete successfully with the private sector. Competition creates a long-term cultural change in the governmental organization if the culture of thinking and operating like a private business is truly fostered. Elimination of internal bureaucracies and clear communication of the organization's vision and expectations leads to empowered employees who accept the challenge of competition.

It is extremely important to know one's competitors and to benchmark the competition. Competition works. However, the effort is intense. In the City of Charlotte, competition has broken down internal bureaucracies and changed the culture of the organization. The results of competition can exceed one's expectations as it did for Charlotte-Mecklenburg Utilities and its operating of the Vest Water Treatment Plant.

REFERENCES

City of Charlotte. Business Support Services (1997a). *Competition Process Handbook*. Charlotte, North Carolina.

City of Charlotte, Charlotte-Mecklenburg Utilities (1997b). *FY98 Utilities Business Plan*. Charlotte, North Carolina.

City of Charlotte, Charlotte-Mecklenburg Utilities (1996a). *Vest Water Treatment Plant Memorandum of Understanding*. Charlotte, North Carolina.

City of Charlotte, Charlotte-Mecklenburg Utilities (1996b). *Vest Water Treatment Plant Request for Proposals*. Charlotte, North Carolina.

City of Charlotte. Charlotte-Mecklenburg Utilities (1996c). *Vest Water Treatment Plant Request for Qualifications*. Charlotte, North Carolina.

City of Charlotte. Charlotte-Mecklenburg Utilities (1997c). *FY98 Utilities Business Plan*. Charlotte, North Carolina.

City of Charlotte. Privatization/Competition Advisory Committee (1994). *Guidelines for Services Contracting and Asset Management*. Charlotte, North Carolina.

City of Charlotte. Privatization/Competition Advisory Committee (1997d). *Guidelines for Services Contracting*. Charlotte, North Carolina.

Sizer, Edward (1998). "A Brief History of Managed Competition in Charlotte." City of Charlotte: Business Support Services. Charlotte, North Carolina.

CHAPTER 14

Key Issues Limiting the Ability of Managed Competition as a Winning Alternative for Communities

Douglas Herbst

The delivery of public services related to water and wastewater are being opened up to competition among the private sector with increasing frequency. In these competitions, the goal of municipal decisionmakers is to determine whether private sector involvement will result in reduced costs and increased service quality along with greater accountability and enhanced risk management.

Simultaneously, another service delivery option, "managed competition," is also being considered. Under managed competition, the existing public workforce is given the opportunity to bid on the same scope of services as the private sector firms. At a city's water or wastewater treatment plant, for example, the existing staff competes with private firms to provide the service of running the plant. Managed competition has been founded on the belief that competition among a wide range of providers will ensure high quality services at the lowest cost.

Private sector firms, in particular, are proponents of competition, not only because the private sector is inherently competitive but also because it has proven to be the best method for achieving operational efficiencies and cost savings. It creates incentives for the efficient delivery of services while preserving public accountability. Competition dictates that the best-qualified and most experienced firm meet a contract's obligations at the lowest practical cost.

In theory, managed competition, where private firms and the public workforce bid on the same job, seems to be a good approach because it opens up the competition

---

Douglas Herbst is senior vice president of Earth Tech, Total Water Management Division and president of the National Council for Public-Private Partnerships.

to let all types of providers compete. However, for any competition to be effective—for it to work in reality and not just in theory—there must be a "level playing field" on which to compete. The same set of rules must apply to all bidders in order for the competition to proceed and the best proposal to prevail.

With managed competition as it is being carried out today, the set of rules are different for the public and private sides, making the creation of a level playing field highly problematic. Only when the rules are equally applied can managed competition offer decision makers an equal basis for comparison between a private sector proposal and that of an existing public workforce. However, this idea is inherently flawed in that it is virtually impossible to apply the same rules to private sector firms and the public workforce.

The ever-changing worlds of business and municipal government are not always fair. For the private sector to cry "foul" in regard to the levelness of the field of competitive play for public procurements may, on the surface, sound simply like sour grapes. If one looks a little closer, it is clear this is not the case. The winner of any procurement process should be the municipality, its citizens and ratepayers. A level playing field is necessary for *the community* to collectively win the competition, regardless of whether it is the public work force or the private sector that are ultimately given the task of delivering the required services.

Privatization, or a public-private partnership, is a public service delivery option. Managed competition is simply another alternative for the delivery of these same services. Therein lies the real issue: How to take two service delivery alternatives and place them in a side-by-side comparison and be sure that the truly right choice was made.

NINE MAJOR HURDLES OF MANAGED COMPETITION

To have an effective, successful competition, there first must be a level playing field on which to compete. In an attempt to get to that level playing field, there are nine major hurdles.

1) Conflict of Interest

First, the issue of conflict of interest must be thoroughly addressed. The same people who are involved in preparing the procurement documents and evaluating proposals should not also submit a proposal or provide direct consultation to those that do. To guard against this, different people are used in all steps of the process; "Chinese Walls" are erected; and third-party independent consultants are retained to prepare and evaluate documents while others assist in proposal preparation. These efforts greatly diminish the potential for conflict of interest. In spite of these efforts, however, perception problems and actual experiences often make it difficult to overcome this issue.

With managed competition, there have been instances where the individuals who prepared procurement documents and evaluated proposals were also allowed to submit proposals. There have also been instances where the public sector workforce had the advantage of reviewing private sector proposals prior to submitting their own proposal. Other times, promises have been made that the best private sector proposals submitted would still be beaten by the public workforce. Cases such as these have left private sector firms apprehensive of taking even the initial step in responding to municipalities' procurement initiatives where managed competition is part of the process.

2) Cost of Service (Addressing "Cross-subsidization")

A major issue concerns the true cost of service. Typically, a price proposal submitted by a private sector firm includes all direct and indirect costs. When evaluating proposals submitted by a public workforce, it is important to determine whether all direct and indirect costs are included as well.

For instance, the proposal should include the cost of services provided by other city departments. These include services such as those provided by a city's purchasing department. If a private sector firm is awarded the contract, the purchasing department will no longer be servicing water or wastewater operations. This represents a direct savings to the department, but some of the cost may now be reflected in the private sector's price proposal. If, on the other hand, the public work force is awarded the project, the purchasing department's services to water and wastewater operations will continue to be an expense, and this cost should be factored into the public workforce's proposal. This concept, referred to as "cross-subsidization," must be thoroughly addressed and the appropriate costs must be included in the proposal submitted by the public workforce.

Other areas for cost evaluation should include insurance, utilities, rent, rolling stock, training, and maintenance. Costs such as personnel and payroll administration as well as other city administrative functions must also be thoroughly evaluated and included. Additionally, outside services related to water and wastewater operations, such as engineering, financial, accounting, and legal services need to be factored into the proposal submitted by the public workforce.

3) Qualifications

Most procurements find qualifications and experience to be extremely important selection criteria. Even if the existing public workforce performed well in the past, it may not be able to adequately compete in qualifications with firms that operate facilities throughout the country or world and have a network of national experts and extensive depth and resources. With this in mind, it can be difficult for a public workforce that operates only one facility to win a place on a shortlist over private sector firms with substantially more experience and qualifications. Therefore, if a public workforce with its limited experience and qualifications makes it to the city's shortlist, then all private sector firms, even those deemed unqualified, should be allowed to submit proposals if there is to be a level playing field for the competition.

In fact, if the qualifications requirements for a procurement were applied equally, the public workforce would probably not qualify and there would be no managed competition. To have a managed competition in most cases, the public workforce must be given a "free pass" to the shortlist. This statement is not meant to be disparaging to public workforces; it is just a realistic statement of what happens when one compares apples to oranges, and it is a common occurrence today in managed competition, where the public workforce makes the short list and qualified national firms often are excluded. Often, highly qualified firms will not even bother to submit proposals in managed competition situations because they do not want to risk their resources and reputations by not making the short list.

4) Accountability

Accountability is the most important feature of a public-private partnership that is noticeably absent in managed competition. It must be thoroughly addressed in order to establish a level playing field so that the community can be the winner of the procurement competition. This is one of the toughest hurdles.

It is difficult to ensure that the existing workforce will perform in accordance with its proposal. Unlike the private sector firm, the public workforce is not subject to termination and its remedies if it fails to perform. With a public-private partnership, the private firm must deliver the contracted services at the contracted price. If shortfalls exist, the private firm must absorb the loss. Also, if objectives in the managed competition proposal are not achieved and the public workforce exceeds the promised price, the shortfall cannot be taken from the cash flow of the public workforce (*i.e.*, out of the employees' paychecks) rather, ratepayers and taxpayers must pick up the tab and absorb the loss.

Because responsibility is less clearly defined under public sector operation, and with no stipulation for penalties for nonperformance, accountability can become diffused and incentives and reasons to perform become weak. With a public-private partnership, however, the private sector is held accountable, and that accountability is backed up by contractual guarantees and other security, along with contractual penalties for nonperformance.

5) Guarantees

Closely related to accountability is the issue of performance guarantees. One of the cornerstones of a public-private partnership is guaranteed performance. The private sector is required to provide guarantees. Securing guarantees from the public sector for the same service, on the other hand, is problematic because the ultimate guarantor of the public workforce proposal is the full faith and credit of the community. The existing workforce cannot provide contract security as a performance bond or letter of credit or provide the guarantee from its corporate parent. The public sector cannot offer similar forms of security to assure performance and an equal basis for comparison to the private sector.

6) Risk

Risk is the possibility that an action may produce an undesirable outcome. The outcome has a value (an economic impact, usually negative), a probability of occurrence, and a consequence to the other party or the "taker." How well risk is managed depends on the extent risk can be allocated to the taker or reduced or eliminated. Risk management is a very crucial area in the area of water and wastewater treatment.

With managed competition, the public sector's proposal places 100 percent of the risk on the community's doorstep. With a public-private partnership, a percentage of risk is allocated, with the actual percentage of risk allocated to the private sector varying from partnership to partnership. Thus a public-private partnership has the added benefit of providing a better risk management profile than could be achieved with managed competition.

Because all risks have an economic consequence, the risk allocated to the private sector should be valued and that value should be added to the public workforce's proposed cost under a managed competition situation. This provision would go a long way toward creating a level playing field. It should be noted that the value of the risk allocation would be significant. A part of this value would be consideration for the value of the guarantees being sought from the private sector firms, in addition to a standard risk-by-risk assessment.

7) Contract Enforcement

When considering contract enforcement issues under managed competition, the contractual relationship between the community and its workforce is uncertain at best. It would be impossible for the community to enter into an enforceable contract with itself. Specific performance could not be enforced. In the case of nonperformance, the entire public workforce could not be terminated as a private firm sector could under a public-private partnership.

It is difficult for a community to allocate risk to its employees and for the public workforce to guarantee its own performance. With a public-private partnership, these issues can be addressed in the contract to the satisfaction of the community. The private sector must perform or it can be subject to contract termination or, at the very least, its contract with the community may not be considered for renewal when it expires.

8) Managed Competition as a "Political Ploy"

It has been reported that, in some cases, municipalities are using managed competition and the "threat" of privatization as an incentive for the public workforce to perform or for unions to give concessions, as if the public employees' livelihoods would be in jeopardy if a private firm assumed control of a public system. In these instances, procurement initiatives were reportedly implemented with the best intentions, only to be used later as leverage or as a bargaining chip with the employees or unions.

9) Labor Concerns

If managed competition is pursued and the community opts for a public-private partnership, it is highly likely that the existing public workforce will become employees of one of the private firms it is competing with. Under managed competition, the public workforce cannot be expected to operate the water and/or wastewater systems, prepare a proposal, provide data and information to the independent preparers of the procurement document, give tours and information to the private sector competitors, and then, if a private firm is selected, be happy about the outcome.

More often than not, managed competition sets up an "us versus them, must win at all costs" environment. This is not a conducive environment in which to place the existing workforce. It can lead to bitterness, frustration and anxiety—not a pleasant atmosphere for public employees if they do not win the competition. The obvious mindset among public employees in the instance that a private sector firm is selected is that they "lost." The public workforce may not only direct hostility to the firm selected but to local decision makers as well.

Private firms have a different approach to the delivery of public services. While this approach can bring increased training, benefits, and opportunity to the existing workforce, significant adjustments are required. Change is difficult under normal circumstances; it would be that much harder if the workforce viewed itself as "conquered."

Many public-private partnerships that have been most successful in the transitioning of employees to the private sector were successful because they dealt with the employee situation early. Time is needed to address the existing workforce's issues and concerns. Prior to the selection of a private firm, the public employees should be comfortable with the transition requirements. Under managed competition, however, the time needed to address employee transition adequately and thoroughly is often not available.

The employee transition to the private sector should not be viewed in a negative manner. The public-private partnership approach would not be quickly and widely growing in use if successful employee transition programs were not part of the process and were not acceptable to the existing workforce. The fact is that the vast majority of employees that make up virtually all private sector firms' workforce are former public employees who are now receiving increased benefits, greater chance for advancement and training, quite content with the change.

One of the myths involved with public-private partnerships is that when a private firm takes over, existing employees are laid off or fired as a way to reduce costs by reducing the workforce. The private sector has by and large been able to provide the same level of service with fewer employees, but traditionally all employees are hired by the private firm and optimum staff size is achieved over time through attrition and reassignment to other open municipal positions.

Additionally, the municipality sets the requirements and can take any position it wishes regarding staffing, from allowing the private sector to operate at the staff size it prefers and offering employment to selected individuals in the existing workforce, to requiring the private firm to offer employment to all existing public staff members with workforce reductions achieved only through attrition. This is a policy decision to be made at the local level and then translated into the requirements of the RFP.

Thus, the public and private sectors are under the same requirements as to staff size. If the requirements are the same for both, the public sector will be under the same pressures to reduce staff in accordance with the requirements of the procurement.

FITTING PUBLIC-PRIVATE PARTNERSHIPS INTO MANAGED COMPETITIONS

With managed competition as it is being carried out today, the set of rules are different for the public and private sides, making the creation of a level playing field highly problematic. Up to this point, managed competition has been shaped to "fit into" public-private partnership procurements. One way to clear the nine major hurdles is to run a different race—create a model that "fits" a public-private partnership into a managed competition.

If the public sector is unable to compete under the same rules as the private sector, then common ground can only be reached under the managed competition rules. The following scenario would be created if a public-private partnership were to be "fit into" a managed competition and the same ground rules were applied to all parties to create a level playing field.

- The selected entity (the "Operator") will not have to enter into a legally binding and enforceable contract; only a non-binding memo of understanding (MOU).
- The cost offered will be only a good-faith reasonable estimate. If there are cost overruns, the Operator will not have to pay for them. The other party to the MOU will pay for them.
- Cost overruns exceeding a specified amount will be deemed a breach of the MOU. The only remedy and penalty for the breach will be the right of the other party to seek an alternative service provider (the "remedy") and preclude the Operator (who has breached) from competing for it again. The same remedy and penalty provisions hold true for any breach related to nonperformance.
- If the MOU is terminated, the Operator has absolutely no liability and will pay no termination damages or penalties to the other party.
- The Operator will not have to take any risk, including the payment of damages or fines due to its actions or negligence.
- No indemnifications are necessary from the Operator.
- The Operator will not be obligated to provide a guarantee for its performance. The other party will act as the guarantor.

It is unlikely that a public decision maker would issue such a procurement or agree to such a transaction, and this scenario has been presented here merely to make a point. However, with managed competition as it exists today this is how the public sector is competing in the public-private partnership arena and more importantly what is being offered in its proposals.

PUBLIC VS. PRIVATE SECTOR OPERATION

Proponents of managed competition cite many advantages that make it the best approach for the delivery of public services. The advantages often cited include: the public workforce knows the systems better than the private competitor; the public workforce's proposal does not include profit and taxes; and the fundamental motivation of the public workforce is public service.

With a public-private partnership, however, the existing public workforce is transitioned or hired by the selected private firm and that knowledge base is not lost. The private sector firm does not have staff "waiting in the wings" to take over the existing positions held by the public workforce. The private firm needs, values, and relies on the existing public employees for their knowledge and expertise.

The private sector includes profit and taxes in its proposal as costs of service and can still perform its services at less cost to the community. The cost savings are brought about because of the higher degree of creativity and innovation brought to bear by the private firm; private sector management and decision-making capabilities; often greater access to expertise and technology; corporate experience and knowledge from the inventory of other facilities the firm operates; profit motive and entrepreneurial spirit; lack of "bureaucratic handcuffs;" incentives offered to employees; crosstraining of employees; and economies of scale.

The private sector also has a strong commitment to public service. In order for a private sector firm to be successful and grow, it must have a good reputation. This reputation is established by providing reliable, high-quality public services.

TWO DIFFERENT INSTITUTIONS

When a public service is under scrutiny and the debate focuses on public versus private sector operation, the instinct is to blame the public workforce that has been running the system and switch to private operation to increase efficiency and lower costs. Communities that have made the switch to private operation for the delivery of their public services have found that the private sector can operate their systems more efficiently and at less cost. But it's difficult to compare public sector operation to private sector operation.

The public sector functions differently than the private sector. Just as managed competition and public-private partnerships are two different alternatives, the public sector and private sector are two different institutions. It is unlikely that the public sector could function like the private sector. The inherent qualities that make up the public sector are difficult to change. The public sector at best can only seek to emulate the private sector, and then the real issue becomes the question of whether this can emulation be sustained over time.

A proposal can be developed by the public workforce, sometimes with the help of consultants, but there is a big difference between what is proposed and the real ability of that workforce to deliver and guarantee performance. There are no assurances that the results and benefits predicted would be attained.

Thus, managed competition is only a price competition: the guaranteed private sector price proposal versus the best estimate proposal of the public workforce, which has no accountability or guarantees for performance and no track record or experience of actually achieving cost savings and/or reductions.

The public-private partnership industry has been successful with the majority of the employees in private sector operations being former public employees transitioned and retained by these firms. Many of these employees become very successful and rise through the ranks of private firms to key management positions. Public-private partnerships are not an indictment of the abilities and capabilities of public workforces. Some people may not perform as well under the constraints and limitations of the public sector. In the private sector, with fewer limits and constraints, the performance of former public employees often improves dramatically.

It is often stated in discussions about managed competition that "open and fair competition between private firms and the public entity will provide the customers with the highest level of service at the least cost. The focus should be on the customers or ratepayers receiving the service." However, with managed competition, it is hard for one to determine how customers and ratepayers can get the highest level of service at the least cost when there are no accountability or guarantees for that service or cost. In the event the public workforce cannot fulfill the terms of its proposal, it is the customers or ratepayers who will ultimately pay for the broken promises.

CONCLUSION

Unless serious shortcomings are adequately addressed to protect the public interest, managed competition will only be a price competition with different rules for the private and public sectors. If low bid is the ultimate objective, with no accountability, no guarantees and no risk allocation, then managed competition is the appropriate choice. Public decision makers need to review all the issues involving the managed competition and public-private partnership approaches when deliberating what is best for their community.

Finally, managed competition is not a way to kill privatization when it is introduced to a community or a concept to be used in a preemptive way to "keep privatization at bay." In reality, managed competition and privatization are alternative public service delivery options. The advantages and disadvantages of each alternative must be thoroughly identified and evaluated. Only then can an informed decision be made as to which alternative may be most suitable for a particular situation and, ultimately, for the community being served.

CHAPTER 15

Learning to be Competitive from the Private Sector

Alan Manning

The utility industry changed dramatically in the 1990s—and utility management is feeling pressure from all sides. Takeovers by contract operations and maintenance (O&M) firms have increased dramatically, while, at the same time, ratepayers are demanding improvements in rates, quality, and service.

Curiously enough, both issues are related, and can be dealt with by taking the same big step: *becoming competitive.* By becoming competitive, a utility will remain in the public sector, *and* will provide a greater value to its customers as a result.

COMPETITION LEADS TO VALUE

Competition is continuing its foray into the utility industry, with private sector companies taking over the operation, maintenance, and administration of water and wastewater utilities or districts. This trend will continue, influenced by many factors—including changing legislation, deregulation of utility services, blurring lines of distinction among utility organizations, and performance of both public and private service providers.

Competition can be healthy, however, by stimulating utilities to rethink and reorganize the way they do business. However, to become competitive, the utility

Alan Manning is chief executive officer of EMA Services.

industry must change from its monopolistic attitude steeped in tradition to a business-driven, competitive mindset. Utility managers are reluctant to change.

RATIONALE FOR CHANGE

Change is always difficult, but it is especially hard in the utility industry. Bureaucracies have evolved over the years. Whenever a problem occurred, rules were created to prevent that problem from being repeated. Over the years, numerous rules, policies, and procedures were written. Later they were no longer needed but had become entrenched, and so they became barriers to competitiveness and efficiency. These barriers cannot be removed entirely—but they can be managed, and their impact on competitiveness can be lessened. These barriers must be attacked in a systematic way, focusing in on those parts that can be affected, and leveraging the *important* changes.

The reality is that competition will sound the death bell for utilities that do not heed the call and take steps to face the challenges—and opportunities—head-on. The great beauty of this challenge, however, is that the utility facing it proactively will become a healthier utility with happier staff, with streamlined organization and processes, and with the technology that makes it possible. By adopting new work practices, organization, and technology, the competitive utility will benefit from:

- Reduced costs—from labor, operations, maintenance, supplies;
- Effective and efficient employees—more knowledgeable and proactive;
- Streamlined management—a simpler, more efficient operation;
- Happy and healthy taxpayers—they know they're getting the best product at the best price.

ADVANTAGE OF PUBLIC UTILITY

Finally, here is the best news of all: almost any public utility can compete with any private utility, because public utilities have an inherent competitive advantage. Privatizers (O&M contractors) have additional costs (*e.g.*, profit, taxes, financing) that public utilities do not have. So even when public and private utilities are equally well run, the public utility will still be more competitive, with the same level and quality of service.

ACCEPTING CHANGE

The question most frequently asked is: How does a utility, regardless of size, come to this agreement—that change is all-important? Permanent, cultural, and behavioral changes take place successfully in an organization only when there is urgency—with identified solutions, developed over a period of time—that is sufficient to overcome organizational resistance. This is shown in Figure 15–1.

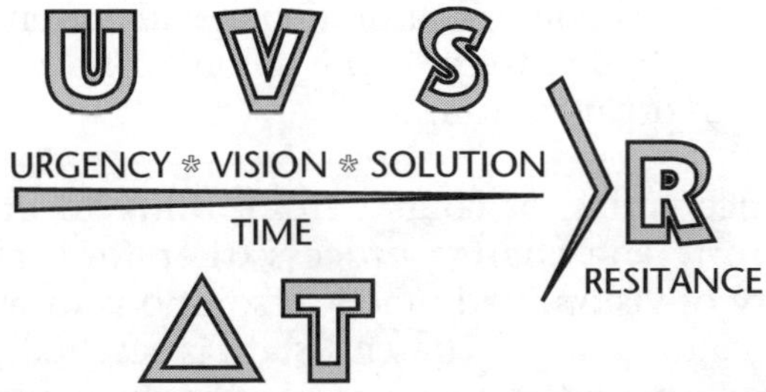

Figure 15–1. Change comes only if urgency and vision have a solution that is strong enough, over time, to overcome resistance to change.

The entire utility, from board, to management, to supervisor, to labor, must realize that they need to change—and must agree that it is the most important item on their agenda. It must be so important that everyone will work together (possibly for the first time in the history of the utility) to achieve change. The desire to change must be greater than any other force within the utility.

The first requisite is an unequivocal sense of urgency throughout the organization. This means that virtually every person in the utility is aware of the need to change and improve, and can explain the reasons *why* the change is being undertaken. The source of urgency can be competition, unhappy customers, funding shortfalls, or unhappy politicians—but there must be a strong and united force demanding change.

The second factor is alignment around organizational vision, a vision of the future that is significantly different from the present. Vision is the ability to genuinely see a future that is different than the present, which means that each individual within the organization must understand what the new environment will look like and feel like. Alignment must be around a new vision of a utility that is business-driven, not monopolistic. This vision must be specific enough that each individual not only understands it, but can explain the new environment and new culture to others—as well as why it is critical to the future of the utility.

The third factor is the mechanism to get from where the organization is now to where the organization needs to be in the future—the "solution" part of the formula. Each person in the organization must understand that there is a process to get from where they are now to where they are going to be. And they need to believe this process will work.

Note that the formula is multiplicative. This means that if any of these components is zero, the chance of succeeding is zero. So there must be urgency, there must be vision, and there must be a practical solution.

THE CHALLENGE OF CHANGE

Several factors make transformational change especially challenging in the public sector. One such factor is history. The autocratic leadership, bureaucratic structures, and civil service mentality that have characterized the water/wastewater industry since its beginning do not change easily.

Another constraining factor is politics—a hidden agenda often conflicts with what is best for the organization. Finally, there is the absence of a profit motive to act as a coalescing force within an organization.

Organizational resistance varies, of course, from utility to utility, and throughout any major change effort. The change process, therefore, must be supported by enough urgency, clarity of vision, and practical solutions to overcome the normally organizational resistance to change. The end state is reached when new work practices have been adopted, the entire staff has new habits, and momentum has been created for continuous improvement.

THE POWER OF BELIEF

Alignment of the organization around these components, over time, allows momentum to be built, which is greater than organizational resistance. If the process of alignment is such that (almost) every individual in the organization understands the urgency, understands the vision, and believes that the solution (the process for change) is realistic, then organizational resistance will be overcome and cultural change will be successful.

The keystone to the change process is helping staff throughout the organization discover, through meaningful involvement in the change process, a better way of doing their work. *A person convinced against his or her will, is of the same opinion still.* Don't tell them, don't show them—but *involve* them in the process—so that they'll discover *why* and *how* to change themselves!

THE PROCESS OF ORGANIZATIONAL CHANGE

Commitment comes from understanding. Understanding comes from involvement. In order to gain the commitment and the alignment of the staff, it is essential that not only do they have an understanding of the new culture, but that they—to a significant degree—actually design the new culture themselves. This is often referred to as the process of discovery, where the people working in teams, in a facilitated manner, with subject matter experts available as resources and advisors, examine their past to chart their course toward new goals.

Buy-in comes from the involvement of every level of staff in every step of the process, and by keeping every employee informed and involved throughout the process. It's the result of frequent and consistent communications at all levels of the organization. It's the result of seeking input from everyone, and it's successful because it involves representatives of all staff levels and categories in making the decisions as to what must be done and how it will be done.

It takes a leap of faith from all stakeholders. The governing board or council must believe that the time, money, and process will be a wise investment in the utility's future; management must believe that the new organization will function smoothly and effectively; supervisors must believe that it will be a better place to work, and staff must believe they will still have jobs and that they will like their work.

A person cannot be "told" that the new way of doing things is better than the old ways. That must be discovered. Figure 15–2 shows the relationship between risk and commitment of the organization and its personnel. As the commitment and alignment of the organization and personnel increases, the risk of success increases. In the beginning, there is a period of time in which alignment is building, but until personnel begin in-depth talk about it, the risk will not be significantly reduced.

Effective communication enables staff to achieve buy-in, concurrence, and commitment. It helps people understand the impact and benefits of a new way of doing work, and prepares them for the changes that will occur. Most importantly, effective communication is a key factor in reducing the fear factor associated with change. Understanding minimizes fear.

FIND, PLAN, AND IMPLEMENT

Once an organization is ready to change, what are the steps? The first step is to determine the difference or "competitive gap" between existing practices and those of the best-run utilities—a work practice comparison. The second step is to identify the opportunities that exist for eliminating that gap, and develop a plan for improvement. And the third step is to implement that plan.

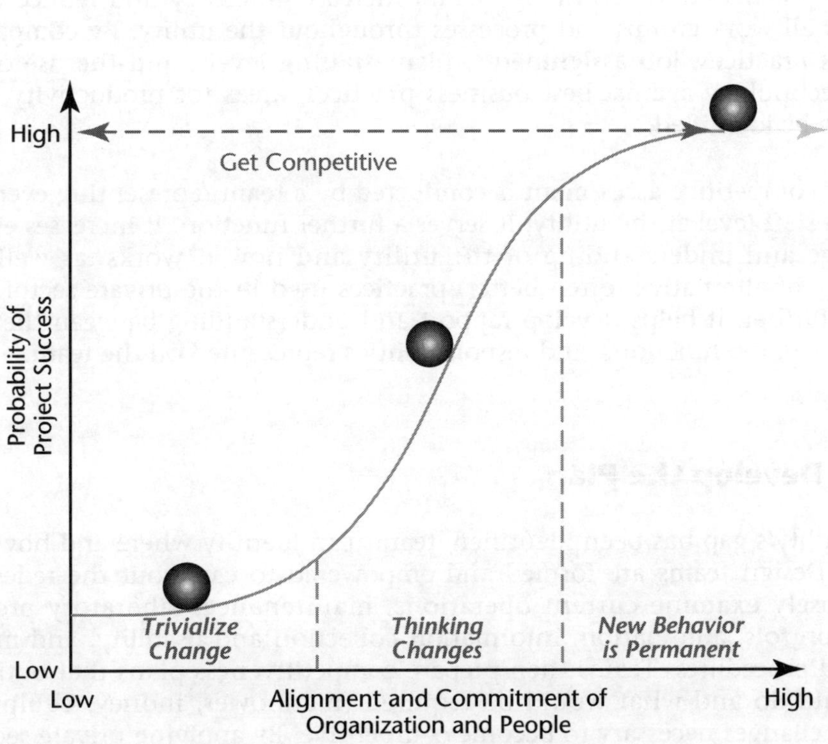

Figure 15–2. As alignment and commitment to change grows, the likelihood of success increases momentum, and at the same time, the utility is becoming more competitive.

Step I: Determine the Gap

The first step to becoming more competitive is to know how the utility measures up against the best, as determined through an evaluation of work practices. This can also be used to determine what the competition would "bid" against a utility.

Studies repeatedly show that public utilities have a competitive gap—the difference between "best practices" in the private sector and how public utilities function. Although some public utilities function very efficiently, the average public utility will have a gap of around 25 percent—which leaves room for significant savings. Most public utilities will be able to change their culture and work practices to successfully become competitive. About 20 percent, however, are so entrenched in culture and tradition, with archaic leadership and supervision, political instability, limited computer literacy, or poor union/management relationships, that they are not likely to be able to change in order to compete.

The assessment begins with the formation of a steering team. The steering team must represent the organization both horizontally and vertically. All facets—labor, union, skilled workers, unskilled workers, managers, and supervisors—must be represented on the steering team. The steering team develops a clear vision of goals, metrics, communication, involvement, and guidelines. It also determines how many design teams will be needed, and how to select team members and leaders.

The steering team begins the assessment by identifying areas of opportunity, where improved practices and technology could increase efficiency and reduce operating costs—for all work groups and processes throughout the utility. By comparing current work practices, job assignments, plant staffing levels, and the use of process control technology against best business practices, areas for productivity improvements can be identified.

Since this competitive assessment is conducted by a team representing every department and staff level at the utility, it serves a further function. It increases everyone's knowledge and understanding of the utility and how it works, as well as their knowledge of alternative (often better) practices used in the private sector. And yet one step further, it helps develop rapport and understanding between the diversity of employees, job functions, and responsibilities represented on the team.

Step II: Develop the Plan

Once a utility's gap has been identified, teams can identify where and how to make changes. Design teams are formed and empowered to carry out the redesign. The teams closely examine current operations, maintenance, laboratory procedures, process controls, automation, information collection and usability, and manpower needs and procedures. Teams then prepare competitiveness plans that outline what they would do and what would be required (manpower, money, equipment) to make the changes necessary to become competitive. By applying private sector paradigms, the savings for each opportunity is identified and ranked by the expected return on investment.

This is often the point where there is a critical need for an outsider, because the team needs to be prompted to think differently (outside of the box). The team also needs help in understanding that an organization *can* change internally, despite old baggage, old history, and old relationships. Therefore, the process and recommendations often require an outside force to provide new thinking, facilitation, and important new directions.

The teams then take their ideas and their concepts to the rest of the organization through a validation process. The idea is that 75 percent of the organization, which is the critical mass, will have had significant input into the design—which is very important in obtaining commitment to new behaviors and new cultures.

For a public utility to successfully compete with the private sector, it must change its philosophies and beliefs, leading to new knowledge and understanding—which in turn will result in new skills for using new and existing tools more effectively. Applying new skills and tools (both human and technological) to traditional philosophies and beliefs, each part must be addressed thoroughly and completely if change is to be effective and lasting.

Implementation

Assuming that the alignment takes place, the question asked is: *How does a utility go about changing its work practices?* The key to changing work practices is shown in Figure 15–3. It is the simultaneous development of new behaviors around three components of the new culture—organization, practices, and technology.

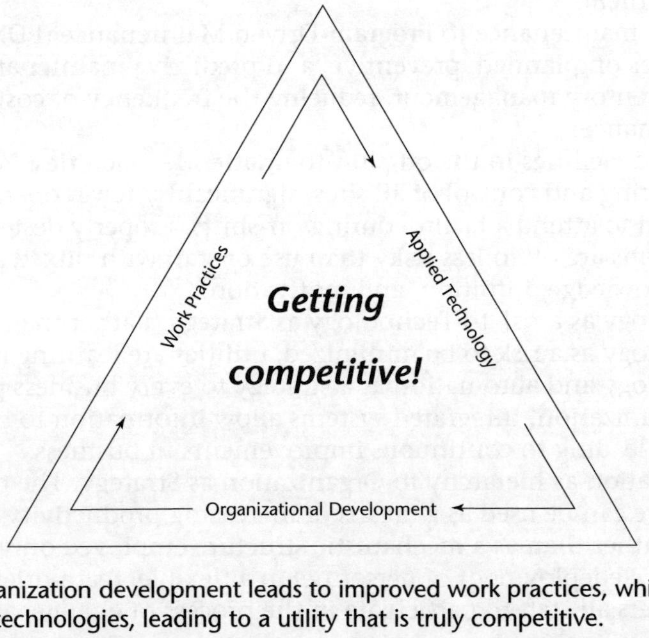

Figure 15–3. Organization development leads to improved work practices, which then leverages the impact of new technologies, leading to a utility that is truly competitive.

1. Organization

The first element that needs to be changed is the way the organization operates—organization development. Perhaps the most significant change will come in the changing roles of management, supervisors, and staff, as the utility becomes "flat" with responsibility liberally sprinkled throughout. Management staff must clearly and continually demonstrate that they support each individual in taking on new roles and responsibilities.

This will lead (often with the help of training) to changes in the way people treat each other, listen to one another, show respect and dignity for each individual, stop judging others and start listening, learn to work in teams, learn to work with less supervision, learn to be good leaders and followers, and understand their shared fate. Organization development is fundamental to any culture change process.

2. Practices

The second component of the triangle is work practices. Teams analyze their current work processes and come up with new schemes to reduce unnecessary steps—or to adopt new approaches, or to completely stop doing things that don't add value. The key to the work practice evaluation is the voice of the customer. The employees and the new culture must ask: *If a customer knew I was doing this particular activity in the way that I'm doing it, would they pay me for doing it?* Another way of thinking about this is: *Would you pay somebody to do what you're doing right now?* This identifies make-work activities, activities that add no value and that must be eliminated. The essence is changing practices from:

- operations and maintenance to Total Productive Operations (TPO). TPO substantially eliminates the segregation of the O&M functions. Operators no longer wait for problems to occur before calling a maintenance expert to resolve them;
- reactive maintenance to Program-Driven Maintenance (PDM). PDM focuses resources on planned, preventive, and predictive maintenance activities and inventory management, reducing the frequency of costly unplanned maintenance;
- attended facilities in the off-shift to unattended facilities. With centralized monitoring and control of all sites, significantly fewer operators are assigned to attend a facility during off-shifts. Properly designed unattended operations are often less risky than use of staff with mixed and variable skill sets, knowledge, intuition, and motivation.
- technology as a risk to Technology as Strategy. Rather than viewing technology as a risk to be minimized, utilities are learning to apply technology and automation as a strategy to every business process within the organization. Integrated systems allow information to be appropriately shared, leading to continuous improvements in business.
- organization as hierarchy to Organization as Strategy. The organization structure can be used as a means of increasing productivity and decreasing costs, rather than as a mechanistic structure employed only to resist change. Redeployment of personnel in a flexible, team-oriented approach empowers all staff to participate in the process of change; and

- work separated by craft to Workforce Flexibility. Increasing the skill set of maintenance and operational personnel will significantly decrease the amount of time spent waiting for other skills or tradesmen to arrive and correct a problem. Skills-based compensation provides the incentive for staff to become cross-trained.

3. Technology

Following development of the new organization, new behaviors, and new work practices, the requirements for technology to support those practices can be identified. Technology from the standpoint of business objectives becomes very simple and straightforward, used strategically to automate the basics. Many bells and whistles and contingencies for downsides are no longer needed because technology will now be used to work smarter and to minimize dangers. Technology becomes the high-leverage enabler of the new work behaviors and practices, the component that allows the change to work successfully.

One of the classic impediments to utilities' progress is not having the right information at the right time in the hands of the right people. Most public utilities spend excessive time accumulating, gathering, formatting, and reporting limited data; technology is now used to obtain more timely and complete (and useful!) information. This not only supports the utility's day-to-day information and planning needs, but it also provides the information needed to make sound management decisions required to become competitive.

Step III: Implementation

The third part of change is the actual implementation of the plan, and this is where things can get difficult as reality sets in and every change is "tested." As teams begin to implement the changes, the team structure is tested as well. Support and encouragement come as teams begin to see "quick wins" from their work, and as incentives provide increased motivation.

Management must remain focused on all organizational aspects, maintaining a steady view of the big picture and a steady course with day-to-day activities. Management must remember to lead from the top, and implement from the bottom, as shown in Figure 15–4. The entire change process typically takes three to five years to complete throughout the utility.

An Uphill Battle

The whole process is precarious at best, since it is the product of human nature—but it is a process, and its success has been demonstrated repeatedly in utilities of all sizes. As was shown in Figure 15–2, it's like rolling a ball up a hill. In the early stages of a change program, the alignment is not as strong because not many people have been involved in a significant way. As the project moves on and more people get involved in design teams and begin learning new ways of thinking and behaving, the alignment and commitment of the organization increases and the project moves toward success.

Figure 15–4. Managers must remember to keep re-focusing on each level of change motivation, and to lead from the top down while implementing from the bottom up.

The middle stage, when the thinking starts to change, is the most dangerous. If the process is not successfully completed, change will not be complete, and a significant amount of momentum will be lost. When people backslide, moving back to their old comfort-zone behaviors, the project fails to achieve its goals and participants lose faith in the process and the future. So it is very important to understand that the process of alignment requires a systematic approach for consistent advancement.

Key Role of Management

Executive leadership is a key component in making transformation successful. It is imperative that top leaders in water/wastewater organizations anticipate the need for transformation, understand the implications of this type of change, create a compelling future vision, and drive their organizations to achieve that vision. Regardless of obstacles, bringing about transformational change is a leadership imperative in the water/wastewater industry. The ability to do so is no less than a core competency.

Leadership not only has the task of creating a future vision, it is also responsible for demonstrating and communicating the change—getting buy-in and support, creating excitement, controlling direction, and institutionalizing leadership throughout the entire utility.

First-line supervisors are a critical part of the change process as well. The supervisor is the primary communicator with the workforce—the major component of any utility organization. The rank and file receive virtually all information and communication through their supervisor, so if change is to occur, the communication-critical point is the supervisor.

The Critical Importance of Vision

One of the most critical ingredients of successful transformational change is a clear, compelling vision of the future. Vision is what makes it possible for people in an organization to picture and grow into the future they are trying to create. Vision empowers, by providing a context for daily decisions and reducing the need for rules. It focuses people on moving *toward* something positive, rather than *away* from something negative.

Creating this vision is a key leadership role and responsibility. In most cases, only the top layer of an organization has the vantage point to see all the environmental cues to anticipate needed changes in strategy, as well as access to the organizational resources needed to initiate such dramatic change. Top management is also in the best position to create a sense of urgency and to demonstrate that change is truly vital to the future of the organization.

In too many water/wastewater organizations seeking to make major change, there is a disturbing absence of this sort of vision. There are a number of likely reasons for this. In many organizations, the concept has been overused; visions and statements have come and gone without significant impact. Some people do not grasp the vital importance of a vision. Others consider the process of developing a vision too time-consuming, complicated, or mysterious; and, to many, the notion seems mushy. Finally, the creative, often intuitive visioning process is outside the comfort zone of many engineer-managers who prefer to deal with facts and figures.

Establishing a vision is a part of the leadership imperative. Even with a clear, compelling vision in place, transformational change is risky and difficult; without such a vision, it is certain to fail.

CONCLUSION

Change is Becoming a Way of Life for Utilities

Public water and wastewater utilities are not the only utilities in the midst of rapid change. We are all familiar with the breakup of AT&T, the world's largest regulated monopoly. We witnessed deregulation of the natural gas industry 20 years ago, and the retail wheeling of natural gas that resulted. Most utilities (especially larger ones) benefited when lower contract prices were negotiated.

Deregulation is well under way in the electric utility industry. Proponents of deregulation in the electric industry predict lower energy costs, similar to the natural gas and telephone industries. Opponents of deregulation say reliability will suffer and prices will go up as it occurred in England with the privatization of electric and water utilities.

This rapid change and deregulation in other utility sectors poses yet another "crisis" situation, which can be viewed as both an opportunity and a danger for public water/wastewater utilities. The opportunity is to develop strategies that will leverage the changes in the electric utility industry to negotiate new long-term contracts at reduced rates. On the other hand, the deregulation of electric utilities is a danger in that investor-owned electric utilities (regulated monopolies) are now eyeing public and investor-owned water and wastewater utilities as attractive takeover targets. Why? Because they have surplus cash and their electric market share is shrinking and there can be significant cost savings by providing several utility services to the same customers.

Utilities must take it upon themselves to become competitive and change before others change them. They will be facing change either way—change for the good of the utility, its employees, and the community—or change that favors private industry.

The advantage to making changes proactively is that substantial savings can be realized and used to:

- Offset future rate increases;
- Reduce debt service;
- Plan needed expansions;
- Offer education/rewards to staff; and
- Contribute to the general fund.

Competitiveness brings out the best in people throughout the utility—it reinvigorates the spirit and the mind and, oftentimes, the heart. Utilities must compete if they are to remain in the public sector. To compete successfully, they must become competitive, which involves many factors.

- The key to getting competitive is a cultural change process.
- Alignment of people and the organization requires a process that is systematic.
- The culture must change, from a monopoly to business-driven.
- Commitment to that process comes from the top.
- Change requires support and participation from both management and labor throughout.
- Whatever changes take place within the organization will enable the change process to happen.
- Organization development is a very important part of the change process.
- Changing work practices, adopting best practices, and stopping work practices that don't add any value are the key to productivity improvement.
- Use of technology is an enabler for new behaviors; technology is designed to support the new way of doing things.
- The formula *Urgency*, *Vision*, and *Solution*, overcoming *Resistance* is a roadmap. The individual, as well as the utility as a whole, must have a sense of urgency, have a sense of the new environment that is being created, and have a sense of how to get from present to future.

A competitive utility is a healthier utility with happier, productive staff that function smoothly and efficiently. The ultimate reason to become competitive is simple: *survival*. Charles Darwin put it succinctly: "It is not the strongest of the species that survives, nor the most intelligent, but the one most responsive to change."

PART IV

Policy Issues

CHAPTER 16

Privatization and Incentives for Efficiency: The Case of Wastewater Treatment

Randall G. Holcombe

The early 1980s saw a flurry of interest in privatizing wastewater treatment facilities, arising from a number of reasons. One impetus was the potential for increased efficiency through private ownership, but other motivations included more stringent U.S. Environmental Protection Agency (EPA) regulations on water quality, changes in federal tax laws in the early 1980s that increased the attractiveness of privatization, and public finance considerations of local governments. In theory, one would expect cost reductions to accompany privatization, but in practice, as an examination of the construction costs of a number of facilities shows, there were no savings. An examination of the actual privatization agreements reveals an explanation for the lack of cost savings. In most cases, the privatization agreements allowed the private contractors to pass most costs through to the governments with which they were contracting. The agreements were written in such a way that there were no incentives for privatizers to act efficiently.

The specific cases of privatization of wastewater treatment discussed here show that privatization does not necessarily result in cost savings, and that the lessons drawn from these wastewater treatment cases can be extended to apply to privatization agreements in general. While there is substantial literature explaining why privatization enhances efficiency, a simple transfer from private to public ownership will not in itself enhance efficiency, and may actually make the activity less efficient. Privatization enhances efficiency because, in general, there are better incentives for efficient behavior in the private sector than in the public sector. In the case of waste-

---

Randall G. Holcombe is professor of economics at Florida State University.

water treatment, private ownership often is established without the accompanying incentives for efficiency.

This chapter first discusses general reasons why privatization might be desirable, and puts the privatization of wastewater treatment in a broader context. It examines the environment surrounding the privatization initiatives in the early 1980s that led to widespread privatization of wastewater treatment facilities. Next, some evidence on efficiency is discussed, followed by an explanation of privatization agreements for wastewater treatment. After examining the specific case of wastewater treatment, the lessons that can be applied to privatization in all areas will be discussed.

RATIONALE FOR PRIVATIZATION

The primary justifications given for privatization are cost savings and improvements in the quality of output. There is substantial literature explaining the advantages of privatization, and while there is little reason to review that literature in detail, it is worth considering the theoretical advantages of privatization in order to place the privatization of wastewater treatment in context.

Privatization has the potential for reducing the cost of publicly provided goods and services. If a private firm contracts to exchange some output for a fee, any gains from efficiency can be retained by the private firm as profit. In addition to general efficiency gains, bidders for privatization contracts tend to build the cost savings into their bids in order to try to get the contract. The general advantages of privatization make it appealing,[1] and a number of specific studies that examine the difference in costs between publicly operated versus privatized facilities of the same type consistently show significant cost savings.[2]

In addition to the general advantages of private ownership, there are several reasons to think that privatization might save costs and result in efficiencies, specifically in the construction of wastewater treatment facilities. A privatized facility would operate by accepting wastewater from a municipality, treating it, and then discharging the treated water. The private firm would build the facility, own it, and operate it. This gives the firm an incentive to be efficient in the construction of the facility, saving costs when it is profitable to do so, but not skimping in areas that would result in higher operational costs down the road. Much more supervision is required during the construction of a municipally operated facility because the contractor who builds the facility will not operate it. The contractor thus has an incentive to skimp in areas that might lead to higher costs later on.

It is common in facilities like this to have design changes during the construction of the facility, or even to begin construction before the design phase is complete. Because of the incentive problems with municipal ownership, the municipality must be much further along in the design phase before construction can begin, and then must adhere very closely to the agreed-upon design in order to prevent inefficient changes by the contractor. Another reason that municipally owned facilities have to be further along in their design is to prevent the contractor from overcharging on any design changes that might be required after the contractor is locked in. This requires more advance planning, a longer design phase, and detailed oversight by the municipality during construction. If the contractor requests design changes, they must be carefully scrutinized.

Consider a proposal from the contractor for a design change once construction has begun. If the facility were to be municipally owned and operated, the contractor would suggest such a change only if it would alleviate construction costs, or if the change in specification would result in additional billing to be paid by the municipality. Thus, some potentially effective design changes might not be suggested if they would not reduce construction costs; further, there is always the potential for changes that reduce construction costs but increase operating costs. The municipality must carefully assess any such requests because the contractor's incentives are at odds with those of the municipality. If the same firm that will operate the facility is also building it, this government oversight is less critical, because faulty construction will impose costs on the privatizing firm, not on the municipality, once the facility is operating. This gives the private firm an incentive to consider life cycle costs—not just construction costs—during the construction of the facility. Because of this incentive issue, efficient design changes are more difficult to make in municipally owned and operated facilities than in privatized facilities. The incentive structure is such that privatized facilities ought to be less expensive to construct and operate.

Closer integration of the construction and operation of the facility offers a number of other potential advantages for privatizing wastewater treatment. Privatized facilities should be able to begin construction earlier, since less of the design needs to be finalized before construction begins, and there should be greater flexibility in design as the project continues. Facilities can be started sooner, can be completed in less time, and can be more efficiently built. For a number of reasons, construction costs should be lower with privatization, and privatized facilities should also be less expensive to operate.

PRIVATIZATION OF WASTEWATER TREATMENT IN THE EARLY 1980s

After considering the potential advantages for privatization, this section turns to an examination of specific factors related to the privatization of wastewater treatment in the early 1980s. Despite the possibility of efficiency advantages explained in the previous section, increased efficiency was not a major motivation for privatization of wastewater treatment. A number of circumstances conspired to produce an environment very favorable to the privatization of wastewater treatment in the early 1980s. First, federal policy toward wastewater treatment created financing challenges because federal funding was cut back at the same time that facilities needed to be built in compliance with the Clean Water Act (CWA). Second, state and local governments found themselves fiscally squeezed in the early 1980s due to the recession and other federal funding cutbacks. Third, federal tax law was changed to make privatization more attractive than it previously had been.

Note that these reasons have little to do with cost savings or improvements in output quality, the major benefits of privatization discussed in the previous section. A more detailed examination of these motivations for privatization can shed more light on the specific forces leading to privatization of wastewater treatment.

Federal Policy Toward Wastewater

One of the key elements in the privatization movement was the challenge of complying with the Clean Water Act. Originally passed in 1972, the CWA had been amended several times prior to the wave of privatization in the early 1980s, and the federal government had become increasingly involved in the financing of local wastewater facilities.[3] In 1981, the CWA was amended again, but this time federal involvement in financing facilities was reduced substantially. EPA grants to local governments for the construction of facilities were cut back significantly, and grants for planning and design were eliminated completely. The reduced financial commitment of the federal government was offset somewhat by a pushing back of deadlines for compliance with clean water standards to 1988, but this deadline was only a few years away, and wastewater treatment facilities can require years to advance from planning to completion.

Federal policy created the necessity in many communities to make substantial investments in wastewater treatment, but at the same time reduced the federal funds that would be available to assist in this investment. Through privatization, these funds could come from private owners, rather than from local government sources.

The Fiscal Climate for State and Local Government

Another contributing factor was the recession of 1982, which made it difficult for local governments to find the money to spend on wastewater treatment when other programs were competing for scarce funds. The fiscal problems of state and local governments tend to be closely correlated to the business cycle. Most states and localities are legally constrained to balance their budgets, but do not do a very good job of planning for cyclical fluctuations. Rather than setting aside any reserves, they tend to spend whatever money is available during the good years, creating very tight budgets during recessions.[4] The result was increasing fiscal distress for state and local governments in the early 1980s, which reduced their capacity to finance additional investments in wastewater treatment.

Gold (1984) and Papke (1983) discuss the fiscal problems of state and local governments during the 1982 recession, giving an indication of the kinds of difficulties they faced. There was much talk about state fiscal crises during the period, but with hindsight one can see that state and local governments had an easier time during the remainder of the 1980s when steady economic growth kept state revenues growing. After the recession of 1991, there was again some talk about state fiscal crises, as described by Matoon and Testa (1992). As this literature describes, state and local governments experience fiscal stress during recessions because they do not set aside adequate reserves to smooth their spending over the business cycle.

This is directly relevant to the wave of privatization initiatives in wastewater treatment that occurred in the 1980s, because the push to construct new facilities due to federal policy came along with the recession that severely limited the ability of local governments to raise revenues for the facilities. In this context, the advantage of privatization was that the private contractor is responsible for financing the facilities, thus reducing the local government's financial concerns.

Federal Tax Law Changes

The early 1980s also saw substantial changes to federal tax law that greatly reduced the tax burden associated with privatization, primarily as a result of federal tax reform in 1981. These changes were reversed later in the decade, especially as a result of the Tax Reform Act of 1986.[5]

One of the more significant changes made in 1981 was the implementation of accelerated depreciation schedules, which allowed the capital cost of privatized facilities to be depreciated much more rapidly than before. Depreciation schedules were made continually less favorable as the decade progressed, culminating in the Tax Reform Act of 1986, which produced depreciation schedules less favorable than those used before the 1981 tax reform.

Another significant tax advantage to privatization was the investment tax credit, which paid 10 percent of the capital cost of a facility through a reduction in the investor's taxes. The investment tax credit existed prior to 1981, but was repealed in 1986. Capital gains were also taxed at a rate more favorable than ordinary income prior to the Tax Reform Act of 1986. Before 1986, corporate income was taxed at a maximum rate of 46 percent, but capital gains were taxed at a 28 percent rate. After the 1986 tax reform, the corporate tax rate was lowered, but capital gains were treated as ordinary income and taxed at the 34 percent corporate tax rate. Thus, prior to 1986 there was a greater tax advantage to investing in capital projects like wastewater treatment facilities, but that advantage disappeared after 1986.

Another factor relevant to privatization initiatives is that prior to 1986, local governments could use tax-exempt industrial development bonds to finance privatized wastewater treatment facilities. The ability to use industrial development bonds was severely curtailed in 1986, to the extent that tax-exempt financing was not practical for privatized facilities.

When all of these changes are considered together, the federal tax climate for privatization became much more favorable in 1981, but by 1986 the tax environment again weighed against privatization (Holcombe, 1990). While one might be tempted to argue that the changes made in 1981 in effect subsidized privatized facilities, this is not the case. In fact, privatization has always had a tax penalty associated with it, because if the facilities are municipally owned and operated, the municipalities pay no tax at all on the facilities. Thus, if the facilities generate any income on which tax must be paid, there is always a tax penalty associated with privatization. This penalty was estimated to be in the neighborhood of 30 percent of the capital cost of the privatized facility after the Tax Reform Act of 1986, and the more favorable tax treatment during the early 1980s served only to put privatized facilities on a more equal footing with municipally owned and operated facilities, but did not create a tax advantage to privatization (Holcombe, 1990).

Changes in the federal tax laws created a window of opportunity for privatization in the early 1980s because they removed some of the substantial tax penalty that is associated with privatizing wastewater treatment facilities. Some tax penalty remained, but the tax cost of privatization was sufficiently small in the early 1980s that federal tax laws did not create a substantial disincentive to privatization as they do today. This is a key point, both as it applies to the privatization of wastewater treatment and as it applies to privatization initiatives in general. Municipalities do not pay taxes, so if some activity is privatized and the private provider makes profits on which taxes are paid, there will be a tax penalty associated with privatization. Thus, in general, tax laws create an impediment to privatization, but one that was lessened in the early 1980s.

Wastewater Privatization Decisions

The decisions that many municipalities made to privatize their wastewater treatment operations in the early 1980s were driven by different factors than those normally cited to justify privatization. The desire to treat wastewater more efficiently was not a big motivating factor, although more effective treatment was needed to comply with the CWA. The desire to save money was not a key motivator either. Municipalities could hope to save money, but typically the facilities are paid for by user fees designed to cover costs, so the budgetary impacts for municipal governments would be limited. Municipal governments could not hope to save some money on wastewater treatment and use the proceeds elsewhere.

The real motivating factors were first, to produce new facilities to comply with the CWA, and second, to find a creative way of financing the facilities at a time when municipalities were finding it difficult to raise money. By privatizing, the private owner would raise the money and be responsible for repayment of debts and for covering operating costs. While it is true that residents of the affected municipalities would be paying the private supplier, the local government would not need to worry about raising money directly for the construction and operation of the facility and would thus have a limited financial exposure. The municipality would simply pay a fee for a service once the facility was online. The favorable federal tax climate made privatization more feasible than it otherwise would have been, creating an additional incentive to explore the privatization option.

COSTS OF PRIVATIZED VERSUS MUNICIPALLY OWNED FACILITIES

This chapter previously discussed the potential for cost savings through privatization, both because of the incentives involved in privatization in general, and because of factors specific to the design and construction of wastewater treatment facilities. Some have attempted to measure the potential savings in the financing of facilities through privatization by looking at actual cases (*e.g.*, Herbst and Katz, 1991). However, this section compares construction cost data for both privatized and municipally owned and operated facilities to generate evidence on the relative costs of privatized versus municipally owned facilities. Details of the analysis are presented in Holcombe (1991), which compares the construction costs of seven privatized facilities with 25 municipally owned and operated facilities.

All of the facilities were similar in that they were completed between 1982 and 1986, and the municipally owned and operated facilities covered the same range of sizes as were present in the privatized facilities. A regression model was used to identify the difference in cost attributable to privatization by using a binary variable to identify the privatized facilities. Other factors that could account for cost differences, such as the starting date, the length of construction, and output measures were included. The design flow of the facility in millions of gallons per day was a variable in the regression, as was the five-day biological oxygen demand of the treated water (BOD5), which is a measure of the quality of treated water.

After accounting for these differences, the regression results show that privatized facilities were substantially more expensive to build. This analysis uses a small sample of only seven privatized facilities. The results are sensitive to the inclusion of the highest-cost privatized facilities. If they are left out, there is no statistically significant difference in the cost of privatized versus municipally owned and operated facilities. If the extreme examples at both ends are dropped (the facilities with the highest and lowest residuals in the regression model), privatization remains the more costly option. The results are sensitive to a few observations, but looking at the data in any reasonable way, there is no evidence that privatization reduces cost, and the evidence suggests that privatization is more expensive than municipal ownership and operation. If these results can be accepted at face value, one would be justified in asking, in light of the substantial theoretical and empirical evidence supporting cost savings from privatization, why no cost savings are evident here.

The Privatization Agreements

A possible explanation for the lack of cost savings can be found by looking at the privatization agreements that have been drawn up for wastewater treatment facilities. This section discusses several actual privatization agreements that were drawn up in the early 1980s within the context of the advantages of privatization reviewed earlier in the chapter. Four of the facilities considered in this section overlap the privatized facilities that were analyzed in the previous section, but three of the agreements are for facilities not analyzed in the previous section. The reason there is not complete overlap is that in some cases privatization agreements were available but cost data were not, whereas in other cases the agreements could not be obtained when cost data were available.

A brief description of each of the seven agreements is found in Holcombe (1991), so this section gives an overview of their terms. While each of the privatization agreements is different, three factors stand out when the agreements are reviewed. First, there is no standardized contract. Second, in most cases the agreement allows the privatizer to pass through most or all of its costs to the local government with which it contracts. And third, some of the agreements contain buy-back provisions allowing the local government, under some conditions, to buy the facility and perform its own operations if the privatization agreement does not prove satisfactory. Each of these aspects will be considered after briefly reviewing the privatization contract terms.

The Privatization Contract

Typically, the private firm in a privatized wastewater treatment agreement owns and operates the treatment facility, but not the collector system that brings wastewater to the facility. This is not universally true, as in some cases the private owner also owns some or all of the collector system as well. Regardless of ownership, it would still be possible for the privatizer to deal directly with individual households and businesses in collecting payment for its services. In practice, however, this is rarely done.[6] Rather, the privatizer deals with a local government, and households and businesses pay the local government for wastewater treatment services in the same way as if the facility were municipally owned. The local government then pays the privatizer for the service of treating the wastewater.

The typical privatization contract, then, is an agreement between a private firm and a local government in which the firm agrees to treat wastewater shipped to it through the local government's collector system in exchange for a fee. The local government collects user charges from local households and businesses, and uses the revenue to pay the private contractor to treat the wastewater. The private firm is responsible for building, owning, and operating the facility. In most cases, if EPA regulations are violated, the private firm is liable for the damages, which means that the firm is responsible for paying any fines levied against the facility. With regard to general operations, each contract specifies how much wastewater may be shipped to the facility, and how much the firm will be paid for its treatment services.

The Lack of Standard Contract Provisions

In many areas of business, there are standard contracts, but all of the privatization agreements examined in this study differed significantly from the standard, even though on occasion the same firms owned the privatized facilities. There are several possible explanations for this. First, privatization of wastewater treatment is relatively rare, so there may have been little incentive to invest in the creation of a standardized contract. Nonetheless, once the first agreement is made, it should be possible to build upon it so that contracts would become more similar, especially to ensure that they contain certain standard provisions. In practice, this does not appear to be the case, based on the seven contracts examined here. Second, each of the seven contracts examined here was from a different state, and differences in state laws might have necessitated differences in the contracts. This explanation does not seem plausible, although it must be recognized that there would be little, if any, overlap between attorneys from the state in question with the attorneys who draw up the agreements. More likely, there was little or no communication among the governments involved, so each privatizing government was unaware of the negotiations the other state governments were undertaking. A third explanation might have to do with differences in local conditions that would necessitate different contracts but, again, it would seem that fundamentally standard contracts could be used with specific changes to reflect local conditions.

The firms that have undertaken privatization initiatives typically are firms experienced in building wastewater treatment facilities, and they have the detailed knowledge and experience to anticipate costs, potential problems, and profit potential from operating the facilities. Local governments are likely to be less experienced, meaning that firms might have a bargaining advantage over the governments with whom they are contracting.[7] This provides another explanation for the lack of standard privatization contracts: It enables the contracting firms to get more favorable terms. If contracts were standardized, local governments could benefit from each others' experiences in developing a continually improving standard agreement. The lack of standardized contracts seems to provide an advantage to the firms in the privatization agreement.

Pass-Through of Costs

In most of the privatization agreements examined, the contract allows the private contractor to pass through most or all of its costs to the municipality. The allowance of cost pass-throughs is an obvious detriment to efficient operation. The method by which costs are passed through varies from agreement to agreement, but in all cases where pass-throughs are allowed, the incentives for efficiency are significantly compromised.

One of the facilities, located in East Aurora, New York, is governed by a simple cost-plus-management-fee contract, which means that all costs of operation are passed through to the municipality. This agreement is a good example of an incentive structure that eliminates the most desirable attributes of privatization. Since all costs can be passed through, there is no incentive for the private owner to operate the facility efficiently, and there may even be perverse incentives if some of the costs provide revenues to the private owner. This type of agreement makes it clear why privatization does not necessarily result in lower costs.

In addition to the lack of cost control, a cost-plus contract like this will require considerable oversight by the municipality to ensure that costs are kept in line. One of the advantages of dealing with a private owner when a service is exchanged for a fee is that the government does not need much oversight, since inefficiencies represent costs to the privatizer. In this case, however, inefficiencies will not hurt the private owner, and might actually benefit the owner. Privatization would not be expected to fare better than municipal ownership when the privatization agreement contains no incentives for cost control and requires considerable oversight on the part of the municipality. As noted above, however, most privatization of wastewater treatment in the early 1980s was motivated by reasons other than the potential efficiency gains.

A privatized facility in Mt. Vernon, Illinois, provides for a fixed fee to the privatizer, but allows most costs to be passed through to the municipality. In addition, if the operation and maintenance costs turn out to be very low, the contract provides for a refund of some of the fee to the municipality. While these provisions make it look like incentives for efficiency were considered, ultimately, the incentives at Mt. Vernon are much like those in East Aurora, and the efficiency potential for privatization is compromised by a pass-through of costs to the municipality. Several other privatization agreements set fees according to a formula, again suggesting an attempt to derive efficient incentives, but these formulas essentially allow the costs of operation and maintenance to be passed through.

One interesting contract governs a privatized facility in Auburn, Alabama. The contract provides for a fixed fee for service, which creates the incentive for efficiency. The privatization agreement extends for 25 years, but has a provision that the fees are renegotiable every five years. The agreement does not provide guidelines for renegotiation, which creates a bilateral monopoly situation. Once built, the privatized facility is locked in to treating Auburn's wastewater, but at the same time

Auburn is relying on the facility for its wastewater treatment. While the fixed fee contract looks like it has taken into account the potential for efficiency incentives, the renegotiation provision effectively subverts those incentives.

Two of the privatization agreements contain efficiency incentives in that costs are not allowed to be passed through, and any cost savings will increase the profit of the privatizing firm. In addition, both of these facilities have other unique characteristics when compared to the other facilities examined in this study. Pine Bluff, Arkansas, has a fixed-fee contract with a privatizer, but the privatizer pumps wastewater to a city treatment plant from other city facilities, and does not treat wastewater. Another agreement, at Culpeper, Virginia, serves an industrial park but not the city itself. Cost-data for these facilities were not available, and the Pine Bluff facility is not directly comparable in any case, so neither of these facilities was included in the cost comparison discussed earlier in the chapter.

Clearly, it is possible to design privatization agreements that do not have cost pass-throughs, and that provide incentives to keep costs down and produce efficiently. However, efficiency incentives are not the norm in privatization of wastewater treatment. Thus, one should not be surprised that the previously reported empirical examination revealed no cost savings to privatization. Private ownership was present, but the incentives for efficient operation that normally accompany private ownership were not.

Provisions to Buy Back the Facility

All of the contracts examined would allow the local government to buy back the facility in the event of a breach of contract. In addition, two of the seven contracts have provisions for the local government to buy back the facility for any reason. The East Aurora, New York, agreement that allows complete cost pass-through also allows the municipality to buy back the facility for any reason on the fifth, eleventh, and each succeeding anniversary of the contract. This creates some incentive for the privatizer to act efficiently despite the cost pass-through provisions of the contract, because if the municipality is dissatisfied with the operation of the facility, it can buy it and operate it itself.

Another privatized facility at Gilder Creek, South Carolina, also allowed for the facility to be bought by the local government for any reason, every five years. In other words, if after five years the local government was not satisfied, they could buy the facility and operate it themselves. If they chose not to buy the facility, the contract would run for another five years, at which time the local government would have another opportunity to buy back the facility. When costs can be passed through, as in both Gilder Creek and East Aurora, the option for the local government to buy the facility for any reason provides a market incentive to the private operator. If the customer (local government) is not satisfied, the contract will be terminated.[8]

Every privatization agreement contains clauses that allow termination of the contract and purchase of the facility by the contracting government, but generally these apply only to cases of breach of contract or some other well-defined reason. This makes termination much harder than it would be if no reason needed to be given. Local governments have little incentive to pursue this course of action, because inefficient activity by the privatizer imposes costs on local residents, not on those in the government directly. If action were taken to terminate the agreement because of breach of contract, this would be tantamount to the local government admitting they had made a mistake to privatize in the first place. The decision to privatize is often controversial, and those who won the battle to try privatization would undoubtedly bear substantial political costs from admitting after the fact that the decision was a mistake. Thus, the breach of contract provision is likely to have little effect.

The inclusion of a provision to allow the local government to buy back the privatized facility for any reason is a good one, because it provides an additional incentive for efficiency, because it assures that the local government is a satisfied customer, and because in so doing it gives the private firm an incentive to be more responsive to the desires of its customers. No contract can cover all contingencies, but a buy-back provision allows for a withdrawal from the agreement without having to assign fault or blame.

GENERAL LESSONS FOR PRIVATIZATION

The examination of the privatization of wastewater treatment in the early 1980s provides some interesting insights into the costs and benefits of privatization. As noted earlier, cost reduction and efficiency gains were not the primary reasons why municipalities turned to privatization, despite the fact that these are the major selling points according to the advocates of privatization. In this case, privatization was desired because it could reduce the financial burden, and especially the up-front costs, of wastewater treatment. In the early 1980s, more stringent EPA regulations combined with recession-induced financial problems for local governments made privatization look attractive, and the federal tax climate in the early 1980s minimized the tax consequences of choosing privatization. The privatization agreements discussed above are consistent with the notion that efficiency gains were not the prime motivators for privatization, because the incentives in those agreements do not promote cost savings or more efficient operation.

One might question why the privatization agreements did not contain better incentives for efficient operation and cost reduction. The fact that other considerations loomed large in the privatization decision should not preclude writing the agreements to enhance efficiency as well. Three possible explanations, each of which might have some effect, will be considered here.

First, in negotiating a contract, the parties to the contract push for what is important to them. Since cost factors were secondary to the governments involved, but since cost pass-throughs would insulate the private contractors from potential financial losses, the cost pass-throughs might have been negotiated in exchange for performance assurances and other contract provisions. Considering the motivations of the parties, the benefits of the pass-throughs to the firms might have been greater than the perceived costs to the governments. This impression might be reinforced when it is recognized that municipal ownership also allows complete cost pass-throughs. If the primary motivation for privatization had been cost savings, cost pass-throughs would have been more unlikely to appear in the agreements, but as it was, they might have been a reasonable compromise in the context of the goals of the contracting parties.

Second, governments do not have much incentive to save costs in any event, since those costs are not borne by the government but are passed through again to the government's customers. A substantial literature on bureaucratic decision-making argues that governments do not have much of an incentive to minimize costs, and this is even truer in the case of wastewater treatment. In areas financed from a government's general fund, money saved in one area can be spent in another, but wastewater treatment is typically financed by a user charge, and cost savings will reduce user fees but will have no impact on the government's ability to spend in other areas. Thus, the incentives for cost savings are weak. This explanation for the lack of efficiency incentives complements the first one, since it helps explain why governments would not be motivated to privatize their wastewater treatment facilities merely to save costs. Other factors would have to be paramount.

Third, the privatizing firms may have had an advantage in bargaining with governments, for several reasons. The privatizing firms were likely to know the wastewater business better than the governments because the firms bidding on the privatization contracts tended to be firms with experience in constructing wastewater treatment facilities. In addition, the firms often worked previously with other governments in privatization efforts; this enabled them to have more direct experience with privatization than the governments they were negotiating with.

In examining the privatization of wastewater treatment in the early 1980s, several factors in the process of creating privatization agreements stand out. Cost savings were not a primary concern for privatizing, and governments have limited incentives to reduce costs in any case. In addition, the firms were likely to have a bargaining advantage with governments in developing the privatization agreements. All of these reasons contributed to the lack of incentives for efficiency in the privatization agreements, and some limited empirical evidence suggests that privatization was no less expensive than municipal ownership, and indeed may have been more expensive. If privatization is viewed as a way to reduce costs and enhance efficiency, then it did not work in wastewater treatment in the early 1980s. However, the real goals of privatizing governments were different: they were looking for a way to build expensive new facilities in difficult times without putting a large financial burden on the government. Privatization did achieve these goals.

Advocates of privatization often show unbridled enthusiasm for privatizing almost everything the government does, but this enthusiasm must be tempered by an understanding of the mechanism by which privatization produces its efficiency gains. It is not private ownership per se that enhances efficiency, but rather the incentives for reducing costs and enhancing the value of output that typically accompany private ownership. The big lesson for privatization in general that comes from examining the specific case of wastewater treatment is that privatization initiatives must be structured carefully so that incentives for efficient behavior accompany private ownership in order for efficiency gains to be realized.[9] Merely transferring an activity from the public sector to the private sector is not sufficient.

The case of wastewater treatment provides a specific example of a potentially widespread problem that could occur with any privatization initiative that combines public finance with private provision. In the privatization of wastewater treatment, the private provider is typically a monopoly seller operating on a cost-plus basis. Since all costs are being reimbursed, there is no incentive to reduce costs, and indeed if the provider's profit is calculated as a percentage of costs, there is actually an incentive to increase costs rather than reduce them. Private ownership can actually make the situation worse than public ownership in this case. Public sector employees have little incentive to reduce costs, but at least they do not have an incentive to increase costs. Private owners under the type of privatization agreements examined above could find it more profitable to increase their costs above the costs of municipal ownership.

Incentives for efficient production occur when producers are rewarded for lowering their costs or increasing the quality of their services. One of the most direct ways of creating these efficiency incentives for public sector output is to privatize. One must remember, however, that it is not private ownership itself that creates efficiency, but the incentives associated with market exchange. Privatization works in most cases because in most cases the arrangement by which the output is purchased is changed from simply covering the cost of an activity under government ownership to paying a market-based fixed price to a private contractor. The case of wastewater treatment illustrates that private ownership can exist without the accompanying market incentives, and can result in increased costs.

The advocates of privatization must make sure that enthusiasm for the concept of privatization does not push governments into ineffective privatization initiatives. Otherwise, poor results in a few cases can be used as an argument against privatization in the cases where it would make the most sense.

NOTES

[1]See Butler (1985) and Savas (1977, 1982) for discussions of the advantages of privatization. A model of competitive bidding for the right to be a monopoly supplier is provided by Demsetz (1968), who shows how competitive bidding can realize efficiency gains.

[2]See, for examples, Ahlbrandt (1973) on fire departments; Blair, Ginsberg and Vogel (1975) on health insurers; Bowman, Hakim, and Seidenstat (1992) on the justice system; Crain and Zardkoohi (1978) on water utilities; Davies (1971) on airlines; and Lindsay (1976) on health care. In addition, Kemp (1991) contains nine chapters reviewing the general experience of privatization in different areas, as well as six chapters reviewing potential problems that show that privatization does not always produce superior results.

[3]See Heilman and Johnson (1992, ch. 2) for a discussion of the history of federal involvement in local wastewater treatment.

[4]There is substantial literature on the incentives that push government to spend as much as possible in the present rather than consider the future. In frequently cited works, Downs (1957) notes that politicians have the incentive to be reelected, which limits their time horizon, and Niskanen (1971) describes the incentives for bureaucracies to maximize their budgets. Much subsequent literature has built on these seminal works, and is reviewed in Stevens (1993).

[5]The major tax factors affecting privatization during this time are discussed in Holcombe (1990).

[6]A facility in Culpeper, Virginia, serves only an industrial park, and deals directly with users rather than going through the municipal government. Still, the privatization agreement was signed by the municipality, and rates are set by agreement with the government, not with the users of the system.

[7]Niskanen (1971) describes in detail why these firms would be likely to have a bargaining advantage in drawing up the privatization agreements.

[8]The Gilder Creek facility actually has been purchased by the government, but for another reason. The contract had specified that if the firm were to sell the facility, the government would have the first right of purchase, and when the owner wanted to sell the facility to another party, the government exercised its option to buy.

[9]Perhaps private ownership is not even necessary for efficiency gains if public sector incentives can be structured in order to produce incentives for efficiency. This is the premise behind Osborne and Gaebler (1992). Examples of incentive restructuring would include voucher systems for public education endorsed by Osborne and Gaebler and first proposed by Friedman (1962).

REFERENCES

Ahlbrandt, Roger (1973). "Efficiency in the Provision of Fire Services." *Public Choice* 16, Fall, pp. 1–15.

Blair, Roger D., Ginsberg, Paul B., and Vogel, Ronald J. (1975). "Blue Cross-Blue Shield Administration Costs: A Study of Non-Profit Health Insurers." *Economic Inquiry* 13, June, pp. 237–251.

Bowman, Gary W., Hakim, Simon, and Seidenstat, Paul, eds. (1992). *Privatizing the United States Justice System: Policy, Adjudication, and Corrections Services from the Private Sector*. Jefferson, N.C.: McFarland & Company.

Butler, Stuart (1985). *Privatizing Federal Spending: A Strategy to Eliminate the Deficit*. New York: Universe Books.

Crain, W. Mark, and Zardkoohi, Asghar (1978). "A Test of the Property Rights Theory of the Firm: Water Utilities in the United States." *Journal of Law & Economics* 21, October, pp. 395–408.

Davies, David G. (1971). "The Efficiency of Public versus Private Firms: The Case of Australia's Two Airlines." *Journal of Law & Economics* 14, April.

Demsetz, Harold (1968). "Why Regulate Utilities?" *Journal of Law & Economics* 11.

Downs, Anthony (1957). *An Economic Theory of Democracy*. New York: Harper & Row.

Friedman, Milton (1962). *Capitalism and Freedom*. Chicago: University of Chicago Press.

Gold, Steven D. (1984). "State Tax Increases of 1983: Prelude to Another Tax Revolt?" *National Tax Journal* 37, March, pp. 9–22.

Heilman, John G., and Johnson, Gerald W. (1992). *The Politics and Economics of Privatization*. Tuscaloosa: University of Alabama Press.

Herbst, Douglas, and Katz, Lanny (1991). "Wastewater Treatment." In Roger L. Kemp, *Privatization: The Provision of Public Services from the Private Sector* Jefferson, N.C.: McFarland & Company.

Holcombe, Randall G. (1990). "The Tax Cost of Privatization." *Southern Economic Journal* 56, No. 3 January, pp. 732–742.

Holcombe, Randall G. (1991). "Privatization of Municipal Wastewater Treatment." *Public Budgeting and Finance* 11, No. 3, Fall.

Kemp, Roger L. (1991). *Privatization: The Provision of Public Services from the Private Sector.* Jefferson, N.C.: McFarland & Company.

Lindsay, Cotton M. (1976). "A Theory of Government Enterprise." *Journal of Political Economy* 84, October, pp. 1061–1078.

Matoon, Richard H., and Testa, William A. (1992). "State and Local Governments' Reaction to Recession." *Economic Perspectives.* Federal Reserve Bank of Chicago, March/April, pp. 19–27.

Niskanen, William A. (1971). *Bureaucracy and Representative Government.* Chicago: Aldine-Atherton.

Osborne, David, and Gaebler, Ted (1992). *Reinventing Government: How the Entrepreneurial Spirit is Transforming the Public Sector.* Reading, Mass.: Addison-Wesley.

Papke, James A. (1983). "The Response of State-Local Government Taxation to Fiscal Crisis." *National Tax Journal* 36, No. 3 September, pp. 401–405.

Savas, E.S. (1982). *Privatizing the Public Sector: How to Shrink Government.* Chatham, N.J.: Chatham House.

Savas, E. S. ed. (1977). *Alternatives for Delivering Public Services.* Boulder: Westview Press.

Stevens, Joe B. (1993). *The Economics of Collective Choice.* Boulder: Westview Press.

CHAPTER 17

The Role of Economic Regulation in Water and Wastewater Utility Privatization

Janice A. Beecher

The debate over public versus private ownership of water and wastewater utilities can be easily oversimplified. Ideologues tend to believe that ownership matters; analysts tend to believe that competition matters. Ownership matters, but not in ways that are entirely understood. Ownership mainly affects how utilities are organized and financed. Ownership also matters because people *believe* it matters. Most managers of public utilities feel strongly that "their" ownership form is superior by virtue of public purpose (in the case of municipalities) or private enterprise (in the case of investor-owned utilities). Competition might actually matter more in terms of actual utility performance, particularly as measured in economic efficiency terms.

Somewhat ironically, the current competition between public and private ownership forms can be more vigorous than the competition between utilities of like ownership. The perception of greener grass on the other side of the fence has led to a form of market contestability in the water industry. Today, an underperforming utility often is likely to have its ownership challenged. The fence might be more protective of publicly owned utilities (because of barriers to entry); municipalities also have a longer reach over the fence (because of powers of eminent domain).

One structural feature that matters at least as much as ownership and as much as competition is economic regulation. Privatization in the U.S. is guided more by pragmatism than by ideology, but a pragmatic plan for privatization should not ignore the actual and potential role of economic regulation. Many of the perceived performance differences between publicly and privately owned water utilities might well

---

Janice A. Beecher is owner and principal of Beecher Policy Research.

be attributed to regulation rather than to ownership or contestability. In fact, many of the hypotheses about differences in performance between publicly and privately owned water utilities probably can be attributed to economic regulation on the private side and absence of economic regulation on the public side. Some of the prevailing hypotheses favoring privatization are summarized in Table 17–1.

TWO MODELS OF PRIVATIZATION

In the United States, two very different worldviews of privatization are identifiable (see Table 17–2). Model I suggests that privatization should be accomplished through competitively bid contracts, with public ownership maintained and no need for state economic regulation. The contract-based model of competition has been somewhat more prevalent in the wastewater industry than in the water industry. Model II suggests that privatization should occur through investor-owned utilities that are subject to economic regulation (either the traditional model or a modified approach).

Importantly, under both models for privatizing the water industry, substantial monopoly power remains. The monopolistic nature of long-term private contracts often is ignored or understated. However, the longer the contract with a private firm, the more the "concession" takes on the characteristics of monopoly.

The potential for abuse of economic power on the part of the monopolist could be a rationale for economic regulation of contract-based privatization, but this usually is not the case. In fact, economic regulation by the state public utility commissions is fairly consistently viewed as a major barrier to Model I privatization. Some privatization advocates have complained openly that regulation constrains profitability while failing to provide the performance incentives of competitive markets (Haarmeyer, 1993).

TABLE 17–1. PREVAILING HYPOTHESES FAVORING PRIVATIZATION
In comparison to privately owned utilities, publicly owned utilities:

Experience more cost overruns during construction projects.	✓
Postpone necessary infrastructure improvements.	✓
Overcapitalize, even more so than regulated private utilities.	✓
Tend to overutilize debt, due to lower financing costs.	✓
Demonstrate higher production and operating costs.	✓
Are less efficient in their procurement and scheduling practices.	✓
Adopt cost-saving devices and innovation more slowly, if at all.	✓
Provide longer periods of tenure to utility managers.	✓
Have greater debt capacity and access to private capital.	✓
Are riskier and realize lower and more variable returns.	✓
Provide subsidies to or receive subsidies from other municipal operations.	✓
Support costs through development fees that keep rates artificially low.	✓
Set rates further from actual costs and with less regard for marginal costs.	✓
Favor voters over nonvoters, business over residential users, and organized over nonorganized political groups in ratemaking.	✓

TABLE 17–2. TWO MODELS OF PRIVATIZATION

Model	Implications
I. Municipal ownership with contracts with private providers	Intense competition for contracts in lucrative markets. Potential for monopolistic behavior under long-term contracts. Limited use of private capital for major projects. Municipal ownership and the competitive market "substitute" for regulation.
II. Expansion of the investor-owned water industry	Monopolistic markets with limited or structured competition under some circumstances. More use of private capital for major projects. Expansion of state regulatory jurisdiction. Potential use of alternative regulatory models.

According to one observer, "A very restrictive feature in many states is the inability to remove the privatization transaction from control of the public service commission of that state. As a result, rates of return are strictly monitored. To enhance privatization, provisions need to be made in state laws to buffer profitability from regulation by the public service commission. Profitability would best be regulated under the terms of the privatization service contract" (Raftelis, 1989).

For privatizers, according to another observer, "the rate of return is the key consideration" (R.V. Anderson Associates, 1990). Regulation of profits clearly seems to run contrary to this motivation. Control is a central issue in the water industry. The water industry is more local in character than other utility industries because of the desire of communities to control growth and development through the control of water service (even though regional utilities might lower costs through economies of scale). Local communities also are reluctant to surrender control over rates and finances to state regulatory bodies. Considerable anecdotal evidence suggests that many of the water and wastewater privatization agreements recently initiated or now underway were structured to avoid economic regulation by the states, to the liking of both privatizers and the cities in which they operate.

While seen as a barrier in the context of Model I privatization, regulation can be seen as an advantage in Model II privatization when investor-owned utilities seek to acquire municipal water utilities. California-American, for example, cited economic regulation as an advantage in its bid to run the water system in the City of Petaluma (Beecher, et al., 1995). Opponents, however, believed that regulation would increase costs and that the community should not surrender control over the destiny of its water system to the state. The city wanted the successful bidder to apply to the California Public Utilities Commission for an exemption from regulation.

Little doubt exists about the potential for abuse of economic power when transferring utility ownership or operational responsibilities from the public to the private sector. Yet the role of economic regulation is rarely addressed in the privatization literature; when the subject of regulation is approached, it usually is treated negatively. Regulation constitutes the continued presence of government oversight and seems antithetical to the goals of divestiture and competitive markets for contracts. In reality, regulatory policies can be used to encourage or discourage privatization activity.

In addition, alternative regulatory models can be used to encourage competition and efficiency while protecting ratepayers.

Some forms of competition are being introduced in the water sector, despite the fact that it is widely understood as the most monopolistic of the utility sectors. As noted above, there is a detectable degree of competition between the public and private ownership forms. Both types of utilities are trying to outdo each other in terms of improving operational efficiency. However, competition is not sufficiently robust to replace some form of oversight via public ownership or economic regulation.

A number of other reasons also argue for continued oversight of the water industry. First, unlike the telecommunications and energy industries, water supply is a *rising-cost* industry. Since the mid-1980s, water prices have increased at a pace faster than the inflation rate. Substantial infrastructure investments will be needed in the water industry over the coming decades. Second, most aspects of water supply (including production, transmission, and distribution) are likely to remain monopolistic and often vertically integrated. Water and wastewater services are highly capital-intensive and demonstrate substantial economies of scale. Third, the strong environmental, public health, and public safety (fire protection) dimensions of the water industry argue for continued regulation, including economic regulation, in the public interest.

ECONOMIC REGULATION

The U.S. water industry is highly pluralistic. More than 55,000 water utilities of various size and ownership characteristics provide water to American communities. However, the uniformity found in the regulation of drinking water quality is not matched by uniformity in price regulation. Most private or investor-owned water utilities are subject to the jurisdiction of the state public utility commissions and the ratebase/rate-of-return form of regulation traditionally imposed on utility monopolies. Public utilities are regulated because they have monopoly power over captive customers. Regulation is viewed as a substitute for competition on the one hand and public ownership on the other.

Commission jurisdiction (over different kinds of systems) and the scope of commission authority (over different kinds of activities) vary substantially from state to state.

Jurisdiction

The states, the District of Columbia, and the U.S. territories all are engaged in the economic regulation of public utilities that provide electric, natural gas, telecommunications, and water services. As of 1995, state commissions regulated approximately 8,537 water utilities, while 28 commissions regulated 2,152 wastewater utilities (see Table 17–3). Water and wastewater utilities are not subject to commission jurisdiction in Georgia, Michigan, Minnesota, North Dakota, South Dakota, or Washington, D.C., primarily because of the limited presence of sizable investor-owned water and wastewater utilities in these areas.[1]

TABLE 17–3. COMMISSION JURISDICTION FOR WATER AND WASTEWATER UTILITIES (1995)

Type of Utility	Water Industry		Wastewater Industry	
	Number of Commissions	Number of Utilities	Number of Commissions	Number of Utilities
Investor-owned	46	4,095	28	1,233
Municipally owned	11	1,547	6	649
Water districts	7	1,300	4	205
Cooperatives	4	1,436	2	50
Homeowners' associations	6	85	1	0
Nonprofits	1	73	1	15
Other	1	1	0	0
Totals		**8,537**		**2,152**

Source: Beecher, et al. (1995). Michigan ceased regulating water utilities following this survey, reducing the number of commissions to 45.

There are many variations in regulatory jurisdiction. Most U.S. water utilities, particularly larger utilities, are publicly owned and therefore not state regulated. When it does apply, commission jurisdiction for publicly owned water utilities tends to be limited to certain issues or triggered only by certain activities. For example, publicly owned systems in Pennsylvania and other states are not regulated unless the city chooses to sell water outside of municipal boundaries. A few states (such as Maine, Montana, New Jersey, Rhode Island, West Virginia, and Wisconsin) have substantial jurisdiction for publicly owned systems. Municipalities can opt out of regulation in Indiana, counties can opt to regulate water systems in Florida, and investor-owned utilities in Kansas are exempt from regulation if they serve only one municipality.

Authority

The commissions do not exercise uniform authority over all of the water utilities under their jurisdiction. In general, investor-owned utilities are the most comprehensively regulated, although specific areas of oversight vary from state to state.

For the different types of water or wastewater utilities under their jurisdiction, the commissions *might* perform the following regulatory functions:

- Determine whether or not a provider is subject to economic regulation.
- Issue certificates of need or "public convenience and necessity."
- Issue certificates for major construction projects.
- Approve service territory boundaries and changes in boundaries.
- Approve financial issuances and loans.
- Approve mergers, acquisitions, and other ownership changes.
- Audit financial accounts and management practices.
- Review utility management prudence.
- Evaluate long-term resource management plans.
- Review conservation and drought management practices.

- Impose metering, billing, and disconnection practices.
- Approve revenue requirements, cost allocation, and rate structures.
- Determine an allowed rate of return.
- Review record-keeping and reporting.
- Resolve consumer complaints.

Commission authority can affect the process of privatization in any number of ways. Establishing an investor-owned utility usually requires certification by the state public utilities commission (in addition to the approval of drinking water quality regulators). Regulatory approval almost always is required for transactions involving a transfer of a regulated utility's assets, including the financial arrangements associated with the transfer and the resulting effect on rates. In some states, there is an exception when the acquisition involves a holding company. Perhaps, most importantly, privately owned water utilities are subject to revenue requirements regulation, by which allowable rates of return are determined.

Ambiguity, Exemption, and Circumvention

A key issue raised by the privatization movement is whether and when the involvement of the private sector in providing utility services constitutes the formation of a jurisdictional utility entity. Statutory definitions of public utilities that once seemed unambiguous can be a source of uncertainty in the context of privatization. For example, Illinois law defines a public utility as follows:

> Sec. 3-105. "Public utility" means and includes, except where otherwise expressly provided in this Section, every corporation, company, association, joint stock company or association, firm, partnership or individual, their lessees, trustees, or receivers appointed by any court whatsoever that owns, controls, operates or manages, with this State, directly or indirectly, for public use, any plant equipment or property used or to be used for or in connection with, or owns or controls any franchise, license, permit or right to engage in:
>
> a. the production, storage, transmission, sale, delivery or furnishing of heat, cold, power, electricity, water, or light, except when used solely for communications purposes;
> b. the disposal of sewerage; or
> c. the conveyance of oil or gas by pipeline.
>
> "Public utility" does not include, however:
>
> 1. public utilities that are owned and operated by any political subdivision . . .
> 2. water companies that are purely mutual concerns . . .

Dealing directly with the issue of jurisdiction, statutes in some states explicitly exempt privatization arrangements from commission authority as long as specified conditions are met (for an example, see Supplement to this chapter). Such exemptions may require a demonstration that local authority is sufficient to protect utility customers from abuses of monopoly power. As noted, privatizers in California must apply to the public utility commission for an exemption from regulation (see Supplement). Otherwise, they may be required to secure a certificate of convenience and necessity from the commission and operate as a regulated public utility. The application to the commission must demonstrate that the contingent application complies with the law.

In addition, the commission requires that the local agency engaged in privatization have and maintain certain authorities and powers. Essentially, the state seeks assurances that local regulatory oversight and enforcement will protect the customers of privatized systems from potential monopoly abuse. Specifically, the commission requires that the local California agency have:

- the exclusive authority to establish all rates and rate changes charged to the public;
- approval powers over any proposal of the privatizer to provide new, additional or alternate service to any other public or private entity or to change the service fee paid to the privatizer by the local agency;
- approval powers over the original design and construction of the project, including any changes in design, alterations, or additions to the project;
- approval powers over any changes in ownership of the party or parties subject to the contingent agreement;
- the authority to impose fines and penalties for noncompliance with any provision of the executed privatization agreement, or for failure to provide the service within the time period agreed to in the agreement;
- the authority to ensure that the project is adequately maintained;
- adequate opportunity to monitor compliance with the agreement and to ensure that the project is operated to meet any applicable federal or state water quality standards or other applicable laws; and
- adequate opportunity to amend the agreement in the event of unforeseen circumstances or contingencies, such as flood, earthquake, fire, or other natural disasters or federal tax law changes.

These provisions reflect a belief that comprehensive and specific local authority can substitute for state economic regulation. In this respect, public partnerships can be "regulated" at the local level, assuming that local governments have the authority, resources, and expertise to carry out this function.

Not every state has explicit statutes or rules governing the privatization process. Few states probably are as explicit as California about the parameters of regulatory exemption. Nonetheless, the basic principles underlying the California exemption procedure may be reflected in the regulatory policies of many states. In other words, potential participants in privatization agreements should be aware of the possibility that their activities may come close to statutory and administrative definitions of utility activities subject to commission regulation. However, legislative proposals recently initiated in some states explicitly provide for regulatory exemptions and thus for a more favorable climate for privatization. Exemption implies a procedure for determining whether or not private involvement in utility operations constitutes the creation of a utility that should be subject to economic regulation.

The strong desire to be exempt from regulation can lead to the deliberate structuring of privatization arrangements to circumvent the regulatory process. Circumvention may become a cause of concern for policymakers, particularly if competition and local governmental authority are insufficient to check the potential abuse of monopoly power. In some jurisdictions, public authorities are being created for the express purpose of establishing public ownership and oversight while immediately delegating operational responsibility to the private sector. The newly created system is not commission-regulated unless state statutes specifically provide for this authority.

Another means of circumvention occurs when municipalities purchase the privately owned systems serving them and immediately contract with a private firm for operations. It is conceivable that a private contractor, motivated by the potential to earn unregulated profits, could persuade local officials to follow this course. The creation of holding companies and unregulated affiliates, as well as certain wholesale arrangements, also may constitute a form of circumvention. For their part, local officials may be tempted to circumvent regulation for the purpose of gaining local control over the utility. Although the point need not be overstated, policymakers cannot afford to be naive about the political and profit motives behind regulatory circumvention.

HOW PRIVATIZATION AFFECTS REGULATION

So far, privatization has had a relatively limited impact on the jurisdiction, authority, and workload of the state public utility commissions. As noted, the bulk of the privatization activity in the United States has been in the area of operation and maintenance contracts. Unless a regulated utility is involved, and the contract is substantial, these agreements are largely transparent to the commissions. Even when a regulated utility is involved, many of the commissions have limited monitoring capability in this area. Commissions might not even be aware of the myriad of privatization activities in their states.

Privatization Activities of Regulated Utilities

Most of the major investor-owned water utilities in the United States are actively engaged in privatization and regionalization through acquisitions of nearby systems. Stringent environmental and public health standards such as those implemented under the Safe Drinking Water Act have provided—despite industry protestations—private water utilities with considerable opportunities for corporate investment and growth. Contiguous systems are especially attractive candidates for acquisition. Some of the state public utility commissions have explicitly encouraged acquisitions by investor-owned utilities as a means of addressing the needs of small utility systems and improving overall industry economies.[2]

The nation's largest network of investor-owned water utilities, American Water Works Company, has demonstrated an aggressive acquisitions strategy. American plans to acquire municipal and other systems that are contiguous to its existing systems. New Jersey-American, West Virginia-American, and Pennsylvania-American are very active in acquiring and privatizing nearby water systems. Since 1983, New Jersey-American has acquired approximately 12 municipal systems. Several large acquisitions by American are pending in commission proceedings throughout the United States. Pennsylvania-American is aggressively pursuing privatization activities throughout the state and has purchased more than 20 systems since the mid-1980s; as of early 1995, the company had plans to close on 15 water and wastewater systems and had several other pending acquisitions.[3] West Virginia-American is involved in the formation of a large regional water supply system.

Several of the larger investor-owned utilities also are actively involved in providing operations and maintenance services, usually through corporate affiliates. American's affiliate, the American Commonwealth Management Service Company, actively markets operations and maintenance services, particularly to water systems having trouble complying with environmental mandates. The private consortium operating the municipal wastewater facility in Indianapolis includes an affiliate of the Indianapolis Water Company. Some investor-owned utilities have formed partnerships with European firms to compete for contracts. All of these activities present new challenges to state utility regulators.

Although 12 state commissions have some authority over publicly owned systems, only a few have comprehensive authority. Only a few specific cases of regulatory oversight were provided. In New Jersey, for example, the board approved an operations and maintenance contract issued by the City of Hoboken after regulators were satisfied with the proposed local oversight process.[4] (For New Jersey policy, see the Supplement to this chapter.) The West Virginia commission must approve both asset sales and contracting.[5] Maine reports having some oversight over sales of assets, but none over contracting.[6] Montana reports that it has the authority to approve privatization agreements, but that legislation under consideration would limit that authority. In most states with jurisdiction over publicly owned systems, the scope of authority is very limited. Pennsylvania regulators see privatization agreements after the fact. Similarly, Texas regulators essentially have no authority over municipal privatization agreements.

Regulatory Implications

Table 17–4 summarizes the regulatory implications of ownership transfers and Table 17–5 summarizes the regulatory implications of service contracts based on the

TABLE 17–4. REGULATORY IMPLICATIONS OF OWNERSHIP TRANSFERS

	Destination	
Origination	**To public ownership**	**To private ownership**
From public ownership	Generally not regulated. In some cases, providing utility service outside of municipal boundaries may be regulated by the state commissions.	A certificate of public convenience and necessity may be required, particularly if the acquisition is made by a newly formed private utility. The transfer of assets and financial arrangements probably require approval as well. Acquisition adjustments require a determination of ratemaking treatment.
From private ownership	The transfer of assets and ownership probably requires regulatory approval. Regulators also may want assurances that the transfer is in the public interest. In most cases regulation will not prove to be a significant barrier to the transfer.	Regulatory approval may be required for both utilities in the transaction. The transfer of assets and ownership probably require regulatory approval. It may be necessary to modify the acquiring utility's certificate of public convenience and necessity. Transfers at the holding company level may be exempt from oversight. Acquisition adjustments require a determination of ratemaking treatment.

TABLE 17–5. REGULATORY IMPLICATIONS OF SERVICE CONTRACTS

	Service Recipient	
Service Provider	**Publicly owned utility**	**Privately owned utility**
Publicly owned utility	Generally not regulated. In some jurisdictions, utility service outside of municipal boundaries may be commission-regulated.	The contract may be reviewed for prudence and financial terms.
Privately owned utility	Subsidiary activities may be regulated to shield captive customers from risks associated with diversification. Prudence of contracts may be reviewed.	Regulatory approval may be required for both utilities in the transaction. The contract may be reviewed for prudence and financial terms. Subsidiary activities may be regulated to shield captive customers from risks associated with diversification.
Service vendor	Generally not regulated, particularly if contractual procedures and local government authority provides sufficient protection. In some cases, the vendor can appear to behave as a public utility entity, which could trigger regulatory intervention.	The contract may be reviewed for prudence and financial terms. Regulators may want to review contractual terms in relation to the obligation to serve, service reliability, and service quality.

current configuration of regulatory jurisdiction and authority in many states. Commission authority probably is best defined when a private or investor-owned utility is directly involved in the transaction, although some states may address other privatization agreements in explicit terms as well. Transferring the assets of a regulated utility often triggers regulatory oversight. In particular, a new or modified certificate of public convenience may be required. Operations contracts may be examined by regulators to varying degrees.

However, broad regulatory powers to review financial transactions, management prudence, and ratemaking may bring some privatization activities under commission scrutiny. New entrants into privatization may be unaware of the potential impact of regulation on various activities. Given the relatively rapid evolution of privatization policy, current state statutes and commission rules should be consulted by potential privatizers. Over time, states will probably continue to fine-tune regulatory jurisdiction and authority in this area.

Changing Roles

In a 1995 survey, regulatory commission staff members expressed a number of concerns about what they saw as emerging issues related to privatization of water and wastewater utilities (Beecher, et al., 1995). They are still valid concerns today.

- Establishing the value of publicly owned assets in the context of acquisitions because of the unavailability of data or incompatible accounting methods.

- Appropriate regulatory and ratemaking treatment of acquisition adjustments.
- Limits on regulatory authority's ability to review purchases and contracts before implementation.
- Delegating total operating responsibility to private contractors or utility subsidiaries.
- Ensuring service quality, reliability, and sound utility management under third-party agreements.
- Protecting ratepayers of regulated utilities from supporting the costs (or assuming the risks) associated with the contract ventures of the utility or its affiliates.
- The creation of unregulated public water authorities that assign substantial responsibility to the private sector.
- The capacity of cities or other public entities to provide adequate economic oversight over monopolistic utility operations.
- The potential burden of privatization on commission staff resources, especially for reconstructing books and records, and processing various applications.

Commission staff members in some states viewed privatization with considerable favor because acquisitions by larger utility systems are considered a potential solution for struggling small systems. However, not all commission staff members view privatization as the ideal solution to small-system viability problems (Curry, 1994). Some regulators also expressed the concern that "privatization looks too good to be true," and that more systematic information is needed about the actual benefits and costs of privatization.

With increasing privatization activity involving investor-owned utilities and other market entrants, commission resources may be affected. Commissions could experience more water and wastewater cases, more frequent cases, and more complex cases (not to mention generic rulemakings, task forces, and special studies). Time spent on regulating the water and wastewater industries cannot be spent on other regulatory matters. Some observers contend, however, that with restructuring and regulatory changes occurring in the electricity, natural gas, and especially telecommunications industries, water utility regulation may rise in status and priority at the state public utility commissions. With further privatization, it is conceivable that regulating water and wastewater monopolies could become a more salient state function. Staff resources for water and wastewater regulation traditionally have been much lower than for other areas of regulation, but over time, this allocation may by necessity change.

HOW REGULATION AFFECTS PRIVATIZATION

Despite the prevailing view of economic regulation as a barrier to privatization, specific regulatory policies can encourage privatization activity, discourage privatization activity, or have no direct or apparent effect. In other words, commission policies vary in the degree to which they increase or decrease regulatory risk for jurisdictional utilities (or would-be privatizers).

Several of the state commissions have conducted targeted studies of water utility regulation and identified specific policies that could play a role in structural and regulatory reform. To foster restructuring and privatization, a 1993 report prepared by the New Jersey Board of Regulatory Commissioners recommended consideration of such incentives as acquisition adjustments, ratemaking and tariff reforms, and forward-looking recognition of capital investments (New Jersey Board of Regulatory Commissioners, 1993). The recommendations also emphasize, however, the need for flexibility, the importance of analyzing impacts on existing and new customers, and the use of appropriate incentives on a case-by-case basis.

A comparative summary of commission policies related to privatization is presented in Table 17–6. These policies are organized into three general areas: regulatory meth-

TABLE 17–6. REGULATORY POLICIES THAT ENCOURAGE OR DISCOURAGE PRIVATIZATION

Policies That Can Discourage Privatization	Policies That Can Encourage Privatization
Regulatory Methods	
• Strict adherence to original costs in valuing utility assets. • Traditional ratemaking methods, including an emphasis on historical costs. • Spatially determined rates for service. • Conventional limits on profitability. • Little or no consideration of changing risk profiles.	• Acquisition adjustments and other acquisition incentives. • Modified ratemaking, including consideration of future costs and cost adjustment mechanisms. • Single-tariff (or "equalized") rates across a regional territory. • Profit-related incentives, including rate-of-return incentives and profit sharing. • Consideration of changing risk profiles in ratemaking.
Regulatory Procedures	
• Complex procedures for regulatory exemption. • Separate regulatory review and approval of asset purchases and ratemaking treatment. • Comprehensive and lengthy ratemaking procedures. • *Ex post* prudence reviews of utility investments. • Lack of coordination with other regulatory agencies.	• Streamlined procedures for regulatory exemption. • Consolidated approval of asset purchases and ratemaking treatment. • Nontraditional and simplified procedures, including alternative dispute resolution. • *Ex ante* approval of utility investments. • Coordination with other regulatory agencies, including one-stop shopping.
Regulatory Scope	
• Extensive review of asset transfers and ownership changes. • Regulatory oversight of affiliates and transactions. • No authority to provide incentives for market-based utility management. • Different regulatory authority for different kinds of jurisdictional utilities. • Comprehensive regulation in all cases.	• Expedited review of asset transfers and ownership changes. • No regulatory oversight of affiliates and affiliate transactions. • Regulatory authority to provide incentives for market-based utility management, including outsourcing through competitive bidding. • Comparable regulatory authority for all jurisdictional utilities. • Limited or more flexible regulation under some circumstances.

ods, regulatory procedures, and regulatory scope. Some policies may be implemented at the discretion of the commissions; others may require legislative authorization. Each approach has advantages and disadvantages. No recommendations are made here, either implicitly or explicitly, for any specific regulatory policy or approach.

Regulatory Methods

Traditional methods of costing and ratemaking in a regulatory context can present a barrier to privatization activity. A leading issue related to acquisitions is the difficulty associated with arriving at an acceptable purchase price for assets. Many states value utility assets at their original cost at the time they were devoted to public use, less accumulated depreciation. In Florida, for example, state laws inhibit the commission's ability to calculate a rate of return for some acquired utility assets due to the lack of historical investment data. Under these circumstances, constructing a ratebase and a rate of return for privatized utilities can be very difficult. Assets of many older utilities may be fully depreciated. As a result, these systems may not be attractive to potential buyers. Buying a system with little or no value in assets would not add to the ratebase of the acquiring utility, thereby limiting the utility's chance to earn a return on its investment. Acquisition adjustments sometimes are recommended to remedy this dilemma and provide an additional incentive for acquisitions by investor-owned water systems.[7]

In 1994, the New York Public Service Commission issued a policy statement regarding the acquisition of small water companies, which was intended to promote mergers and consolidation in the water supply industry.[8] These incentives may be particularly useful in prodding the acquisition of an otherwise undesirable system by a larger investor-owned utility.

States that provide for "above-the-line" ratemaking treatment allocate the cost of the adjustment to ratepayers; "below-the-line" treatment means that shareholders pay for the adjustment. Some important policy issues are raised by acquisition adjustments, which are perceived as contrary to traditional ratemaking practices. In addition, they may introduce unintended incentives. For example, the presumption that an adjustment will be approved by regulators could dampen the acquiring utility's drive to buy a system at the lowest possible price.[9] In fact, adjustments can cause an artificial inflation of the purchase price. In addition to acquisition adjustments, the state commissions also can provide other incentives for beneficial acquisitions, such as a bonus on the rate of return.

Other costing and ratemaking methods can be designed either to encourage or discourage privatization. The traditional emphasis on historical costs in general ratemaking, for example, can be a deterrent, while forward-looking ratemaking is preferred by utilities. Another example would be the use of a future test year in determining revenue requirements, which can reduce the shortfall between costs and revenues. Regulated utilities also tend to favor the use of more-or-less automatic cost-adjustment mechanisms or "pass throughs," although these methods can reduce the incentive for utilities to control certain costs. For some multisystem utilities, another preferred approach is single-tariff pricing (that is, a pricing structure that provides for cost averaging across multiple systems rather than spatially determined rates). Averaging mitigates against rate shock for customers and revenue instability for utilities, and is relatively simple to administer. Single-tariff pricing can encourage economically efficient industry consolidation and regionalization through privatization.

How regulators treat profitability is a central issue in privatization. Nontraditional alternatives to ratebase/rate-of-return regulation have been proposed to provide utilities with efficiency incentives by enhancing earnings potential. A possible option for publicly and privately owned utilities is to design privatization agreements that include profit-sharing provisions. Finally, the emergence of privatization and competition will affect industry risk profiles. Investor-owned utilities will probably want regulators to take note of these trends in determining allowed rates of return.

Regulatory Procedures

Various regulatory procedures can present a barrier to privatization, including the initial process to determine whether or not an arrangement constitutes the creation of a jurisdictional public utility. Commissions that must make this determination can adopt expedited procedures to grant or deny exemptions from regulation. An expedited exemption from regulation will greatly reduce uncertainty to the parties engaged in the privatization agreement.

Commissions also can facilitate privatization by combining certain regulatory determinations into a single proceeding. In particular, acquisitions by investor-owned utilities may involve one proceeding to approve the transfer of assets and a modification of the purchasing utility's certificate, and a later proceeding to determine the associated ratemaking treatment of costs. This can present a dilemma in the use of acquisition adjustments. If utilities assume that an adjustment is forthcoming, negotiating strategies may be affected. If the adjustment is disallowed or does not meet expectations, the entire transaction may be at risk. Commissions can reduce these uncertainties through consolidated proceedings.

Regulation is perceived as a barrier to privatization in part because the ratemaking process can be lengthy and cumbersome. Regulatory lag, or the time between the initiation of a proceeding and its resolution, can be costly to utilities as well as inconvenient. The process can be particularly difficult for small water and wastewater systems with limited resources. Regulation also can be very adversarial. Some nontraditional methods have emerged to address these issues. Many states use simplified procedures for smaller systems, including simplified filing and reporting requirements. Another modern tool of administrative law is alternative dispute resolution, by which parties can negotiate a resolution of their differences. One procedure that can be particularly frustrating to utility managers is an after-the-fact review of management prudence, which can lead to disallowances of certain expenditures. This can occur for major construction projects, as well as other utility capital and operating activities. One method for reducing the need for *ex post* review is to institute an *ex ante* review of utility investment plans. Electric utilities, for example, have lobbied for preapproval so that the regulatory risks associated with capacity additions can be reduced.

Finally, another potential problem with the regulatory process is the lack of coordination among the various state agencies responsible for quantity, quality, and economic regulation. Regulated utilities may be required to comply with multiple and overlapping requirements for certification, permits, planning, and other concerns. A coordinated system of state regulation could facilitate private involvement in the water and wastewater sectors by simplifying regulatory requirements and reducing regulatory risks. For multistate regional systems, particularly in the northeastern area of the nation, coordination among state regulators could be beneficial as well.

Regulatory Scope

The scope of regulation also can affect the level of interest in privatization. Most of the state commissions that regulate water and wastewater utilities have authority to approve transfers of utility assets or changes in ownership. Generally, commissions want assurances that such changes will not be contrary to ratepayer interests. Expedited regulatory processes can encourage the development of markets for utility assets.

A significant amount of privatization activity involves corporate affiliates of privately owned utilities. Corporate affiliations in the water supply, wastewater treatment, and stormwater management fields are becoming increasingly complex. Based on experience in electricity regulation, many of the state commissions have substantial authority to regulate holding companies and affiliate relationships, including: general power and authority regarding reorganization; specific authority over holding company formation; specific authority over setting up subsidiaries; financial input regarding reorganizations; access to books and records; and authority to review, investigate, or approve affiliate contracts and agreements (Rosenthal, 1994). Commission authority to review affiliate relationships is based on the importance of protecting captive ratepayers from risks associated with unregulated activities. This authority may present a deterrent to privatization ventures by investor-owned utilities.

Commissions have differing authority and inclination to provide incentives for specific utility management practices. Most commissions have some authority to review utility management prudence. In theory, regulators can use this authority to encourage utilities to seek out least-cost practices. Under some circumstances, outsourcing certain functions through competitive bidding might be desirable. At the same time, commissions may want to reconsider some policies that can provide disincentives for efficiency. For example, automatic cost adjustments (especially in a rising-cost environment) can shield utilities from risks that they otherwise would have to assume in a competitive marketplace. Regulated and unregulated firms without the ability to simply "pass through" costs probably are more disciplined about cost control.

The scope of commission authority can be adjusted over time to meet changing needs. First, the scope of commission authority can be adjusted to provide comparable oversight to all jurisdictional utilities. Specifically, it has been argued that publicly and privately owned water utilities should be similarly regulated, at least in the states that have jurisdiction over both types of utilities. Some believe that comparable regulatory treatment of utilities with different ownership forms would help create a level playing field for competition among utilities. Second, commissions might want to introduce certain limited or more flexible regulatory approaches, such as performance benchmarking. As competitive markets mature, regulatory requirements could be relaxed. Alternatives to traditional ratebase/rate-of-return regulation might be considered as well.

POTENTIAL BENEFITS OF REGULATION

The prevailing view of regulation as a barrier to privatization, and the considerable energy spent on avoiding the regulatory process, denies the potential benefits of regulation in the expansion of private involvement in the water and wastewater utility industries. Economic regulation historically has been well justified for capital-intensive industries with rising infrastructure costs and limited opportunities for competition. Regulation can have a very stabilizing effect on utility finances. Regulation provides some assurances that legitimate utility revenue requirements will be met. Regulation does not guarantee a profit, but it does set an allowable rate of return that many jurisdictional utilities are able to achieve. Regulation requires utilities to accept a degree of regulatory risk, but shields them from other forms of risk, including risks associated with municipal politics at the local level and potentially ruinous competition at the global level.

Economic regulation by the states offers certain advantages over alternative methods of societal control, including public ownership under some circumstances. The state commissions have resources and expertise focused exclusively on economic regulation of utility services. Most cities do not have comparable expertise and resources. Commissions demonstrate economies of scale and scope in regulation when compared to decentralized oversight by local governments. While their expertise is grounded in traditional ratebase/rate-of-return methods, the commissions also have responded to the economic and technological changes affecting the utility industries, including emerging competition. The commissions are well equipped to consider complex issues, such as economic efficiency and ratepayer equity. State regulation generally is less parochial, less political, and less driven by expedience than local regulation.[10]

Without significant safeguards, local contracting and oversight can be prone to corruptive influences. It also may be somewhat easier for the states to make politically unpopular decisions. Commission regulation can be more flexible and less arbitrary than regulation imposed through legislative or judicial means. Commission regulation also can be more effective in protecting consumers than antitrust law or other public policy options. New roles for regulatory agencies, such as dispute resolution for contractual agreements, might prove very beneficial as well. In general, state regulation can be used to further various state policy goals, such as efficiency pricing, integrated resource planning, and universal service.

It can be argued that privatization and economic regulation share the common goal of establishing managerially sound and financially viable water and wastewater systems. Strategic use of acquisition and other regulatory incentives already has had a considerable influence on the restructuring of the water industry. Modern public utility regulation ideally encourages utilities to meet least-cost and efficiency goals, and use marketlike methods in the process (for example, competitive bidding). The regulatory process provides policymakers with various tools and incentives for guiding utility performance. In fact, it may be easier to reward investor-owned utilities than unregulated utilities for implementing efficiency and other desirable measures. In theory, regulation can be an agent of privatization by providing positive incentives for the expansion of investor-owned systems. Moreover, regulation can provide a level playing field for emerging markets (or "structured competition," as discussed later on). In sum, appropriate and effective regulation can provide appropriate incentives to privatize while also protecting customers from potential abuses of persistent monopoly power in the water industry.

A Global Perspective

The American dilemma about privatization and regulation may be somewhat unique. In the global experience, the divestiture of public utility monopolies generally is not envisioned without a well-designed system of economic regulation. World Bank analysts strongly encourage close attention to the capacity of governments to regulate when nations choose to privatize state-owned enterprises. The capacity to regulate may vary across governmental systems, and within systems in terms of the strength of different governing institutions (for example, the executive branch versus the judiciary).

In the privatization movements of Europe and Latin America, the formation and legitimization of economic regulation appears to be a high priority during the course of utility privatization. However, the traditional U.S. model of utility regulation often is not the model of choice. In Great Britain, for example, privatization of the nation's major utility industries (energy, telecommunications, and water) was accompanied by the creation of a system of economic regulation using price caps as a central mechanism of control (Beecher, 1997).

ALTERNATIVE REGULATORY MODELS

Despite the potential advantages of economic regulation in the context of water utility privatization, the prevailing regulatory models can be improved. The recent movements to restructure the telecommunications and energy industries are intrinsically related to the emergence of alternatives to the traditional modes of economic regulation.

As noted, many versions of privatization emphasize the importance of local control. Most of the unregulated privatizers seem to envision a market defined by public ownership and competitive contracts for operations with no state regulatory involvement. Methods of oversight are specified in the terms of the contract between the private vendor and the local government. The logical conclusion of this model is the virtual demise of state economic regulation, as well as the investor-owned water and wastewater industries in their present form.

This vision of the future, which negates the role of private ownership of utility facilities, seems contrary to many of the goals of privatization. Specifically, it undermines the goal of using private capital to independently build and operate major capital facilities. It also suggests that communities now served by investor-owned utilities would purchase these systems, which given the current fiscal circumstances of most cities, seems highly unlikely.

A strikingly different model for implementing privatization is to transform *all* municipal water systems to private systems (so that all water systems would be subject to the same regulatory and taxation policies), and replace ratebase/rate-of-return regulation with price-cap regulation (or a similar approach) to foster efficiency and competition (Haarmeyer, 1993). This model essentially suggests the divestiture of all publicly owned systems, much like privatization in Europe. As in total public ownership, total private ownership would create a level playing field for the industries. Total privatization also would create considerable opportunities for consolidation, but the resultant private monopolies would require some form of economic regulation. Given the politics of local control, total divestiture seems very unlikely. In sum, no extreme model of privatization (either total public ownership with contracts or total divestiture) is likely to prevail.

Although some agreement can be found on the need for overseeing private utility monopolies, not everyone agrees that the current system of ratebase/rate-of-return regulation by the states is the best form of oversight. Just as long-held assumptions about vertically integrated utility monopolies have been challenged, so have the regulatory models that traditionally were deployed to protect consumers from the potential abuse of utility monopoly power. Several alternative regulatory models are under consideration in the context of emerging competition within the telecommunications, natural gas, and electricity industries. Emerging regulatory models probably have applicability for the water and wastewater sectors as well. Some of the available approaches are summarized in Table 17–7.

Price caps are used in the regulation of utilities in Great Britain. Under the price-cap model, the focus of regulation is on setting maximum prices; utilities have strong incentives to keep costs down in order to enhance profits. Supporters of price-cap regulation contend that it would help the water supply industry take full advantage of market forces. Another regulatory approach involves the use of performance benchmarking, which is designed in part to replicate or augment competitive markets. Utilities operating within specified financial and operational benchmarks could be subjected to less stringent regulatory requirements. Benchmarks could be used in a system of incentive regulation to reward utilities for exceptional performance. In general, the consideration of incentive regulation is appropriate in the context of rapidly changing needs.

TABLE 17–7. A GLOBAL PERSPECTIVE: RECOMMENDED APPROACH BASED ON INSTITUTIONAL AND MARKET CONDITIONS

Institutional Conditions	Market Conditions	
	Competitive	**Noncompetitive**
High capacity to regulate with a "friendly" market	Privatize	Ensure or install appropriate regulation environment prior to considering privatization.
High capacity to regulate with an "unfriendly" market	Privatize with attention to competitive conditions.	Consider privatization of management arrangements; establish a market friendly framework and an appropriate regulatory environment prior to considering privatization.

Source: Adapted from Sunita Kikeri, John Nellis, and Mary Shirley. *Privatization: The Lessons of Experience* (Washington, DC: The World Bank, 1992).

Structured Competition

Future regulatory models should not necessarily be encumbered by traditional views of economic regulation. Given the potential for increased privatization and some limited forms of competition, a structured competition model may be appropriate for regulating water and wastewater industries. Structured competition could help the industries realize efficiencies through competition without putting ratepayers at undue risk. A structured competition model could consist of the following actions (Beecher, 1995):

- Precertify providers as to technical, managerial, and financial fitness to provide water service, with binding assurances as appropriate.
- Establish a reasonably level playing field by imposing comparable standards and requirements on all regulated water and wastewater utilities (and possibly all publicly and privately owned utilities).
- Encourage market-based utility management practices (such as competitive bidding) and remove disincentives for operational efficiency (such as cost pass-throughs).
- Institute profit-sharing or other mechanisms to distribute efficiency savings between privatizers and ratepayers.
- Provide state commissions with the authority to resolve disputes between private contractors and public entities.
- Streamline the regulatory process so that regulated firms can respond to changing market conditions and competitive opportunities.
- Consider registering or certifying major providers of public utility services to ensure that they meet minimal qualifications.
- Facilitate capacity building at the local level for cities engaged in privatization activity.
- Simplify economic regulation and relax regulatory requirements when competition provides sufficient protection from monopoly abuse.
- Use benchmarking to monitor utility performance along critical indicators and trigger regulatory intervention as needed.
- Coordinate the privatization policies of different governmental agencies and design consistent requirements and standards.
- Establish mechanisms for expedited dispute resolution among parties to a privatization agreement.

Regulators may play fundamentally different roles in overseeing restructured and more competitive industries. A leading example is the idea of using the commissions as a "court of last resort" (prior to and ideally instead of litigation) for resolving disputes between privatizers and cities. Early and competent dispute resolution should facilitate the development of competitive markets. A somewhat more invasive model would be to empower the commissions to approve contracts before they are implemented, which could provide ratepayers with additional protection and help avoid later disputes. The commissions could readily adapt their expertise and resources to these new functions.

Although surely controversial, the commissions also could play a role in registering or certifying privatizers to ensure that they have adequate financial, managerial, and technical capability to be entrusted with utility operational responsibilities. For example, registered or certified providers would meet minimal requirements in such areas as insurance coverage and the number of certified plant operators at their disposal. The commissions could provide rudimentary monitoring and oversight, such as a review of annual financial reports, and initiate inquiries only under certain circumstances (such as a dispute or a petition by ratepayers). Registration or certification also could be very beneficial to local officials when prequalifying bidders for major contracts.

Regulators will face numerous challenges in structuring competition. Perhaps the greatest challenge will be trying to create a more level playing field for competition. The mix of utilities with different ownership forms, the creation of corporate affiliates to provide various services, and the presence of many unregulated competitors makes this task especially difficult. As currently structured, the industry itself presents a significant barrier to economic efficiency. Regulators can continue to *react* to this structure or they can begin to design and implement regulatory models that will actively *change* it. Regulation should not stand in the way of economic industry restructuring that will allow the water and wastewater industries to take full advantage of economies of scale and scope. The concept of structured competition seems an appropriate regulatory model for introducing the competitive spirit to the water and wastewater industries without forcing ratepayers to incur unnecessary or unacceptable risks.

OBSERVATIONS

As noted above, economic regulation strives for many of the same goals as privatization; the two concepts are not as ideologically incompatible as they might appear. While regulation may be perceived as a barrier to privatization, not all regulators share that view. In fact, some members of the National Association of Regulatory Utility Commissioners (NARUC) have actively supported the idea of privatization because of its potential role in restructuring the U.S. water industry. According to one regulator:

> *[P]rivatization can lead to corporate structures of a size which can transcend local political boundaries, attract necessary capital and achieve operating efficiencies to an extent that exceeds the capabilities of most governmental authorities. From a regulatory perspective, therefore, if one believes in the goals of the new standards and the legitimacy of the heightened expectations which NARUC has certainly supported, then we should welcome privatization as not only beneficial but essential to achieving these goals (Foran, 1991: 23).*

Competitive markets are considered the best "regulator" of economic behavior. Competition for market share and profits drives firms toward efficiency. Customers, including municipalities, are protected by their freedom to choose. Yet pure competition for all aspects of public utility services probably is not an achievable goal, given the realities of market failure associated with monopoly services.

In the foreseeable future, the water and wastewater industries will continue to combine elements of public and private ownership, monopoly and competition, and state and local regulation. A critical issue for policymakers is how to gauge when competition and/or local oversight are sufficient to reduce or remove economic regulatory requirements. The mere presence of a service contract is not sufficient evidence of a competitive market. Competition and the prospect of deregulation pose a chicken-and-egg problem: some will argue that competition cannot emerge as long as economic regulation is intact, while others will argue that regulation cannot be removed until competition is well established.

Importantly, the emerging competition in the water and wastewater industries is profoundly different from the manifestations of competition in the other major utility industries because of the aggressive presence of private vendors who have several competitive advantages over investor-owned utilities. Not the least of these advantages is the reality that these vendors are unregulated profit-seekers in a relatively monopolistic enterprise governed by a relatively immature competitive market. Investor-owned water and wastewater utilities will not compete effectively with unregulated private vendors unless substantial institutional changes are implemented, including changes in the regulatory regime. Thus, the commission role in regulating the water and wastewater sectors will evolve by necessity and affect the future of the investor-owned industries in fundamental ways. Without regulatory reform, privatizers will continue to seek exemptions from regulation or try to circumvent the regulatory process altogether. Given persistent monopoly power, the result may do ratepayers more harm than good and deny society the opportunities that structured competition can bring.

SUPPLEMENT

Policy Governing Water Privatization Projects (Kentucky)

Kentucky Acts

107.700. Legislative policy—The General Assembly declares that the policy of this Commonwealth is to assure its citizens adequate public services, at reasonable cost, and that such services are essential to the maintenance and general welfare of the citizens of this Commonwealth and industrial base. However, the cost of constructing, owning, and operating capital facilities to meet the demand for those public services is becoming increasingly burdensome to cities, counties, urban-counties, and improvement districts and it is desirable that innovative financing mechanisms be made available to assist the communities of this Commonwealth in developing drinking water, water, and wastewater projects at reasonable cost. Private sector ownership and operation of capital facilities providing public services can result in cost savings to communities contracting for those public services. It is therefore in the best public interest of the Commonwealth and its citizens that cities, counties, urban-counties and improvement districts be authorized to cause such services to be provided by private enterprise, and to contract with private owners or private owner/operators for providing the service to the public. (Enact. Acts 1986, ch. 456, Section 1, effective July 15, 1986.)

107.720. Privatization contracts—Notice—Hearing—Competitive bidding—Assignment of contract—(1) Any political subdivision may enter into a privatization contract with a private owner or private owner/operator to accomplish the transfer of any political subdivision owned drinking water, water or wastewater project or the designing, construction, operation, maintenance, financing of cost or any combination thereof, of a drinking water, water or wastewater project pursuant to the provisions of KRS 107.720 to 107.760.

(2) Prior to a political subdivision entering into a privatization contract pertaining to its drinking water, water or wastewater project, or any portion thereof, the governing authority shall cause notice of its intention to adopt an ordinance to accomplish such privatization to be published pursuant to KRS Chapter 424. . . . [continues]

(3) Notwithstanding whether the political subdivision has adopted the provisions of KRS 45A.345 to 45A.460, the privatization contract, the service agreement or any other purchase by the local government in connection with a privatization contract under KRS 107.720 to 107.760 may be made or awarded by competitive bidding, competitive negotiation, or negotiation.

(4) The privatization contract or the service agreement may be assigned by either party to secure the performance of any obligation in connection with the financing of construction or operation of a drinking water, water or wastewater project. (Enact. Acts 1986, ch. 456, Section 3, effective July 15, 1986)

107.730. Service agreements—Notice—Hearing—In connection with a privatization contract, a political subdivision, if authorized by its governing body, may enter into one (1) or more service agreements with a private owner or private owner/operator pursuant to which the private owner or private owner/ operator will provide one (1) or more service to or for the benefit of such political subdivisions. The service agreement shall provide for the purchase by the political subdivision of all or any part of the capacity, capability, or output of the facilities used to provide the applicable service, and the charges or rates for such services, and shall contain such other terms and conditions as the political subdivision and the private owner or private owner/operator may agree including, without limitation, a covenant by the political subdivision to maintain rates sufficient to pay debt service incurred in connection with the financing of construction of a drinking water, water or wastewater project. The service agreement, the privatization contract, the charges and rates for services, and private owner or private owner/operator shall not be subject to the jurisdiction of the Kentucky Public Service Commission or any successor regulatory agency unless and only to the extent that the private owner or private owner/operator sells any part of the services or output of the facility to a person or entity other than the political subdivision which is a party to the service agreement.

. . . . . [continues]

(Enact. Acts. 1986, ch. 456, Section 5, effective July 15, 1986.)

Procedure for Exemption from Regulation (California)

California Public Utilities Code
10013. (a) Subsequent to signing a contingent franchise, license, or service agreement with a local agency, a privatizer shall apply to the commission for a determination that the proposed privatization project is not a public utility within the meaning of Section 216 and is therefore exempt from commission regulation. When a privatizer files an application with the commission, the privatizer shall include the information the commission requires to make a determination in accordance with subdivisions (b), (c), (d), and (e).

10013. (d) The commission may determine that a privatization project is not a public utility within the meaning of Section 216, and is therefore exempt from commission regulation if it finds that the franchise, license, or service agreement both demonstrates that the local agency retains sufficient jurisdiction to protect the public interest and adequately addresses all aspects of the provision of service which would otherwise be subject to commission regulation.

In making its determination, the commission shall determine whether the local agency has complied with Section 54253 of the Government Code. The decision of the commission shall be final and conclusive in the absence of any subsequent changes.

10013. (e) In making a determination pursuant to subdivision (c), the commission shall review the franchise, license, or service agreement to ensure that the agreement grants the local agency, at a minimum, all of the following:

(1) Exclusive authority to establish all rates and rate changes charged to the public.

(2) Approval over any proposal of the privatizer to provide new, additional, or alternative service to any other public or private entity or to change the service fee paid to the privatizer by the local agency.

(3) Approval over the original design and construction of the project, including any changes in design, alterations, or additions to the project.

(4) Approval over any changes in ownership of the party or parties subject to franchise, license, or service agreement.

(5) Authority to impose fines and penalties for noncompliance with any provision of the executed franchise, license, or service agreement, or for failure to provide the service within the time period agreed to in the franchise, license, or service agreement.

(6) Authority to ensure that the facility is adequately maintained.

(7) Adequate opportunity to monitor compliance with the agreement and to ensure the project will be operated to meet any applicable federal or state water quality standards or other applicable laws.

(8) Adequate opportunity to amend the agreement in the event of unforeseen circumstances or contingencies, such as flood, earthquake, fire, or other natural disasters or federal tax law changes.

Regulatory Review of Privatization Contracts (New Jersey)

New Jersey Statutes
58:26-23

Section 5. e. A contract entered into pursuant to P.L.1995, c.101 (C.58:26: 26-19 et al.) shall include provisions addressing the following:

(1) The charges, rates, fees or formulas to be used to determine the charges, rates, or fees to be charged by the public entity for the water supply services to be provided.

(2) The allocation of the risks of financing and constructing planned capital additions or upgrades to existing water supply facilities.

(6) The employment of current employees of the public entity whose positions or employment will be affected by the terms of the contract.

58:26-25
Section 7. c. In its review of a contract, the board shall apply the following criteria in determining whether to approve the contract:

(1) The private firm entering into the contract has the financial capacity and technical and administrative experience to ensure continuity of service over the term of the contract and that the standards and requirements contained in the application documents concerning the financial, technical and administrative capacity of the private firm are necessary and sufficient to protect the public interest.

(2) The terms of the contract are not unreasonable. In determining whether the terms of the contract are not unreasonable, the board shall review the fees and charges to be charged or assessed under the contract to determine that they are reasonable to the public entity, taking into consideration all of the obligations undertaken by the private firm and all the benefits obtained by the public entity. In making this determination, the board shall not use the traditional rate based rate of return methodology.

(3) The franchise customers of a public utility participating in a contract are protected from the risks of the proposed contract and that they are not subsidizing the contract. If a private firm is not a public utility, the board shall ensure that under the terms of the proposed contract the users of water outside of the jurisdiction or service area that will receive water supply services under the contract are also protected from the risks of the contract and that water users outside the jurisdiction or service area are not subsidizing the contract through increased charges, rates or fees for the supply of water.

(4) The contract contains the provisions required by paragraphs (1), (2) and (6) of subsection e. of section 5 of P.L.1995, c.101 (C.58: 26-23). Upon approval of a contract as proposed or as revised in response to a conditional approval, the jurisdiction of the board over the contract shall terminate until or unless the contract is amended to change the formula or other basis of determining charges contained therein.

NOTES

[1]Michigan gave up jurisdiction shortly after the 1995 survey.

[2]A few state commissions (including Connecticut, New Jersey, Pennsylvania, and West Virginia) have the authority to impose mandatory takeovers of problem systems by nearby regulated systems.

[3]Correspondence with Charles W. Johnston dated February 3, 1995.

[4]The Board's authority to review public-private partnerships was affected by legislation passed in mid-1995.

[5]West Virginia's statute, §24-2-12, prohibits a utility from selling its assets or contracting for services without state consent for reasons of protecting the public interest.

[6]See Title 35A §§ 708, 1101, and 1102 of Maine's Statutes.

[7]Pennsylvania passed an acquisition policy for this purpose in 1990.

[8]State of New York Public Service Commission, "Statement of Policy on Acquisition Incentive Mechanisms for Small Water Companies," Case 93-W-0962 issued August 8, 1994.

[9]This issue was raised in a 1995 case involving the purchase of Indiana Cities Water Company by the Indiana-American Water Company.

[10]It is not suggested, of course, that state commission regulation is entirely free from parochialism, politics, or the desire for expedience.

REFERENCES

R. V. Anderson Associates, Ltd. (1990). *Private Sector Participation in Provision of Halifax-Dartmouth Wastewater Treatment Services 1 and 2.* A study prepared for the province of Nova Scotia, Canada. October, p. 15.

Beecher, Janice A. (1995). "Regulatory Alternatives for Water Utilities: A Comprehensive Framework." *NRRI Quarterly Bulletin*, March.

Beecher, Janice A. (1997). "Water Utility Privatization and Regulation: Lessons from the Global Experiment." *Water International*, March.

Beecher, Janice A., Dreese, G. Richard, and Stanford, John D. (1995). *Regulatory Implications of Water and Wastewater Privatization.* Columbus, OH: The National Regulatory Research Institute.

Curry, Fred L. (1994). "Public-Private Partnerships." *NAWC Water*, Summer: 16-19.

Foran, Paul (1991). "Internationalization and Privatization: New Challenges and Opportunities in a Changing Water Industry." NARUC Bulletin 48-1991, December, p. 23.

Haarmeyer, David L. (1993). *Privatizing Infrastructure.* Los Angeles: Reason Foundation.

Kikeri, Sunita, Nellis, John, and Shirley, Mary (1992). *Privatization: The Lessons of Experience* Washington, DC: The World Bank.

New Jersey Board of Regulatory Commissioners (1993). Water Industry Challenges and Opportunities, A Report of the Water Management Task Force. May 26: pp. 43–46.

Raftelis, George A. (1989). *Arthur Young's Guide to Water and Wastewater Finance and Pricing.* Chelsea, MA: Lewis Publishing.

Rosenthal, Beth (1994). *State Regulation of Utility Diversification.* Washington, DC: Edison Electric Institute.

CHAPTER 18

H_2Ownership: Privatizing Waterworks in Theory and Practice

Steve H. Hanke and Stephen J.K. Walters

Privatizing a natural monopoly may strike many as a fool's errand. But when conditions of cost and demand are such that it is cheapest to organize production within a single firm—conditions that certainly apply in the case of water supply—privatization can replace a public monopoly with a private one. Many argue that in such circumstances the oft-touted efficiencies associated with privatization will not eventuate, or will be dissipated by the regulatory regime required by privately owned monopolies. Thus, proposals to privatize public utilities can be met with suspicion, if not outright hostility. In Britain in the 1980s, for example, when the Thatcher government announced plans to privatize water supply, 75 percent of poll respondents—and even 57 percent of Conservative voters—opposed the program.

Nevertheless, the 10 public water authorities in England and Wales were privatized in late 1989 and, despite considerable debate about the results, water privatization efforts since then have gained momentum worldwide. Existing water infrastructure has been sold to private investors in several Eastern European, Latin American, and Pacific Rim countries, and sales are under consideration in many more. In the U.S., where investor-owned water companies currently serve about 15 percent of the market, Atlanta, Buffalo, and several other cities have decided to privatize their systems.

This momentum probably owes more to mundane budgetary considerations than to theories of management efficiency propounded by privatization scholars. Often, the infrastructure requirements of a water system become overwhelming, making an

Steve H. Hanke is professor of economics at Johns Hopkins University and Stephen J.K. Walters is professor of economics at Loyola University in Baltimore, Maryland.

infusion of private capital attractive to political officials. Alternatively, where the subsidies embodied in under-priced municipal water have become unsustainable, a well-timed privatization program can provide convenient political cover for necessary rate increases. On occasion, too, costs of compliance with new regulations make private ownership attractive for either political or fiscal reasons.

Such motivations can be dangerous; their presence increases the risk that a water privatization program will fail to enhance public welfare by the maximum extent possible. We submit that the preferred model for waterworks privatization belongs to the French, who have long employed several creative strategies to obtain the benefits of private water supply without the fuss and failures involved in standard regulation of natural monopoly.

In succeeding sections of this chapter, we discuss these issues in fuller detail. First, we lay the foundation for this discussion by outlining the many organizational problems of naturally monopolistic markets. Next, we describe a mechanism that is uniquely capable of bringing the benefits of private ownership to such markets without the normal problems associated with monopolistic conduct or regulatory attempts to forestall such conduct. We then enumerate the crucial lessons from both the French and British experiences with water privatization. We conclude with some suggestions about the implementation of a waterworks privatization program in the U.S.

NATURAL MONOPOLY: MULTIPLE PROBLEMS

An industry qualifies as a natural monopoly if production may be done most efficiently by a single entity. This will generally be the case if unit costs decline as output rises (*i.e.*, there are massive economies of scale in the production of the good).

The water supply industry is a straightforward illustration. A pipe's capacity is roughly proportional to its circumference. If the cross-section area of a pipe is doubled, the circumference less than doubles; it follows that doubling the volume of water to be transmitted between any two points less than doubles the cost, and average transmission cost is everywhere declining. Within any single market, therefore, it would involve wasteful duplication of facilities to have more than one water utility.

Of course, a firm that is the sole producer of a commodity enjoys significant market power, and the fear has long been that natural monopoly would charge prices well above costs. Hence an apparent unhappy choice: shall we have competition that is wasteful, or monopoly that is exploitative? The typical answer has been for government to recognize and license a single producer in markets identified as natural monopolies, and then to regulate the prices (or rates of return) received by these firms to forestall monopolistic conduct.

Rate regulation, however, entails numerous problems of its own. Since most rate regulation involves enforcing some sort of cost-plus pricing rule, the general tendency is for regulated firms to be less than aggressive about controlling their costs. Sensing this perverse incentive, some regulatory authorities have become involved in de facto management of utilities, ruling on the desirability of various expenditures. It is difficult, however, for even a well-intentioned regulatory authority to determine which are legitimate costs and which are not. There is also a strong possibility that the authority will be "captured" by the firms it is supposed to regulate (*i.e.*, will serve the interests of the utilities instead of the broader interests of consumers).

Such issues pose another difficult question for policy makers: If natural monopolies must be regulated so closely by some governmental authority, might it be better to have public ownership of the enterprise? Instead of creating a layer of bureaucracy to supervise the pricing and production decisions of managers acting on behalf of private shareholders, this option inserts the government as owner and allows elected officials to hire managers to achieve the firm's objectives.

However, public ownership poses serious problems as well. Since public enterprises issue no ownership shares whose value will rise or fall depending on the firm's performance, public managers' incentives to aggressively control costs will be weak, perhaps weaker than those of managers of private, regulated firms. Owners of private firms can generally improve their managers' incentives by tying a significant portion of their compensation to firm profits or some other measure of performance. Also, the market for private shares acts as a manager: If management fails to perform, share prices will fall and the firm will become a takeover target. The threat of new owners gives management yet another performance incentive. In public firms, these devices usually are unavailable. Managers of public enterprises, therefore, are often given less managerial autonomy than their private-sector counterparts, and often must manage subject to a set of bureaucratic rules aimed at limiting their opportunities to pursue their own objectives at the expense of the firm. Such rules, of course, may simply cause public managers to take more circuitous routes to enhance their well-being (*e.g.*, a cap on a manager's salary may cause the manager to spend company resources on a larger support staff or more comfortable surroundings).

Studies comparing the efficiency of public and regulated private water utilities show mixed results. Crain and Zardkoohi (1978) found that regulated private waterworks in the U.S. enjoy significant cost advantages over their public counterparts, owing primarily to their 35.9 percent higher output per employee. In two cases where water utilities changed ownership forms during the period for which they had data, Crain and Zardkoohi found that where a public firm was privatized, output per employee increased by over 25 percent; where a firm went from private to public ownership, output per employee fell by over 40 percent. Such findings are in accord with an earlier study by Morgan (1977).

Recent studies are more equivocal. One, for example, concluded that public water utilities are more efficient than private utilities on average but show greater variance in performance (Bhattacharyya, Parker, and Raffiee, 1994). Another found private water utilities more efficient at small scales, but public firms comparatively more efficient when the scale of operations is large (Bhattacharyya, Harris, Narayanan, and Raffiee, 1995).

This ambiguity is not surprising. There is ample evidence that economic regulation can lead private firms to overinvest in capital, pad payrolls, and otherwise inflate costs (Joskow and Rose, 1989). Occasionally, then, it will be hard to distinguish the performance of public and regulated private firms. That has led economists to search for ways to improve regulation (Joskow and Schmalensee, 1986), or to change its fundamental nature.

THE FRANCHISE BIDDING APPROACH

What we would like are the advantages of private ownership without the problems associated with standard rate regulation. One approach (Demsetz, 1968) revived an idea originated over a century ago by Edwin Chadwick (1859): a system of competitive bidding for privately owned utility franchises. Chadwick recognized that those markets most cheaply served by a monopoly need not be afflicted with monopolistic conduct so long as there is meaningful competition for the rights to the monopoly franchise. Though competition within a market may be impossible, competition for the market may be feasible. So long as there is vigorous bidding for the utility franchise, the best of all possible worlds—no wasteful duplication of facilities, no monopolistic overcharges, no perverse incentives stemming from regulation—can be had. But how would such a system work?

The key to the franchise bidding approach to privatization of a natural monopoly is that bidding for the monopoly franchise should not be in terms of a sum to be paid for the franchise, but in terms of the prices that the franchisee would charge and the services the franchise would provide the public on award of the right to be the exclusive seller. If the franchise were merely awarded to the bidder willing to pay the highest price for this exclusive right, competition would drive bids up to an amount equal to the present value of expected future monopoly profits in the market. This would filter these profits through the franchisee to whatever authority granted the monopoly franchise in the first place, but the net result would still be overpricing of the product.

By contrast, Chadwick's proposal is that an auction be held in which the franchise is awarded to whichever bidder promises the best combination of price and quality to consumers. Here competition would drive bid prices down to competitive levels for each possible level of service quality. So long as this bidding process is open and competitive, and so long as inputs to the production process are available in open and competitive markets, there would be no need to fear monopolistic results even though a single firm would indeed be granted rights as an exclusive seller.

Theory is not necessarily reality, however, and scholars, such as Richard Schmalensee (1979), have expressed reservations about such franchise bidding. One set of problems relates to the bidding process itself. Selecting a winner (*i.e.*, determining an optimal price structure and mix of products) may be exceedingly complex, requiring the kind of expertise in the franchise-granting authority that one normally associates with a regulatory commission. In addition, there is no guarantee that bidding will be truly competitive. In particular, significant numbers of new firms may be reluctant to bid on a franchise that has expired when the previous franchisee is also in the bidding (since the previous supplier is almost certain to be better informed about actual cost and demand conditions than are the rivals).

Another set of problems relates to the likely behavior of the winning bidder during the term of the franchise contract. First, if the contract is for a reasonably long term, there must be some formula to allow for rate changes as costs, demands, and technologies change over time—or renegotiation must be allowed. If a formula approach is impractical and renegotiation allowed, the need for some sort of agency similar to a regulatory commission again becomes apparent. Such an agency will also be needed to police the franchise contract, since the agreement will not be self-enforcing. Another problem is that, as the end of the contract approaches, the franchisee may stint on maintenance and underinvest in new assets, leaving its successor to cope with any resulting problems.

These are important but not intractable problems. Three aspects (the difficulty of selecting a winning bidder, the difficulty of specifying or renegotiating contracts, and the need to police the contract) require the existence of some sort of "buyers' agency" to represent consumers. Critics of franchise bidding have asserted that such an agency would, after all, be reduced to performing the same tasks that traditional regulatory commissions now perform with the same difficulties and potential for inefficiency or abuse.

This is not necessarily the case, however. The degree of technological complexity and the swiftness of technological change in the relevant industry are crucial variables here. Selecting a winning bidder may be difficult where technology has created a myriad of potential service options. Where it is possible to specify a limited number of service standards, as with water supply, awarding the franchise may not be troublesome at all. Where the pace of technological change is not too rapid—as, again, with water supply—it may be quite easy to agree on some sort of formula for rate increases, and the possibility of mid-contract renegotiation may never arise. Enforcing the contract also will be facilitated in industries where the number of specified service standards is relatively limited. Note also that the enforcement role of the buyers' agency is fundamentally different from that of the traditional regulatory commission. The buyers' agency in franchise bidding is merely enforcing the contract; there is no need for this agency to try to authenticate (as public utility commissions must) the costs, which is a much more complex and difficult task.

Doubts about the competitiveness of the bidding process are based largely on a misperception about existing presumed advantageous position vis-à-vis new bidders. It is true that a current franchisee will know more about the market than a prospective franchisee, but this in no way implies that bidding for franchise renewal must be less than fully competitive. Knowledge is an asset like any other; it must be created at a cost. That the current franchisee already has acquired the asset and paid the associated cost is irrelevant. Bids of current franchisees will incorporate a return to the knowledge asset that has been created, while bids of prospective franchisees will incorporate the capital cost of acquiring the asset. Any bidder (including prospective franchisees) with a comparative advantage in creating this asset cheaply will be well situated to submit a winning bid.

The issue of a franchise holder's tendency to stint on maintenance also relates to the technological complexity of the industry. If the fixed assets of the operation may feasibly be inspected and evaluated by the authority, then the problem may be addressed by having the franchisee post a bond (to be forfeited and used to replenish any assets found deficient) as part of the contract. This would be similar to the process every tenant must follow in renting an apartment: a "security deposit" is posted as part of the lease agreement and forfeited if the apartment is depreciated more than normal. Obviously, the more complex the production process the more difficult such an approach would be to implement in practice.

It appears, however, that the nature of the water supply industry is well suited to the franchise bidding approach. The technology of water supply is well known and relatively static, and specifications about service standards and quality are readily formulable. All the critics' qualms about the practicability of franchise bidding recede in such a context. Additionally, the benefits of such a system would be considerable. Giving the winning bidder a monopoly franchise would ensure that the firm is able to exploit all possible economies of scale in the provision of service. Requiring that the bidding be in terms of price and service standards would prevent the firm from using its market power to overcharge consumers. Granting the franchise to private owners would harness the incentives of these owners to control costs efficiently in order to maximize profits.

To implement the system, the local government need only create a buyers' agency to conduct the auction and devise the contracts for the construction, maintenance, or operation of the facilities. Once the franchise is granted, enforcement of the contract can itself be privatized: an accounting firm, for example, could be retained to audit the franchisee and confirm that the terms of the contract have been observed. To assure that the auditor has the appropriate incentive to closely monitor compliance, the contract might include a "bounty" that would be payable to the auditor in the event noncompliance were proven.

Once in place, the franchisee, unlike private utilities laboring under rate regulation, would have every incentive to control costs, adopt new technologies, and so forth, since every dollar of cost saved would be an extra dollar of profit earned. If the firm's managers were not attentive to cost control, the profits would fall, share prices would decline, and the firm would become a ripe target for takeover by owners seeking to reap the gains which would result from turning out (or better motivating) the inefficient management.

THE FRENCH EXPERIENCE

It is quite likely that practice preceded theory in this area. Indeed, in his original article Chadwick noted that the principle of awarding franchises to the bidder promising to charge the lowest price to consumers already had been undertaken in France in a variety of activities, including water supply. The first franchise contract for water distribution in France dates back to 1782, when the Perier brothers (note the spelling—this is not related to the firm that now markets bottled water) were given exclusive distribution rights in Paris for a period of 15 years. The charge was about one cent per liter. The Perier firm was later nationalized, however; by 1854, the price of water in Paris had quintupled. In other population centers (*e.g.*, Avignon, Calais, Le Havre, Cannes) private water franchises were common from the early 1850s onward, though many municipalities opted to build and manage their own systems. Over time, however, many of these municipalities (including Paris) have turned to private companies to manage their systems. In the last 50 years, the market share of

France's private water suppliers has grown from less than 20 percent to more than 70 percent.

Privatization of water supply in France generally has taken one of two basic franchise forms. In a concession, a private company is entrusted with the construction (or, possibly, overhaul or modernization) of the facility as well as with its operation. Such a system is especially advantageous when the municipality lacks funds for such a major capital expenditure. The concessionaire advances all capital for construction and all working capital, assumes full responsibility (and risk) for monitoring, management, and maintenance of facilities, and is paid through the direct collection of payment from users. The contract is usually signed for a long-term period—generally 30 years—to enable amortization of the original capital outlay. The contract sets the price of the water sold with a formula including a fixed and variable component. For example, a user may pay a fixed monthly fee for access to a supply pipe of a certain diameter, plus a variable charge based on the number of cubic meters consumed.

In the *affermage* system, the expenses for the initial installation of major civil works are borne by the local community; the private firm then manages the completed facilities and provides working capital. Such systems are popular where municipal financing can be provided at preferential interest rates. The contract contains detailed specifications regarding maintenance or upgrading of facilities. All electromechanical, hydraulic, and metering equipment is the operator's responsibility, while civil works, water collection, and facility expansion are the responsibility of the municipality. Pipe renewal may be the responsibility of either party. As with a concession, a formula fixes the price of water. Often, this formula contains a surcharge, which the operator remits to the municipality for debt service.[1] It is standard to include in all franchise contracts clauses specifying (a) the standard and quality of services, (b) the minimum quantity to be supplied to each individual consumer, (c) the main's pressure, and (d) procedures for renewal of the contract; these are in addition to the aforementioned clauses relating to maintenance responsibilities and a pricing formula.

The French experience demonstrates clearly that, at least in regard to waterworks, the franchise bidding approach is a practical and workable system. The technology of water supply and transmission is such that bidders may be selected, contracts drawn up and enforced, and maintenance and expansion conducted efficiently and at acceptable cost. While no comparative studies are available to compare the costs of private and public water supply systems in France, the pronounced trend toward greater privatization (in a country not generally well disposed to the private sector) suggests that enhanced cost-effectiveness is a signal virtue of the private systems.

The French experience is, moreover, not unique. In Spain, private water companies account for nearly as large a share of drinking water requirements as they do in France. Starting as early as 1914, French concessions branched out and set up systems in Italy, Brazil, Peru, Morocco, Indonesia, Kuwait, and a dozen or more African nations (though some of these were expropriated during the 1960s). In Santiago, Chile, two private concessions long have operated in separate areas of the city, and the Chilean Congress and Senate recently approved regulations to allow privatization of up to 65 percent of the nation's water/wastewater authorities. Australia's water concession contracts have been held up as models of precision and sophistication (Haarmeyer, 1994).

THE BRITISH EXPERIENCE

Despite the success of franchise bidding in France and elsewhere, the promoters of "H_2Ownership" in Britain turned their backs on this system when implementing their water privatization program in 1989. The 10 new water companies they created, which include three of the seven largest private water suppliers in the world, are all regulated utilities in the traditional sense. Production facilities are owned by private shareholders (and, in fact, French water companies have taken significant ownership interests). To prevent the private firms from charging monopolistic prices for water, they are subject to rate regulation by a government authority. To address the possibility that a subject firm might inflate costs or ignore opportunities for innovation, however, the British have made the water companies subject to a relatively new type of rate-setting formula.

Under ordinary economic regulation, a subject firm realizes few benefits from reductions in operating expenses, wages, or capital costs, for when these reductions produce higher-than-permissible profits regulators simply cut rates. Indeed, it is ironic that scholars often cite "regulatory lag"—a term that refers to the fact that it commonly takes a while for the bureaucratic gears to grind out requisite rate changes—as a force for good. The idea is that if regulated firms get to keep the gains resulting from enhanced efficiency for a few months or years, their incentive to be efficient will be improved.

The British have institutionalized this virtue of regulatory lag. Prices of the regulated water suppliers are tied to changes in the consumer price index (CPI). Regulators fix prices initially, and then allow prices to rise by the percentage change in the CPI plus or minus some amount, *k*. This approach, dubbed "rate indexing" or "price capping," can greatly improve the firm's incentive to operate efficiently. The *k*-factor usually is negative, to account for some normal or expected amount of productivity enhancement within the firm. If the level of actual productivity gain is less than this norm, the firm's profits fall; if productivity gains exceed this norm, the firm gets to keep the difference as extra profit. Since subject firms are punished for inferior performance and rewarded for superior performance, they should have reasonably strong incentives to invest wisely, innovate optimally, and avoid "gold-plating." In fact, where price caps have been employed in the U.S., they have generally been successful in reducing costs.

In Britain, the *k*-factor initially was added to the CPI-factor for most British water suppliers. In the case of Thames Water, for example, the *k*-factor was set at 4.5 percent; thus, if the annual increase in the CPI was 10 percent, Thames could raise rates by 14.5 percent. The government fixed positive *k*-factors for water suppliers because (a) it was widely recognized that the industry's infrastructure had decayed significantly under government ownership, and significant new capital spending was required, and (b) new environmental controls specified by the European Community imposed additional capital requirements on the newly privatized firms. Review and revision of *k*-factors was scheduled for every five to 10 years.

The result has been a public relations nightmare for the privatized firms. Editorials scolding them for real increases in water and sewer charges—usually linked to sky-high profits and generous dividends for shareholders—have been a staple of the

British press for several years now. Perhaps this is part of the reason the privatization occurred in the first place. Policymakers knew of the impending need for capital infusions and rate hikes, and it does not strain the imagination to suppose that they also realized there would be political advantage in letting multinational companies take the blame for this dirty work.

The real issue is not whether the privatization-cum-regulation program is winning applause from the public, or even whether it has improved upon the previous state of affairs. The crucial question is whether the British program has worked as well as known alternatives might have worked. We have our doubts; we see the high industry profits that have been so problematic for public relations as *prima facie* evidence that (a) the privatized and regulated firms are, indeed, significantly more efficient than their publicly managed forebears, and (b) the British privatization program has nevertheless failed to maximize social welfare.

Within three years of the privatization, water companies' operating profits rose by more than a third (Solanes, 1995), average share prices doubled, and dividend payments rose by 40 percent (*Economist*, 1994). Soaring profits need not be a sign of enhanced efficiency, of course. In this industry, higher profits might be obtained by depreciating the capital stock, stinting on product quality, or degrading the environment. The bureaucracy charged with regulating the privatized firms, the Office of Water Services (OFWAT), seems confident that none of these opportunistic strategies have been employed. On the basis of its periodic audits, OFWAT concluded that there has been a "significant catching up on maintenance of older assets" (OFWAT, 1995a: 46), that "although there are occasional fluctuations in levels of [service] quality, the overall trend has been one of steady improvement" (OFWAT, 1995b: 4), and that there has been "gradual improvement in performance" (OFWAT, 1995b: 32) with respect to one oft-used measure of environmental impact, the frequency of untreated sewerage discharges into the sea.

There is, in fact, ample anecdotal evidence that the newly privatized water suppliers' high profits were at least partly the result of their aggressive efforts to shed some of the excess labor and underutilized capital and land they had accumulated while government-owned, and of their use of innovative maintenance and construction techniques. For example, within a year of privatization, Thames Water's staff was nearly 20 percent below its peak levels; the company made a killing selling its expensive offices in London and rural parcels that had stood idle for many years; its hundreds of small research laboratories became a handful of larger labs (*Economist*, 1990). Under a new engineering director, Thames doubled the speed with which it digs new tunnels, and derived substantial savings by designing and constructing much of its pipe-installation equipment and supplies in-house; savings also resulted from capital expenditures that reduced the fraction of output that is lost to subterranean leaks by almost a third within two years of privatization (Kay, 1991).

Such savings are to be applauded, of course. By economizing on the amounts of labor, land, and capital required to accommodate demand at the regulated prices, the British privatization program has almost certainly enhanced welfare relative to the pre-1989 status quo. But Britons have gotten only half a loaf. They are enjoying the welfare benefits associated with a decline in the industry's cost curves (relative to those of publicly owned firms). They are missing, however, potential additional welfare gains stemming from the lower prices that would result if these lower costs were combined with greater competition, with the kind of "competition for the market" that is so common in France.

In essence, the British water companies' supra-competitive profit rates make it clear that their prices are also supra-competitive. Had there been franchise bidding for the right to serve the British markets (following the French model), there is every reason to believe that the observed downward shift in production costs would have been combined with lower prices to consumers and more normal, competitive returns to investors. This would have (a) increased the welfare gains from exchange in this market, as lower prices enabled increases in quantity demanded, and (b) enabled consumers rather than shareholders to enjoy the lion's share of the welfare gains stemming from the industry's enhanced efficiency. Of course, OFWAT has been neither deaf to public opinion nor blind to the potential welfare gains from lower prices. The *k*-factors negotiated and fixed in the 1989 privatization have regularly been revised downward, some prior to the five-year review date (under the terms of an "Interim Adjustment" mechanism), others later.

Clearly, then, converting public monopolies into regulated private ones has generated substantial resource savings for British consumers. However, though the short-term results have been favorable, there is reason to worry about the long run. Ironically, one problem is that the newly privatized firms have been so successful at cutting costs—and, therefore, at netting profits—that they have provoked cries for more stringent regulation. Thames Water, for example, announced pretax profits of £212 million on revenue of £835 million in 1991. Shortly thereafter, the water companies' chief regulator announced that he would revise the industry's *k*-factor if profit rates did not converge to more "sensible" levels.

There is reason to believe that this is no idle threat. In other privatized industries, *k*-factors have been revised significantly in order to reduce firms' profits. British Telecom's price cap, for example, initially was set at CPI minus 3 percent; this was later adjusted to CPI minus 4.5 percent and is now CPI minus 6.25 percent. For British Gas, the price cap has gone from CPI minus 2 percent to CPI minus 5 percent.

The danger here is manifest. As the *Economist* has editorialized, "what, after all, is the point of striving to be efficient . . . if the resulting higher profits lead only to the regulator raising [*k*]?" (1991a: 55) The more frequently regulators revise the *k*-factor to capture for consumers the gains from firms' enhanced efficiency, the more this system begins to resemble ordinary rate-of-return regulation, with all its flaws. In such a system, private managers will soon realize it is advantageous to (occasionally) pad costs or postpone realizable efficiencies in order to avoid rousing the sleeping regulatory dogs. For once, the regulators are awake and determined to punish efficiency, the trouble begins. The *Economist* has summarized recent trends in Britain as follows (1991b: 20):

> Privatization was supposed to free firms from piecemeal, day-to-day interference. Yet regulation is itself beginning to evoke echoes of pre-privatization quarrels in Whitehall. Managers are continually trying to second-guess regulators; investment is becoming a chancy business; shareholders worry that they have bought a lottery ticket.

Clearly, this is not the way privatization is supposed to work. Unless regulators in Britain learn that the high profits shown by the newly privatized water suppliers are a good sign rather than a bad one, they risk discarding much of the efficiency gain resulting from privatization.

CONCLUDING REMARKS: SOME LESSONS

Policymakers often become interested in privatization during periods of crisis. In normal times, the interest groups whose members benefit from the status quo (*e.g.*, unionized workers, government employees, consumers of subsidized goods) generally will be strong enough to neutralize those groups holding out the prospect of enhanced economic efficiency in support of a privatization program.[2] Let there be a fiscal crisis necessitating a large tax hike, however, or (as in Britain) looming regulations requiring massive new capital outlays and unpopular rate increases, and privatization proposals will gain at least a respectful hearing. In our view, it is at this point that privatization advocates need to exercise caution. If privatization is offered primarily as a means of delivering an infusion of cash to government coffers, the risks are great that the ultimate program will fail (again, as in Britain) to deliver maximum welfare gains to consumers. Auctioning off public assets to the highest bidder and simply subjecting the winner to economic regulation, even in its most sophisticated, modern form will not deliver all of the potential benefits of a more carefully constructed privatization program.

We offer the following paradigm for waterworks privatization in the relatively common case of a "mature" municipality (or larger jurisdiction) with the great bulk of its water supply and waste treatment facilities in place. The French *affermage* system is well equipped for such circumstances. First, the municipality must specify the amount the franchisor would need to pay in order to lease the facilities.[3] In addition, the municipality specifies any improvements or enhancements (*e.g.*, installation of meters or pollution-control equipment) to these facilities it wishes the franchisor to make, or simply specifies the condition in which the facilities must be left when the lease expires.[4]

Then, the municipality puts the franchise up for bids. Given bidders' calculations of the lease payment they will make, the costs of maintaining or improving the facilities as specified in the franchise contract, and the operating costs it expects to incur to meet the volume and quality specifications of the contract, they will specify a price structure to manage the facilities and serve consumers' water and wastewater needs.[5] The bidder who can do this most efficiently will offer the lowest bid price(s) and win the franchise.

Notice that the price structure offered by the bidders need not be as simplistic as those that public firms typically specify. Given some latitude to be creative about pricing, private suppliers would have an incentive to adopt sophisticated time-of-day or time-of-year pricing regimes aimed at solving peak-load demand problems. In addition, private suppliers would have an incentive to adopt nonuniform pricing aimed at recovering the higher costs of serving customers in sparsely settled areas. Such seasonal and/or zonal rate regimes can significantly reduce wasteful overconsumption of water (Hanke, 1972). This, in turn, will alleviate some of the pressure on local governments to embark on expensive and environmentally risky capacity-expansion projects, and will also reduce the cost and ecological impact of wastewater treatment. Both results will be applauded by environmental groups (Baker, 1991).

We do not mean to suggest, of course, that privatizing public water authorities will be politically easy, or that devising and administering the aforementioned contracts will be simple. We are under no illusions on this score. We think it is clear, however, that the evidence on the application of the French model demonstrates that privatizing a natural monopoly such as water supply is possible, and, indeed, produces ample benefits in numerous dimensions and to a wide variety of constituencies. It is time to apply the French model more widely, and bring the benefits of private provision of water supply to many more of the world's people.

NOTES

[1]A variant of the *affermage* system, called *gerance*, involves roughly the same relationship between municipality and operator, but with a more limited set of responsibilities for the private firm. The firm is paid on the basis of a tariff list agreed upon by the local authority. Yet another variant, *regie interessee*, involves management of a public authority by a private firm, which shares in the revenues or profits. In theory, the municipality retains overall management; the firm is an agent of the municipality and is paid a percentage of revenues (to which may be added productivity bonuses or a share of net profits). This system leaves greater authority with the municipality, but retains the advantage of access to the technical services of a company with specialized knowledge and abilities.

[2]Note, however, that techniques exist to overcome the opposition of entrenched interest groups (Hanke and Walters, 1990). The associated costs generally will reduce, but not eliminate, the social gains from privatization.

[3]The French have developed a sophisticated but workable system for making such calculations, which involves consideration of economic depreciation, amortization, and valuation of equipment. For a good introductory review of the French approach to such matters, see Mothes (1980).

[4]This may be in as little as a year or two, or as long as 10 or 12 years. Clearly, however, longer-term contracts must involve some price-escalation formula, which may be similar to the "CPI-minus-*k*" system in use in a regulated setting—though lacking the possibility that regulators will revisit the formula in order to punish efficiency.

[5]If the public assets the franchisor will manage include any facilities that are idle, underutilized, or merely have a higher-valued alternative use, the franchisor may also be able to derive revenue by sub-leasing these facilities during the term of the franchise contract. Of course, any anticipated revenues from such sub-leases would enable the franchisor to submit a lower bid in the auction.

REFERENCES

Baker, N. C. (1991). "Private Firms Rescue Ailing Wastewater Plants." *Environment Today* 2, No. 4, May, pp. 29–31.

Bhattacharyya, A., Parker, E., and Raffiee, K. (1994). "An Examination of the Effect of Ownership on the Relative Efficiency of Public and Private Water Utilities." *Land Economics*. Vol. 70, No. 2 May, pp. 197–209.

Bhattacharyya, A., Harris, T. R, Narayana, R., and Raffiee, K. (1995). "Specification and Estimation of the Effect of Ownership on the Economic Efficiency of the Water Utilities." *Regional Science and Urban Economics*. Vol. 25, No. 6 December, pp. 759–784.

Chadwick, E. (1859). "Results of Different Principles of Legislation and Administration in Europe of Competition for the Field as Compared with Competition Within the Field, of Service." *Journal of the Statistical Society of London*. Vol. XXII, pp. 381–420.

Crain, W.M. and A. Zardkoohi (1978). "A Test of Property-Rights Theory of the Firm: Water Utilities in the United States." *The Journal of Law and Economics*. Vol. 21, No. 2, October.

Demsetz, H. (1968). "Why Regulate Utilities?" *Journal of Law and Economics*. Vol. 11, No. 1 April, pp. 55–65.

Economist (1990). "Of Wealth and Water." 317, No. 7675, October 6, pp. 69–70.

Economist (1991a). "Can the Regulators Hold Them?" 319, No. 7709, June 1, pp. 25–26.

Economist (1991b). "Rules for Behemoths." 321, No. 7729, October 19, p. 20.

Economist (1994). "Overflowing: Water Prices." 332, No. 7873, July 23, pp. 24–28.

Haarmeyer, D. (1994). "Privatizing Infrastructure: Options for Municipal Systems." *Journal of the American Water Works Association*. 86, No. 3, March, pp. 42–55.

Hanke, S. H. (1972). "Pricing Urban Water." In Mushkin, S. I., ed., *Public Prices for Public Products*. Washington, DC: The Urban Institute.

Hanke, S. H., and Walters, S. J. K. (1990). "Privatization and Public Choice: Lessons for the LDCs." In Gayle, D. J. and Goodrich, J.N., eds., *Privatization and Deregulation in Global Perspective*. Westport, CT: Quorum Books.

Joskow, P., and Rose, N. L. (1989). "The Effects of Economic Regulation." In Schmalensee, R. and Willig, R., eds., *Handbook of Industrial Organization*, Vol. 2. Amsterdam, Oxford, and Tokyo: North-Holland.

Joskow, P., and Schmalensee, R. (1986). "Incentive Regulation for Electric Utilities." *Yale Journal on Regulation* 4, No. 1, Fall, pp. 1–49.

Kay, J. (1991). "Thames Water Makes Waves." *Management Today*, 20, No. 3, Autumn, pp. 437–453.

Morgan, W. D. (1977). "Investor Owned vs. Publicly Owned Water Agencies: An Evaluation of the Property Rights Theory of the Firm." *Water Resources Bulletin* 13, No. 4, August, pp. 775–781.

Mothes, J. (1980). *Faux Comptes et Mecomptes des Enterprises*. Paris: Dunod.

OFWAT (1995a). "1994–95 Report on the Financial Performance and Capital Investment of the Water Companies in England and Wales." Birmingham, UK.

OFWAT (1995b). "1994–95 Report on the Levels of Service of the Water Companies in England and Wales." Birmingham, UK.

Schmalensee, R. (1979). *The Control of Natural Monopolies*. Lexington, MA: Lexington Books.

Solanes, M. (1995). "The Privatization of Public Water Utilities." *CEPAL Review* 56 August, pp. 153–167.

Appendix

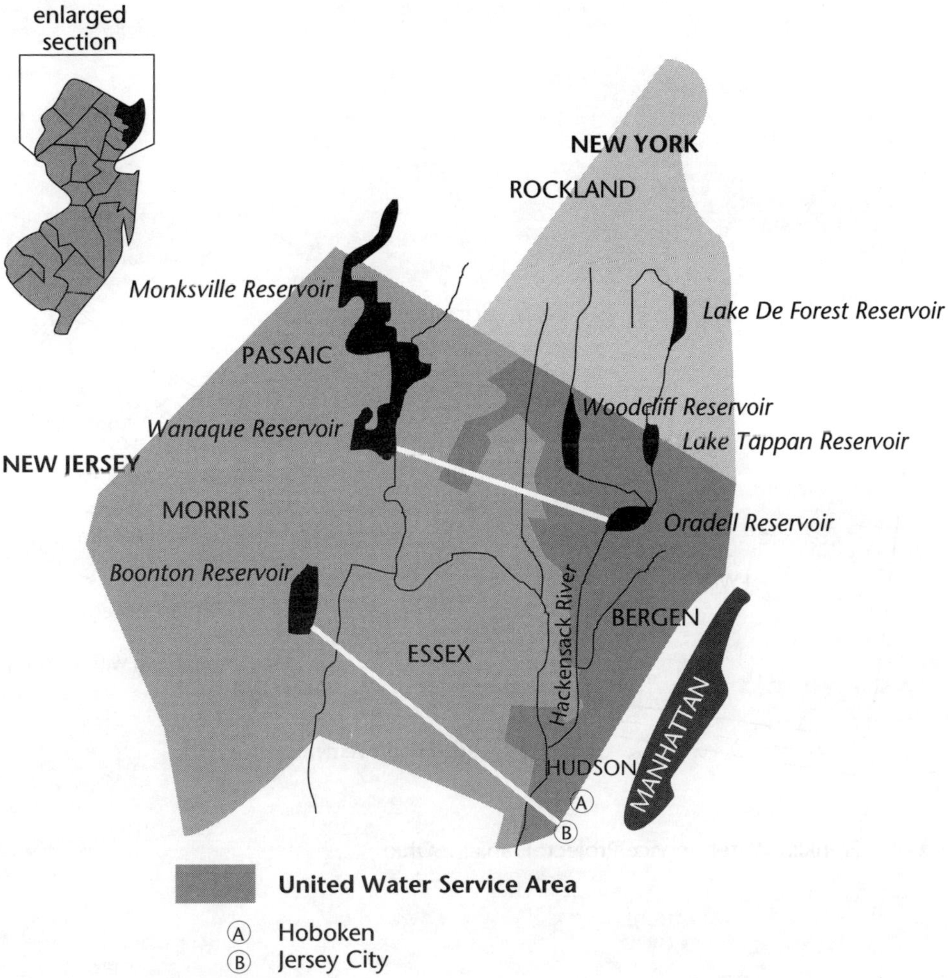

Figure A–1. United Water Supply and Service Territory

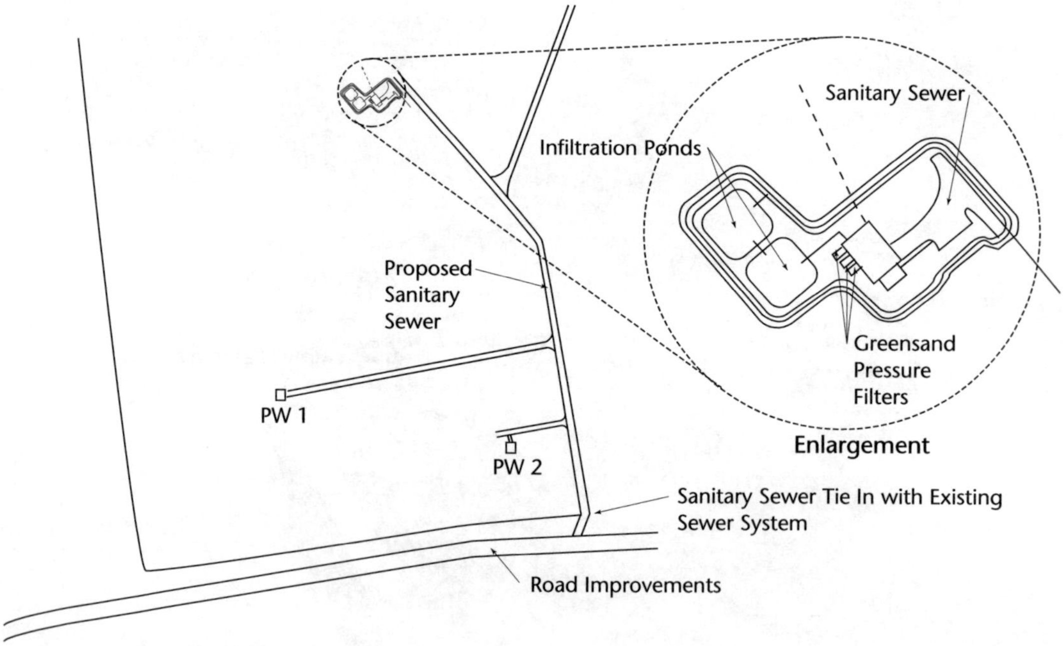

Figure A–2. Franklin Water Service Project, Franklin, Ohio

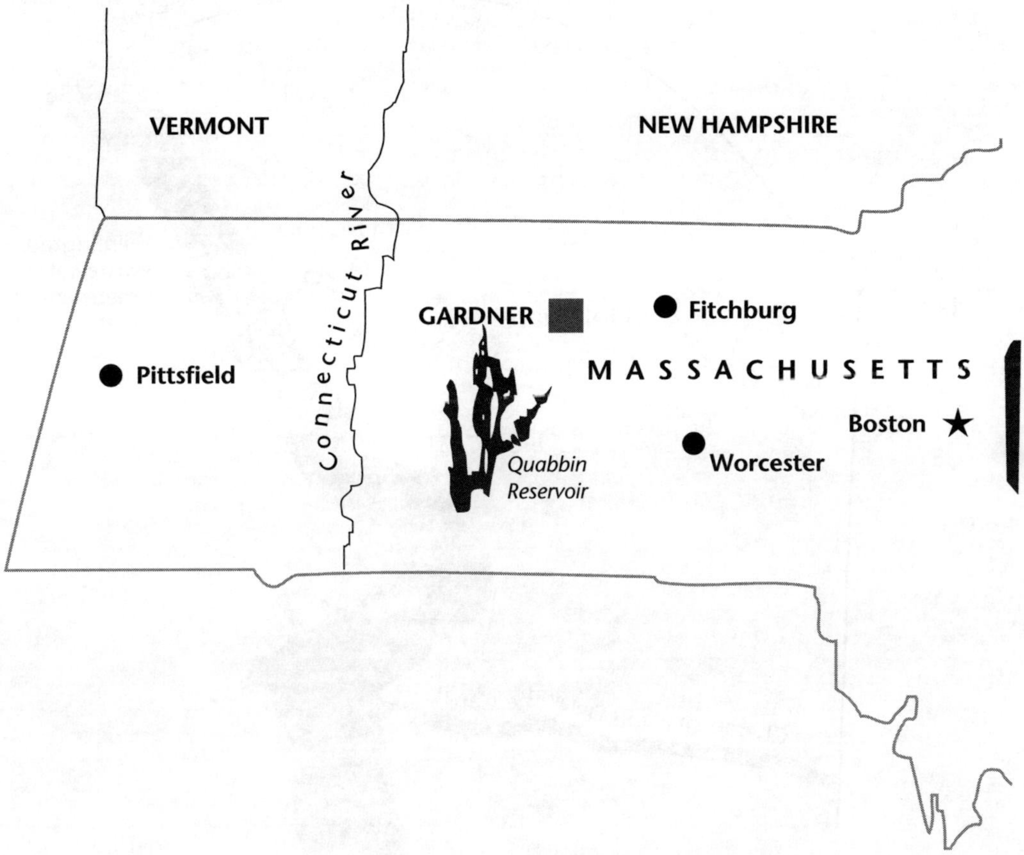

Figure A–3. Gardner, Massachusetts

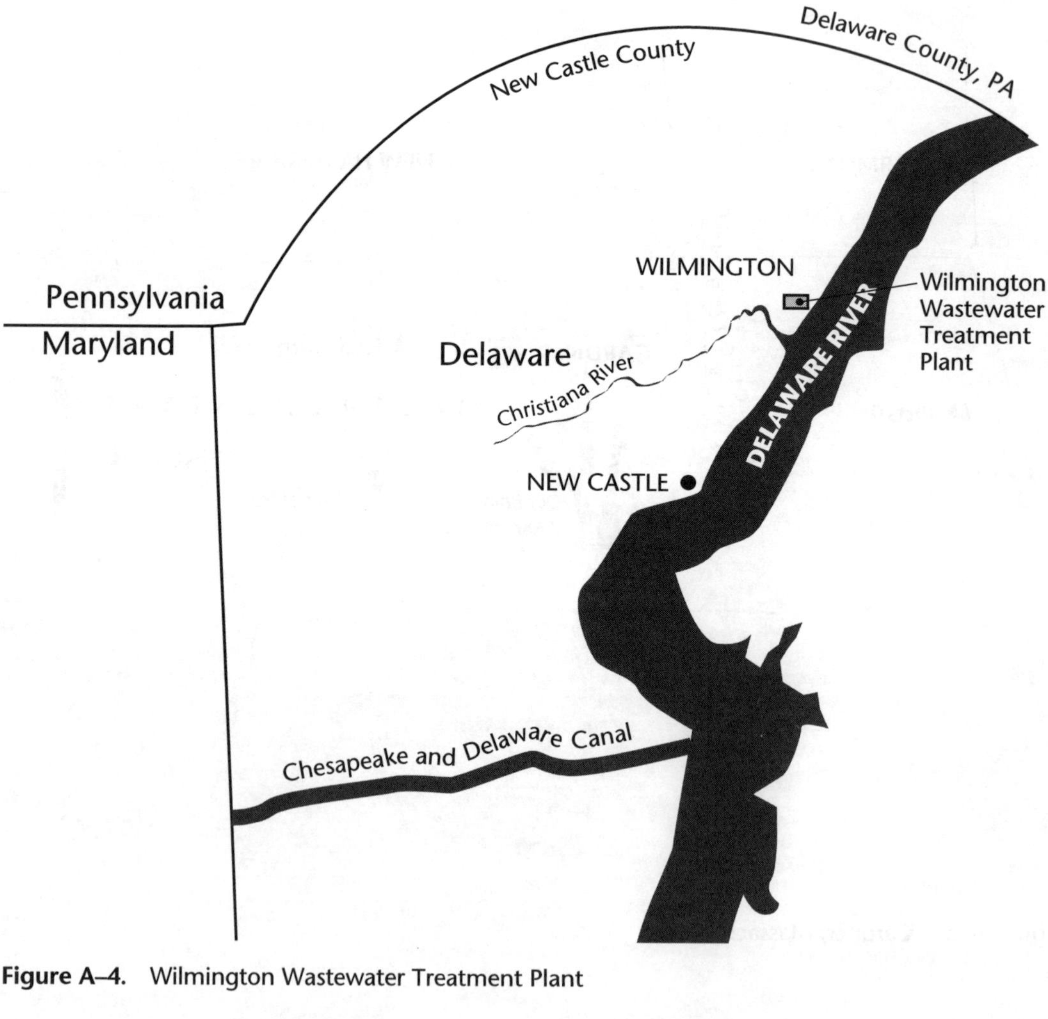

Figure A–4. Wilmington Wastewater Treatment Plant

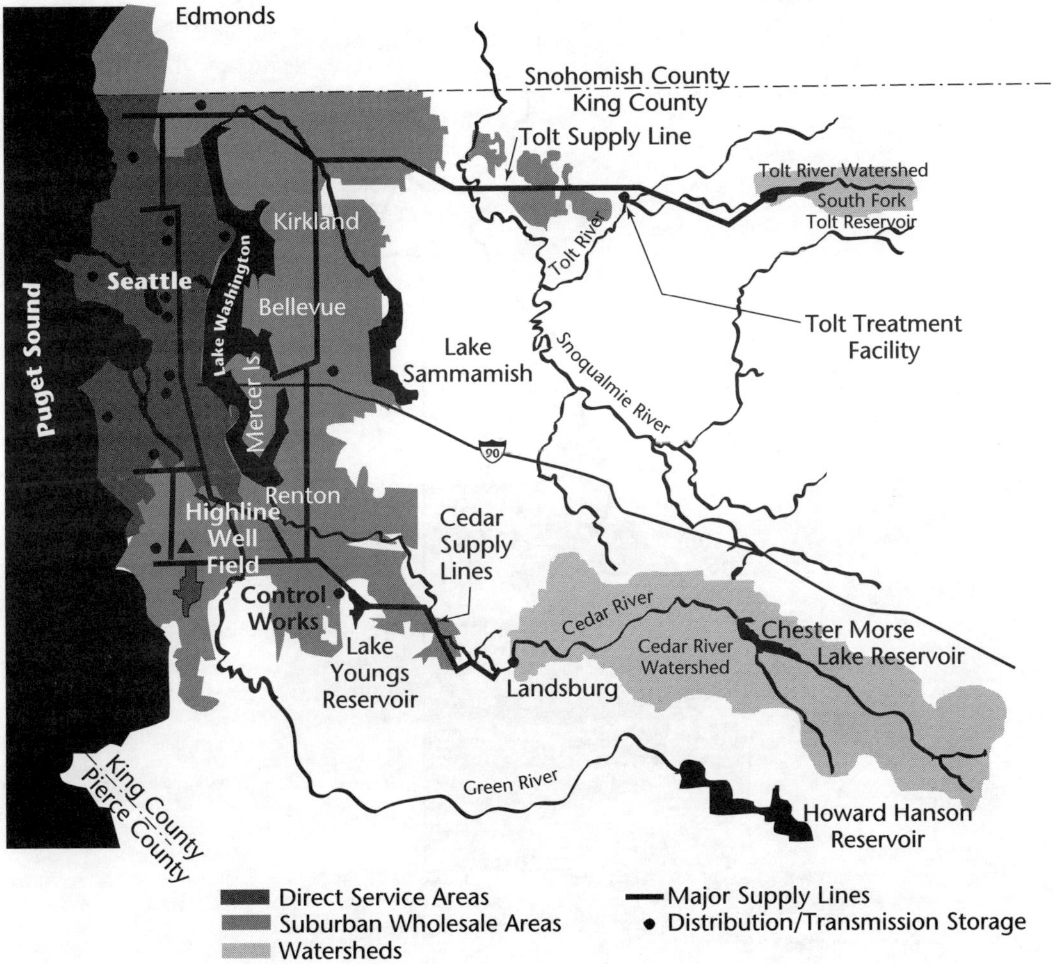

Figure A–5. Seattle, Washington Water System

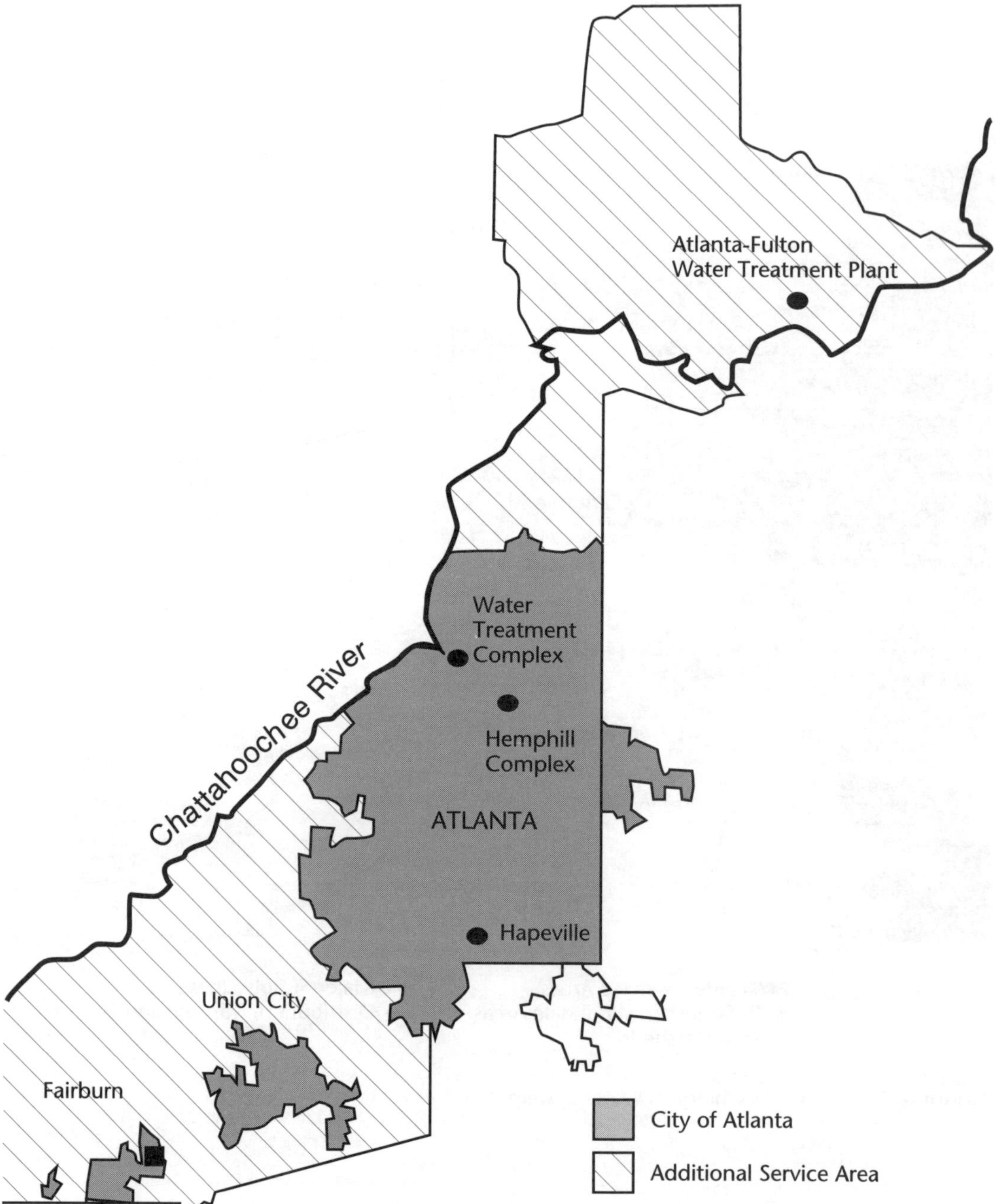

Figure A–6. Atlanta, Georgia

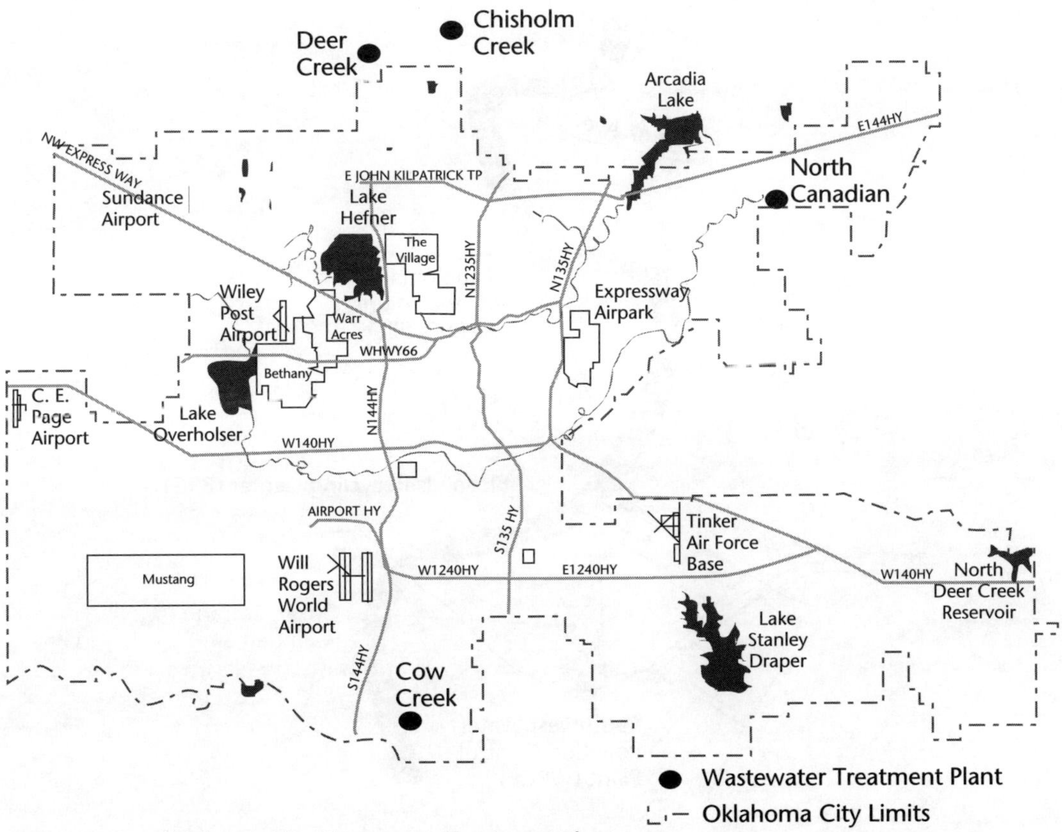

Figure A–7. Oklahoma City Wastewater Treatment Plants

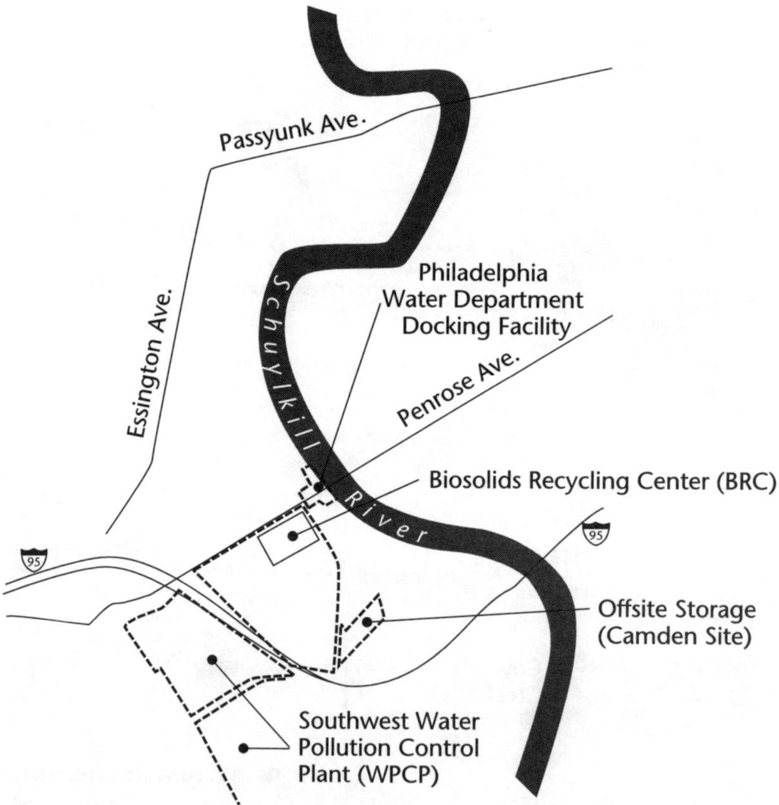

Figure A–8. Philadelphia Water Department

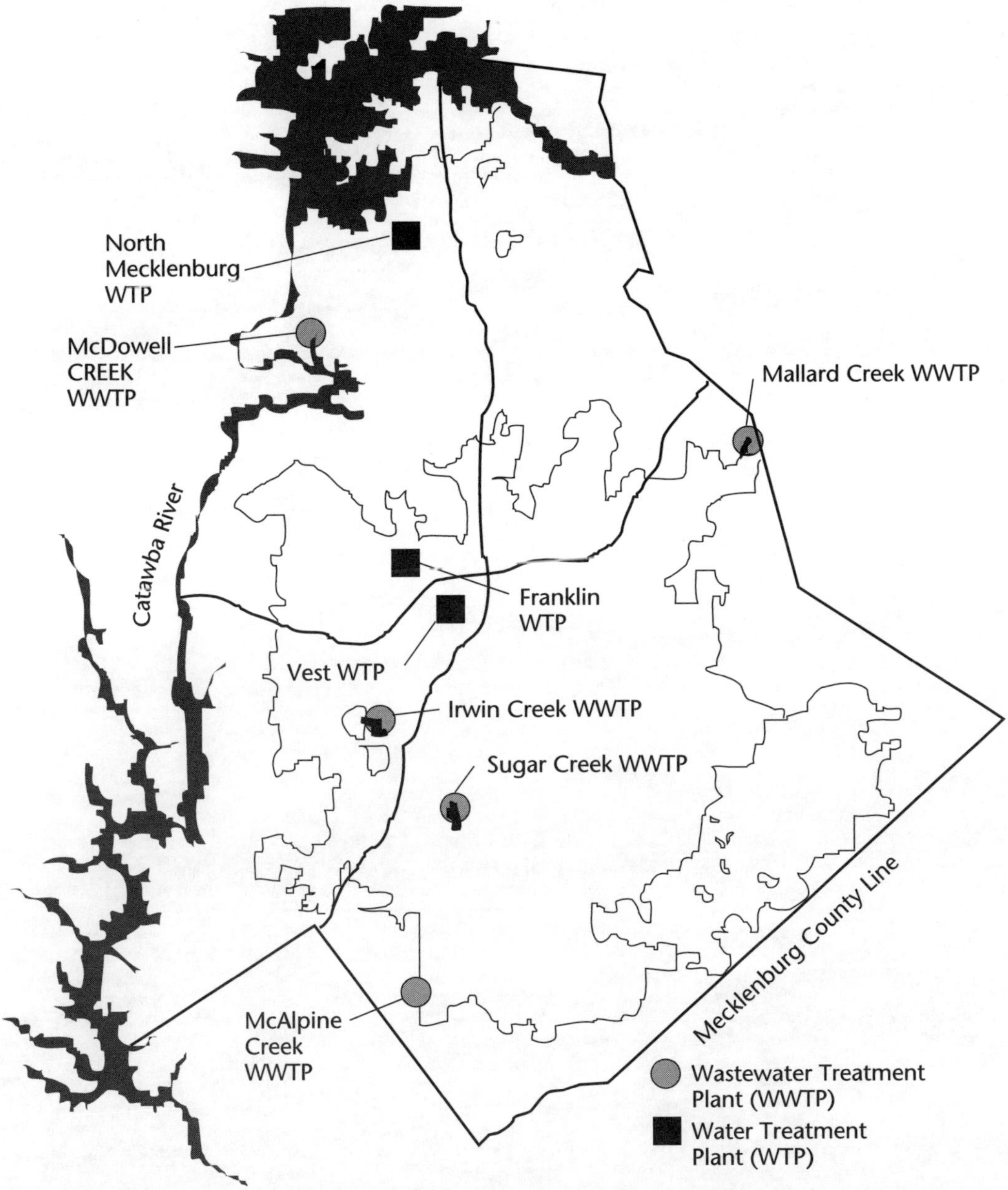

Figure A–9. Charlotte-Mecklenburg Utilities, Charlotte, North Carolina

Author Biographical Information

EDITORS

PAUL SEIDENSTAT is an associate professor of economics at Temple University where he previously was director of the graduate program in economics. He has written numerous books and articles in the field of privatization. In 1999, he edited *Contracting Out Government Services* (Praeger). A specialist in public finance, he has been principal investigator on several research projects for federal government agencies and has served in local governments as finance director and financial advisor. He is currently doing research in the area of state and local government finance and management. Seidenstat and Simon Hakim are co-editors of *Privatizing the United States Justice System* (McFarland), *Privatizing Correctional Institutions* (Transaction), *Privatizing Education and Educational Choice* (Praeger), and *Privatizing Transportation Systems* (Praeger).

MICHAEL NADOL is director of finance for Philadelphia and has formerly served as chief financial officer for the Philadelphia Water Department, one of the nation's largest municipal water and wastewater utilities. He is an adjunct professor for the Fels Center for Government at the University of Pennsylvania.

SIMON HAKIM is professor of economics at Temple University. He is director of the Privatization Research Center at Temple University and a co-editor of the series, *Privatizing Government Assets and Services* (Praeger). He has published over 30 scientific articles and has edited five books. He has conducted funded research projects for governmental agencies and private companies and also has written extensively on the analysis of criminal behavior, police operations, and the privatization of various types of institutions.

CONTRIBUTORS

JANICE A. BEECHER, a public utilities expert, is owner and principal of Beecher Policy Research, a major utilities consulting firm. She holds a Ph.D. in public administration and has written extensively in the field of public utility privatization.

GURU P. BOSE is manager, southwest operations of the Philadelphia Water Department. He is a registered professional engineer and holds a masters degree in business administration from Drexel University. He received the City of Philadelphia's 1995 Public Management Award.

KIMBERLY S. EAGLE is the assistant business manager for Charlotte-Mecklenburg Utilities. She has a master's degree in public administration. She has written several articles about public finance issues.

JAMES GOLEMBESKI is plant manager of the Biosolids Recycling Center of the Philadelphia Water Department. He is a registered professional engineer and received the City of Philadelphia's 1995 Public Management Award.

DAVID HAARMEYER is a consultant for Stone and Webster Management Consulting, Inc. in Boston. He previously was an economic consultant for the World Bank and Global Environmental Fund. Prior to that he was a senior policy analyst at the Reason Foundation and spent three years in New Zealand working with the government's Economic Development Commission. He has authored numerous studies and articles on infrastructure privatization.

STEVE H. HANKE is professor of economics at Johns Hopkins University and is the Director of the Institute of Applied Economics and the Study of Business Enterprise. In the early 1980s, Hanke served as a senior economist on President Reagan's Council of Economic Advisors. During the 1990s, he held advisory posts with various foreign governments, including Yugoslavia, Lithuania, Argentina, Bulgaria, and Indonesia. He has written numerous books on privatization, economic development, capital markets, and currency reform. He received a B.S. in Business Administration and a Ph.D. in Economics from the University of Colorado at Boulder.

ROGER HARTMAN, Ph.D., P.E. is owner and principal of THC, Utility Management Specialists in Boulder, Colorado. He is an independent consultant with 25 years of experience in the municipal water and wastewater utility industry. He holds a Ph.D. in environmental engineering and is a professional engineer. He has written extensively in the field.

SCOTT HASKINS is Resource Management Branch Director at Seattle Public Utilities, where he oversees watershed management, water quality and supply, planning, project development, and community programs for water, solid waste, drainage and wastewater. He also has provided the city of Seattle's management oversight for the Tolt design-build-operate project. Previously, he served as Finance Director and Deputy Director of Seattle Water. He has been the Chair of the Western Regional Water Utility Benchmarking Group since 1996. He has given numerous papers and presentations on utility management, assessment and contracting issues at national and international water symposiums and conferences. He holds a Master of Public

Administration and a Bachelor of Arts Degree in Political Science from the University of Washington.

DOUGLAS HERBST is senior vice-president, Earth Tech, Total Water Management Division and is president of the National Council for Public-Private Partnerships. Herbst was one of the pioneers who developed the privatization concept for the water and wastewater industry. As a municipal advisor, he has advised state and local governments on the various aspects of privatization, and as a private sector service provider has negotiated and structured numerous privatization (public-private partnerships) transactions.

RANDALL G. HOLCOMBE is professor of economics at Florida State University. He is an expert in the field of public finance. He has authored many publications relating to public finance issues.

KIRK HUMPHREYS is mayor of Oklahoma City, Oklahoma. He previously was a successful businessman, having built a national distribution company and real estate development firm.

ELIZABETH S. KELLY, P.E. has been with Seattle Public Utilities for 10 years. She manages large capital improvement projects for the department, using conventional and alternative contracting procedures. She managed the procurement phase of the Tolt Treatment Facilities design-build-operate project, and currently is responsible for ensuring successful implementation of all aspects of the project. She is also managing the procurement process for the Cedar River Treatment Facilities DBO project.

JANINE M. LANDOW-ESSER is an environmental attorney and a partner in the Chicago law firm of Holleb & Coff in Chicago. Her specialty is environmental and OSHA law. She previously was at the U.S. Department of Energy and served as one of the U.S. representatives to the International Energy Agency. She has served as chairperson of the Environmental Law Committee of the Chicago Bar Association.

ALAN MANNING is chief executive officer of EMA Services. He has conducted many seminars for public water companies and has written extensively on the subject of making public utilities more efficient.

MELISSA A. MANUEL is an environmental attorney with Holleb & Coff. She specializes in environmental and OSHA law.

ROBERT MARKOWITCH is the regional director for Earth Tech for New Jersey, New York, and New England. Markowitch has more than 20 years of experience in all aspects of managing wastewater treatment facilities and business development/client service.

KEITH OLDEWURTEL is executive vice-president, Earth Tech, Total Water Management Division. Oldewurtel has managed numerous projects for municipal and industrial clients and has 21 years of experience in the negotiation of contracts for water and wastewater projects.

REMEDIOS del ROSARIO is commissioner for Atlanta's department of water. With over 25 years experience in municipal water and sewer management, she has served as an officer of the American Water Works Association Research Foundation and the Association of Metropolitan Water Agencies.

PAUL D. REITER is director of Strategic Policy for Seattle Public Utilities and has more than 20 years' experience in utility management, quantitative and economic analysis (focusing on integrated resource planning), strategic policy development, and public finance. He has a master's degree in economics from the University of Oregon. He is a former adviser to the mayor of Seattle on utility issues and was also a private consultant in utility economics with experience in least-cost supply and conservation development for electrical and solid waste utilities.

JOHN SALO, a professional engineer, is a senior vice president of Brown and Caldwell. He has over 30 years of experience in environmental engineering and specializes in water and wastewater utility management consulting. He is actively involved in legislative and regulatory issues for the Water Environment Federation and the American Water Works Association.

DAVID SEADER is a director of the water industry practice of Pricewaterhouse Coopers. He has over 30 years of experience in infrastructure project analysis, planning and development; he writes and speaks frequently on public-private partnerships.

KASH SRINIVASAN, a professional engineer, is commissioner of public works for the City of Wilmington, Delaware. He has also held the positions of director of the water division, and wastewater facilities manager for the department over the past 19 years.

ROBERT E. THIELE, JR. is assistant vice president of United Water Services, Inc. and is responsible for the management of the Jersey City and Hoboken, New Jersey contracts. He is a professional engineer and has over 20 years of experience in the water industry.

WILLIAM E. TOFFEY is a biosolids water manager for the Philadelphia Water Department. He is a professional engineer.

STEPHEN J. K. WALTERS is professor of economics at Loyola University in Baltimore, Maryland. He has authored a number of publications dealing with privatization and public finance.

Index of Abbreviations Frequently Used

ACO	State Administrative Consent Order
AWWA	American Water Works Association is the largest drinking water industry association
BAFP	Best and final proposal
BOO/BOT	Build-Own-Operate/Build-Operate-Transfer
BRC	Biosolids Recycling Center
CIP	Capital improvement program
COLAs	Cost of living allowances
CPI	Consumer price index
CSO	Combined sewer overflow
CWA	Clear Water Act establishes a national program for the protection of the nation's waters
DBFO	Design-Build-Finance-Operate
DBO	Design-Build-Operate
EO	Executive Order
EPA	United States Environmental Protection Agency
IDB	Industrial development bonds
IRS	Internal Revenue Service

ITC	Investment tax credits
LDE	Suez-Lyonnaise des Eaux
M&E	Metcalf & Eddy
mgd	Millions of gallons per day
Moody's	Moody's Investors Service
NBV	Net book value
NPDES	National Pollutant Discharge Elimination System program regulates the discharge of pollutants into U.S. waters
O&M	Operation and maintenance
OM&M	Operation, maintenance, and management
OSHA	Occupational Safety and Health Administration is a federal government regulatory agency within the U.S. Department of Labor
PDM	Program-driven maintenance
PMSI	Parsons Municipal Services, Inc.
POTWs	Publicly owned treatment works
PPI	Producer price index
PUC	Public utility commission
RFP	Request for proposal
RFQ	Request for qualifications
SCADA	Supervisory, Control, and Data Acquisition system enables personnel from the city's public works department to access live operating data about the plant from their offices and other remote locations.
SDWA	Safe Drinking Water Act requires stringent environmental standards governing water and sewage
SOQ	Statement of qualifications
TPO	Total productive operations

Index

-D-